CW01572403

Recovery of Gray Wolves in the Great Lakes
Region of the United States

Adrian P. Wydeven
Timothy R. Van Deelen · Edward J. Heske
Editors

Recovery of Gray Wolves in the Great Lakes Region of the United States

An Endangered Species Success Story

 Springer

Editors
Adrian P. Wydeven
Wisconsin Department of Natural Resources
Park Falls, WI 54452
USA

Timothy R. Van Deelen
Department of Forest and Wildlife Ecology
University of Wisconsin – Madison
Madison, WI 53706
USA

Edward J. Heske
Illinois Natural History Survey
Champaign, IL 61821
USA

ISBN: 978-0-387-85951-4   e-ISBN: 978-0-387-85952-1
DOI: 10.1007/978-0-387-85952-1

Library of Congress Control Number: 2008940849

springer.com

We dedicate this book to *Pamela Sue Troxell* (January 8, 1959–November 9, 2007). Pam Troxell coordinated the Timber Wolf Alliance, Ashland, WI from 1994 until her untimely death in 2007. Pam worked not only to educate people about wolves and wolf conservation but also had a great gift for bringing people together over wolves and other environmental issues. In April 2007, Pam was honored with the Silver Eagle Award given by the US Fish and Wildlife Service (USFWS), and in May 2007 she and her coworkers received the USFWS's Cooperative Conservation Award for their efforts at promoting the recovery of gray wolves in the Great Lakes region. We are grateful for having known Pam and for her love of life, people, wolves, and wild places, and we thank her for helping to make the recovery of this species successful.

Preface

In this book, we document and evaluate the recovery of gray wolves (*Canis lupus*) in the Great Lakes region of the United States. The Great Lakes region is unique in that it was the only portion of the lower 48 states where wolves were never completely extirpated. This region also contains the area where many of the first modern concepts of wolf conservation and research where developed. Early proponents of wolf conservation such as Aldo Leopold, Sigurd Olson, and Durward Allen lived and worked in the region. The longest ongoing research on wolf–prey relations (see Vucetich and Peterson, Chap. 3) and the first use of radio telemetry for studying wolves (see Mech, Chap. 2) occurred in the Great Lakes region.

The Great Lakes region is the first place in the United States where "Endangered" wolf populations recovered. All three states (Minnesota, Wisconsin, and Michigan) developed ecologically and socially sound wolf conservation plans, and the federal government delisted the population of wolves in these states from the United States list of endangered and threatened species on March 12, 2007 (see Refsnider, Chap. 21). Wolf management reverted to the individual states at that time. Although this delisting has since been challenged, we believe that biological recovery of wolves has occurred and anticipate the delisting will be restored. This will be the first case of wolf conservation reverting from the federal government to the state conservation agencies in the United States.

In the process of wolf recovery, we have learned much about wolf biology and ecology, endangered species management, carnivore conservation, landscape ecology, depredation management, and social aspects of wildlife conservation. Our book traces wolf recovery in this region and highlights lessons learned by conservationists during the recovery process.

The concept for this book grew out of a well-attended symposium held at the annual meeting of The Wildlife Society in Madison, Wisconsin on September 29, 2005. Many of the authors of the chapters in this book presented portions of their material at that conference. The chapters also cover a broader and more complete range of information than was possible in a half-day symposium. To that end, we recruited additional authors to contribute chapters in the book. These authors are professionals who are or were directly involved in major portions of research and conservation of wolves in the Great Lakes region. Authors represent federal, state, and nonprofit conservation agencies, and universities in the region.

Our goal was to produce a semitechnical book on wolf recovery that is both rigorous with respect to science and policy and accessible and interesting for the lay reader. The story of wolf recovery in the Great Lakes region is one of international significance for conservationists, and wolves themselves are controversial, charismatic, and fascinating on many levels. Each chapter presents a thorough review of the pertinent literature. Some chapters also present new data or new perspectives and interpretations. To maintain scientific rigor, each chapter was reviewed by at least two professionals who are specialists in the relevant fields.

Contributing authors represent a remarkable breadth in professional expertise, and cover topics ranging from ecology to policy to cultural, social, and historic significance of wolves. Moreover, the authors address wolf recovery from diverse perspectives that range from important ecological theory developed and applied by academicians in some of the region's best universities to the on-the-ground, muddy-boot realities of local management pioneered by dedicated conservationists working for public and private agencies. Indeed, we are especially proud of this cross-disciplinary collaboration because it parallels the cross-disciplinary collaboration between researchers, managers, and private conservationists that facilitated wolf recovery in the Great Lakes region.

For the purpose of this book, we assume wolves in the Great Lakes region are mainly gray wolves of the subspecies *Canis lupus nubilus*. However, we recognize some recent research suggests that a mixture of gray wolves (*Canis lupus*) and Eastern wolves (*Canis lycaon*) may exist in the region (see Nowak, Chap. 16).

We thank the reviewers of chapters of the book including the following: Ed Bangs, Robert Beschta, John Bissonette, Luigi Boitani, Paolo Ciucci, Dwayne Etter, Jacqueline Frair, Steve Fritts, Todd Fuller, Tom Gehring, Jon Gilbert, Bob Haight, Paula Hollahan, Mike Jimenez, Paul Krausman, Dolly Ledin, Mark Lenarz, John Linnell, Patty Loew, Angela Mertig, Steve Nadeau, Lisa Naughton, Rolf Peterson, Bill Ripple, Colleen Sculley, John Shivik, Carolyn Sim, Doug Smith, Gus Smith, Richard Tedford, Jörn Theuerkauf, and Dirk Van Vuren. We'd like to thank Springer for publishing this book, Janet Slobodien for encouraging us to produce the book, and Frank McGuckin, Senthil (Balasubramaniyam Senthilkumar), Elizabeth Thompson, and Thomas Brazda for producing the book.

Foreword

I will always remember the morning of January 12, 1995. A light snow swirled across the road as we passed beneath the great stone arch marking the entrance to Yellowstone National Park. Safely inside, we unloaded the crates, transferred them to a horse-drawn sleigh, and then carried them through waist-deep snow into the fenced release enclosure. We stepped back and watched as the wolves emerged, looked about, sniffed the morning air, and loped across the snow, brushing along the fences, ready to break out to freedom.

Within 10 years the original 31 wolves had multiplied to over a hundred. The Yellowstone ecosystem came alive as wolf packs roamed the park whittling the elk herds down. Coyotes declined, allowing populations of red foxes and ground squirrels to recover. The numbers of hawks, ravens, owls, and eagles increased. As overgrazed aspen and willow stands rebounded, beavers and nesting songbirds proliferated.

With such success, many of us believed we had finally arrived at the threshold of a new era in which people and wolves could at last learn to live peaceably together on the western landscape. Today, however, the high optimism of that time is fading as the west once again lapses back into the wolf wars of old.

The Bush administration, deferring to western ranchers, proposes to withdraw federal protection. Western politicians are once again stoking antiwolf hysteria. The Governor of Idaho has announced, "I'm prepared to bid for that first ticket to shoot a wolf myself." Wyoming proposes to classify the wolf as an unwanted predator, inviting residents to kill on sight, whether by shooting or deliberately running them down on highways. Alaska has reinstituted aerial shooting.

As the situation in the western US deteriorates, I now turn to Michigan, Wisconsin, and Minnesota, the western Great Lakes states where a success story is unfolding that is the subject of this fascinating and very important book. Wolf management in these three states, while hardly free of conflict, has proceeded in an atmosphere of state and federal cooperation, strong research, and expansive education and political leadership. And in the process these states have innovated and demonstrated important lessons for the future of all endangered species in every region of the country.

The western Great Lakes experience is so different from the Rocky Mountain west that it hardly seems possible that we are talking about the same subject, in the same century, in one and the same country. And the questions are: why the differences? And what can we learn from this Great Lakes success story?

One might begin by looking to history. The essays in this book tell us that wolf hysteria is as old as the history of settlement, whether in New England, the Midwest or the Rocky Mountains. The big attitudinal change came in the 1960s. In Michigan, Wisconsin, and Minnesota, legislatures enacted strong endangered species legislation. Scientists and environmental groups stepped forward to advocate, making the case for protection and restoration. Aldo Leopold of Wisconsin became an iconic figure for his pioneering research and advocacy.

Out west, these transformational changes were slow to emerge and remain anemic to this day. Political leaders, deferring to public-land ranchers, continue to vilify wolves and all other predators, real and imagined. In my state of Arizona, the posthumous voice of Aldo Leopold goes unheard – in the very state where he described the dying wolf on Escudilla Mountain in the most eloquent and oft-quoted words in all of environmental literature.

Looking to the future, there are many lessons that this book imparts to the west and other regions of the country. The first is that good wolf management begins with good science. Old myths must yield to facts established by quality research. And this book summarizes an astonishing amount of that research into the complex biology, population dynamics, predator–prey relationships, and habitat needs of these animals.

Wolf management is of course not just about wolves; it is equally about us, and our history and culture and attitudes toward nature, what sociologists refer to as the "social construct" that shapes our attitude toward wolves. The social science essays illuminate many issues often overlooked by those of us most interested in the wild-lands and biodiversity side of the equation. While the concept of carrying capacity of the land is quite familiar, most of us are less familiar with the parallel concept of "social carrying capacity" of the people who reside in the region and whose acceptance is essential to wolf survival.

Another theme that should be heard more often out West is the manner in which the Great Lakes state agencies and environmental groups have cooperated to innovate methods of wolf management. The use of interstate agreements, state-federal cooperation, the work with nongovernmental organizations, and efforts at public education should be more widely understood and implemented in other parts of the country, not only for wolf management but also for management of other large predators and endangered species generally.

Yet another important lesson that emerges from the experiences recounted in these chapters is the continuing requirement for political leadership. How many areas should be set aside as parks, wilderness, and protected areas; issues of land use; how to manage wolves in the rural interface between wild and urban areas; and acceptable levels of lethal control: all are in the end political issues. It takes all of us – scientists, advocates, citizens, and political leaders – working together to hash out and formulate what one author refers to as "the rules of co-existence."

The success stories recounted in these chapters are by no means complete. With the recent removal of Midwestern wolves from the federal endangered species list, there will inevitably be pressure for the three states to retreat from their success. The need for advocacy, research, education, and political leadership will continue.

For all these reasons I hope this book will be widely read and referenced, not just in Michigan, Wisconsin, and Minnesota, but even more urgently in the public-land states of the West and indeed throughout the world as other countries awaken to the possibilities of protecting and restoring not only wolves but also many other top predators, and other species now at the brink of extinction.

Bruce Babbitt   Secretary of US Department of the
June 23, 2008   Interior (1993–2001)

Contents

Contributors

Dean P. Anderson
Landcare Research, Lincoln 7640, New Zealand
andersond@landcareresearch.co.nz

Eric M. Anderson
College of Natural Resources, University of Wisconsin Stevens Point,
Stevens Point, WI 54801
eanderso@uwsp.edu

Karlyn Atkinson Berg
Wolf/predator conservation consultant to HSUS, Bovey, MN 55709
karlyn@uslink.net

Dean E. Beyer, Jr.
Michigan Department of Natural Resources, Northern Michigan University,
Department of Geography New Science Facility, Marquette, MI 49855
dbeyer@nmu.edu

Peggy Callahan
Wildlife Science Center, Forest Lake, MN 55025
peggy@wildlifesciencecenter.org

Murray K. Clayton
Departments of Plant Pathology and Statistics, University
of Wisconsin – Madison, Madison, WI 53706
clayton@stat.wisc.edu

Peter David
Great Lakes Indian Fish and Wildlife Commission, Odanah, WI 54861 USA
pdavid@glifwc.org

Glenn D. DelGiudice
Minnesota Department of Natural Resources, Grand Rapids, MN 55744 USA
glenn.delgiudice@dnr.state.mn.us

Michael W. DonCarlos
Minnesota Department of Natural Resources, St. Paul, MN 55155 USA
Mike.DonCarlos@dnr.state.mn.us

John Erb
Minnesota Department of Natural Resources, Grand Rapids, MN 55744
john.erb@dnr.state.mn.us

Wayne Hall, Jr.
Wisconsin Department of Natural Resources, Sandhill Wildlife Area,
Babcock, WI 54413, USA
wayne.hall@wisconsin.gov

James H. Hammill
Department of Natural Resources, Crystal Falls, MI 49920, USA
jimhammill@hughes.net

Ellen Heilhecker
New Mexico Department of Game and Fish, Reserve, NM 87830, USA
elleheilhecker@yahoo.com

Edward J. Heske
Illinois Natural History Survey, Champaign IL 61821 USA
eheske@uiuc.edu

Holly Jaycox
Wolf Park, Battle Ground, IN 47920 USA
wolfmagazine@wolfpark.org

Randy L. Jurewicz
Wisconsin Department of Natural Resources, Madison, WI 53707, USA
Randle.Juewicz@wsconsin.gov

Paul W. Keenlance
Biology Department, Grand Valley State University, Allendale, MI 49401 USA
keenlancep@mail.ab.edu

Bruce E. Kohn (retired)
Wisconsin Department of Natural Resources, Rhinelander, WI 54501 USA
wolfie2@newnorth.net

Donald H. Lonsway
USDA/APHIS/Wildlife Services, Ironwood, MI 49938 USA
donald.lonsway@aphis.usda.gov

Kerry A. Martin
Nelson Institute for Environmental Studies, University of Wisconsin – Madison,
Madison, WI 53706 USA
kerryamartin@gmail.com

Keith R. McCaffery (retired)
Wisconsin Department of Natural Resources, Rhinelander, WI 54501 USA
keithmccaffery@wisconsin.gov

L. David Mech
US Geological Survey, University of Minnesota, St. Paul, MN 55108 USA
david_mech@usgs.gov, mechx002@umn.edu

Curt Meine
Aldo Leopold Foundation/International Crane Foundation, Prairie du Sac,
WI 53578 USA
curt@savingcranes.org

David J. Mladenoff
Department of Forestry and Management, University of Wisconsin – Madison,
Russell Labs, Madison, WI 53706 USA
djmladen@facstaff.wisc.edu

Michael E. Nelson
USGS Northern Prairie Wildlife Research Center, Kawishiwi Field Laboratory,
Ely, MN 55731 USA
michaelnelson@usgs.gov

Ronald M. Nowak
Falls Church, VA 22043 USA
RON4NOWAK@cs.com

William J. Paul
USDA/APHIS/Wildlife Services, Grand Rapids, MN 55744 USA
bill.j.paul@aphis.usda.gov

Rolf O. Peterson
School of Forest Resources and Environmental Science,
Michigan Technological University, Houghton, MI 49931 USA
ropeters@mtu.edu

Sarah D. Pratt
Department of Forest and Wildlife Ecology, University of Wisconsin – Madison,
Madison, WI 53706 USA
sdpratt@wisc.edu

Ronald L. Refsnider (retired)
US Fish and Wildlife Service, Federal Building, Fort Snelling, MN 55111 USA
ron_refsnider@fws.gov

Brian J. Roell
Michigan Department of Natural Resources, South Marquette, MI 49855 USA
roellb@michigan.gov

Thomas P. Rooney
Department of Biological Sciences, Wright State University, Dayton,
OH 45435 USA
thomas.rooney@wright.edu

David B. Ruid
USDA/APHIS/Wildlife Services, Rhinelander, WI 54501, USA
david.ruid@aphis.usda.gov

Kevin Schanning
Northland College, Ashland, WI 54806 USA
KSchanning@northland.edu

Ronald N. Schultz
Wisconsin Department of Natural Resources, Woodruff, WI, 54568 USA
ronald.schultz@wisconsin.gov

Theodore A. Sickley
Department of Forest and Wildlife Ecology, University of Wisconsin – Madison,
Madison, WI 53706 USA; National Geographic Society, Washington,
DC, 20036 USA
tsickley@gmail.com

Andrea Lorek Strauss
International Wolf Center, Ely, MN 55731 USA
astrauss@umn.edu

Peggy Struhsacker
National Wildlife Federation, Montpelier, VT 05602 USA
pstrusacker@nrdc.org

Richard P. Thiel
Department of Natural Resources, Sandhill Wildlife Area, Babcock,
WI 54413 USA
richard.thiel@wisconsin.gov

Adrian Treves
Nelson Institute for Environmental Studies, University of Wisconsin-Madison,
Madison, WI 53706 USA
atreves@wisc.edu

Pamela S. Troxell (deceased)
Timber Wolf Alliance, Sigurd Olson Environmental Institute, Northland College,
Ashland, WI 54806 USA

David E. Unger
Division of Natural Sciences, Alderson Broaddus College, Philippi,
WV 26416 USA
dunger@uky.edu

Timothy R. Van Deelen
Department of Forest and Wildlife Ecology, University of Wisconsin – Madison,
Madison, WI 53706 USA
trvandeelen@wisc.edu

John A. Vucetich
School of Forest Resources and Environmental Science,
Michigan Technological University, Houghton, MI 49931 USA
javuceti@mtu.edu

Jane E. Wiedenhoeft
Wisconsin Department of Natural Resources, Park Falls, WI 54452 USA
jane.wiedenhoeft@wisconsin.gov

Robert C. Willging
USDA/APHIS/Wildlife Services, Rhinelander, WI 54501 USA
Robert.C.Willging@usda.gov

Adrian P. Wydeven
Wisconsin Department of Natural Resources, Park Falls, WI 54452 USA
adrian.wydeven@wisconsin.gov

Chapter 1
Early Wolf Research and Conservation in the Great Lakes Region

Curt Meine

1.1 Introduction

The history of wolf research and conservation in the upper Great Lakes is only one chapter in the epic story of evolving relationships between people and land in North America. It is, however, an especially significant chapter. The rapid pace of Euro-American settlement and environmental transformation from the early 1800s to the mid-1900s led (among other impacts) to the near extirpation of the wolf from the region. During this same period, however, the American conservation movement arose in response to reckless resource exploitation. Shifts in conservation science, policy, and philosophy allowed the wolf to be understood within a broader ecological and ethical framework, preparing the way for the recent recovery of the species in the region. In this way, the fate of the wolf in the Great Lakes has reflected broader trends in the history of conservation.

Since its historic low point in the mid-1900s, the wolf population of the Great Lakes region has recovered due to two overriding factors: ecological conditions of the landscape have been conducive to the population's growth and expansion; and the knowledge, values, and actions of the region's people have provided space – on the ground and within our human society – for such growth and expansion to occur. The natural and cultural history of wolves in North America and around the world has been well told in both popular and professional publications (e.g., Lopez 1978; Mech and Boitani 2003). Other chapters in this volume provide accounts of the history of wolves in the Great Lakes states. This chapter provides a brief overview of early wolf research and conservation efforts in the region.

1.2 Wolf Research and Conservation in Historical Context

The science of wildlife ecology and the practice of wildlife management fully emerged only in the late 1930s. This was several decades after conservation first found traction as a public concern and a policy goal during the presidency

A.P. Wydeven et al. (eds.), *Recovery of Gray Wolves in the Great Lakes Region of the United States*,
DOI: 10.1007/978-0-387-85952-1_1, © Springer Science+Business Media, LLC 2009

of Theodore Roosevelt (1901–1909), and several decades before the modern environmental movement reconfigured the older conservation movement in the 1960s and 1970s. More recently, interdisciplinary fields such as conservation biology, landscape ecology, environmental history, and environmental ethics have transformed the scientific foundations of conservation as well as our understanding of its social and cultural context (Knight and Bates 1995; Meine 2004). Through all these changes the wolf has served as an indicator species, telling us much about the status not only of our ecosystems but also of our scientific knowledge and our evolving conservation ethic. To track the history of the wolf in the Great Lakes is to follow a trail through the heart of American conservation history.

Knowledge of wolf biology, ecology, behavior, and populations in the Great Lakes of course predates the arrival of modern Euro-Americans (see David, this volume). Humans and wolves shared the Great Lakes regional landscape for at least ten postglacial millennia, the populations of both responding to changes in climate, flora, and fauna. The shifting presence of (and relationships among) native communities influenced the numbers and distribution of wolves and other wildlife species (Hickerson 1970). How, where, and to what degree that influence played out over time may never be known with precision. However, the basic fact that wolves and humans coexisted for so long suggests that, however native people conceived their ethic, it was sufficient to accommodate large predators within the land community. Wolves and other creatures were meaningful inhabitants not only of the land but also of the myths, stories, and traditions of the people. For the Ojibwe and other tribes of the region, wolves and people belonged to, and were connected within, the same moral community (Callicott and Nelson 2004).

As Euro-American explorers, missionaries, trappers, and settlers moved into the mid-continent, economies based on the commodification of nature transformed the region's human-land relationships (Cronon 1991). In the increasingly humanized and privatized landscape, wolves and other large predators began their long slide toward near-extirpation. The loss of viable wolf populations proceeded along with the widespread disruption and depletion of the region's other natural assets: its pine (and later hardwood) forests, wetlands, grasslands and savannas, its game populations, and Great Lakes and inland fisheries.

The arrival of Euro-Americans also brought a more beneficent force to the landscape. A strong tradition of education and scientific inquiry was activated in the new institutions of the emerging states – museums, public schools and universities, scientific organizations, and historical societies. These provided important cultural underpinnings for the nascent conservation movement. Even as the region's resources were diminishing, the response was being born. The deforestation of the Great Lakes pineries provided strong impetus to the national forestry movement. Under Roosevelt and his chief forester Gifford Pinchot, this burgeoned into a broader conservation movement in the first decade of the 1900s. Michigan, Wisconsin, and Minnesota, having experienced the profligate exploitation of their land resources, emerged as national leaders in the movement.

The utilitarian cast of the early conservation movement reflected its pre-ecological origins and its narrow economic premises. The persecution of predators was already deeply embedded in a culture that viewed them mainly as "vermin" and "varmints." Beginning in 1914, predator control became national policy. Congress directed the US Bureau of Biological Survey (BBS) to undertake formally what until then had been a sideline for federal foresters, range managers, and game wardens (Dunlap 1988). The removal of wolves and other predators was akin to the suppression of fire, the damming of rivers, and the plowing under of the prairies: all served the goal of rational management of nature for human benefit through maximum efficiency and the sustained yield of resources. Science of a sort informed that rational management approach. But a new kind of science – one that stressed the diversity, functioning, and interrelationships inherent in natural systems – would soon start to call into question the aims of purely utilitarian conservation. And it would lead at least some conservationists to reevaluate the role and value of wolves and other predators.

1.3 Shifting Policies and Emerging Insights

The concerted campaign against predators that began in 1914 accelerated into the 1920s. The BBS was highly successful in its efforts to remove remnant populations of large predators from the mountains and rangelands of the American West. The Bureau found itself less devoted to its prior mission of research, and increasingly in the business of predator eradication. Among scientists (both within and outside the agency) a split developed between those who regarded predators first and foremost as agents of destruction and those who viewed them as creatures of scientific interest and value. This tension erupted into open dispute in the mid-1920s, pitting leading members of the American Society of Mammalogists (ASM) against BBS administrators and field agents (Dunlap 1988). Into the 1930s the annual meetings of the ASM served as the arena for pitched debate over the perpetuation of large carnivores in the American landscape.

Through these same years, one of the most consequential events in the annals of wildlife management played out on the Kaibab Plateau, on the north rim of the Grand Canyon (Young 2002; Binkley et al. 2006). After the Kaibab was designated a national game reserve in 1906, deer hunting was curtailed and livestock grazing restricted, and the BBS set about removing the plateau's wolves, mountain lions, coyotes, and bobcats. The deer herd swelled. By the early 1920s foresters were reporting the damaging effects of the superabundant deer on forest and range vegetation. The irruption of the Kaibab deer herd occurred before the science of wildlife ecology existed. Reliable techniques of game censusing and range assessment had yet to be developed. However, the informal evaluations and visual inspections of the rangers, foresters, and biologists were convincing enough. Among professional resource managers and the public alike, the Kaibab episode became a starting point for reconsideration of the role of predators.

Historian Thomas Dunlap (1988, p. 43) noted that "the ecology of large predators and their prey presented technical problems that would discourage researchers for

... years. Wolves ranged widely and swiftly over enormous areas of forested wilderness land. Even counting them was a formidable task; discovering their relations to other species and to the environment was even harder." The thin science involving wolves of the Great Lakes region as of the late 1920s demonstrates the point. Published research, such as it was, consisted of descriptive accounts, local reports of occurrences and extirpations, records within site-specific mammal inventories, and the occasional anecdote involving noteworthy wolf behavior. Reflecting its focus on "economic mammalogy," the BBS had published Vernon Bailey's "Destruction of Deer by the Northern Timber Wolf" in 1907 – and little of scientific relevance on the topic since (Bailey 1907).

Even Aldo Leopold in his obscure but critical *Report on a Game Survey of the North Central States* (1931), which included a chapter on predators, had essentially nothing to report on the subject of wolves. Leopold's vehement antipredator stance, evident when he served with the US Forest Service in the Southwest, had shifted. He began keeping his own office file on wolves and coyotes in the mid-1920s. He knew most of the main actors in the ASM/BBS debates over predator policy. He was also acquainted with many of the biologists and foresters who were trying to understand the cascading ecological effects on the Kaibab Plateau. By the late 1920s, Leopold was publicly expressing tolerance for predators. As the chief author of a new American game policy report, a key document in the development of professional wildlife management, Leopold recommended that "no predatory species should be exterminated over large areas" and that "rare predatory species ... should not be subject to control" (Leopold 1929).

Although Leopold's *Game Survey* contained scant reference to wolves, his fieldwork in preparing the report provided support for his evolving attitude toward predators. He wrote after his survey of Missouri in 1930: "Predators show no alarming trends. All past and present ideas about predator-control seem inadequate. A rational policy must be built up on a foundation of scientific facts yet to be determined" (quoted in Meine 1988, p. 274). As the field of wildlife management emerged in the 1930s, Leopold, his students, colleagues, and contemporaries supplied the first layers of that foundation.

1.4   "... A New Appreciation of Carnivores and the Role They Play"

By the early 1930s, university and agency biologists had begun to probe more deeply into the phenomenon of predation. Paul Errington, a University of Wisconsin graduate student whom Leopold began advising in 1929, had begun long-term research on northern bobwhite quail populations. This led him to focus on the impact of predation relative to other factors affecting quail productivity. In addressing that issue, Errington called into question long-held assumptions about the supposed destructive effect of predators on prey populations (Errington 1934). His research showed that predation was only one of many factors that together determined a prey population's fortunes. Errington's study quickly became a cornerstone in the scientific study of predation, and initiated his own career-long focus on the topic (Errington 1967).

As Errington was studying quail in the Midwest, Olaus Murie was beginning to contribute his voice and field experience to the predator debate. Murie was a BBS biologist who in 1927 had begun research on predators in Jackson Hole, Wyoming – specifically, the impact of coyote predation on elk populations (Murie 1935). His findings put him at odds with many in the BBS. For years to come, Murie would be a staunch critic of his own agency's predator control policies. After reading the predation chapter of the *Game Survey* report, Murie wrote to Leopold about his own field studies: "Personally, I have felt that too much attention has been given to the predatory animal factor. …I do not find the coyote a bad fellow at all. As far as the elk are concerned, he is not nearly as big a factor as several other things" (quoted in Meine 1988, p. 286).

The personal communications with Errington, Murie, and other trusted informants provided Leopold with the material he needed to define more clearly his own take on predation. He expressed it most clearly in his book *Game Management* (1933), the first text in the new field. In his discussion of predation, Leopold appealed to all parties in the debate to acknowledge the complexity of the issue and to maintain an attitude of "fairness" and open-minded curiosity. "There is only one completely futile attitude on predators," Leopold wrote, "that the issue is merely one of courage to protect one's own interests and that all doubters and protestants are merely chicken-hearted" (Leopold 1933, p. 252). In *Game Management*, Leopold offered few references to wolves, reflecting the still-thin body of solid information. He alluded to the wolf's breeding potential, its capacity for recolonization, and, significantly, the positive influence that "normal depredation" by wolves may have on deer distribution. Ever aware of the need for deeper research, Leopold noted that "many possible predator influences [are] as yet beyond our vision" (Leopold 1933, p. 247).

The evolution of Leopold's attitude toward predators advanced quickly after the mid-1930s, through experiences in two very different landscapes: the intensively managed forests of central Europe and the semiarid woodlands of northern Mexico's Sierra Madre (Flader 1974). Traveling across Germany and Czechoslovakia in 1935, Leopold examined the long and intertwined history of forestry and game management in the mid-continent – and the ills that resulted from their inability to reconcile competing resource management goals. "We Americans," he later wrote, "in most states at least, have not yet experienced a bearless, wolfless, eagleless, catless woods. We yearn for more deer and more pines, and we shall probably get them. But do we realize that to get them, as the Germans have, at the expense of their wild environment and their wild enemies, is to get very little indeed?" (Leopold 1936). In Mexico, by contrast, Leopold experienced prey populations thriving amid normal predator populations, reflecting (as he would later phrase it) "a biota still in perfect aboriginal health."

Over the next decade, Leopold continually elaborated and refined this new concept of "land health." It became the focus of his scientific research, a companion concept to his "land ethic," and an all-encompassing measure of conservation success (Newton 2006). Central to the concept was an understanding of the role of predators in the healthy functioning of ecological communities. For Leopold and his like-minded colleagues, the fate of the wolves in the Great Lakes region would provide a critical example – and test – of the concept in the years to come.

First, however, the "foundation of scientific facts" needed further bolstering. The next significant contribution to that foundation came through the work of Sigurd Olson on wolves and coyotes in the Superior National Forest. Like Leopold, Olson's view of predators shifted dramatically from outright hostility to appreciation to advocacy (Backes 1997). His early antipathy toward wolves began to change when he decided to pursue graduate studies. After an opportunity to study under Leopold fell through, Olson signed on with Victor Shelford at the University of Illinois. Shelford, a pioneering animal ecologist, a leading voice for the protection of natural areas, and an occasional participant in the ASM/BBS debates, guided Olson in his unprecedented study.

Olson undertook his field research in northeastern Minnesota in December 1930. Previously a proponent of wolf-control measures, Olson was skeptical of the effectiveness of poisoning and trapping due to the continual influx of wolves into Minnesota from the north. By the time he completed his thesis – "The Life History of the Timber Wolf and the Coyote: A Study in Predatory Animal Control" – Olson had come to question not only the wisdom of control techniques but also the cultural stereotype of predators that had motivated the control programs.

In 1938 Olson published two articles based on his thesis (Olson 1938a, b). "Organization and range of the pack" appeared in the journal *Ecology* and "A study in predatory relationship with particular reference to the wolf" in *Scientific Monthly*. In the introduction to the latter, Olson emphasized the ecological and aesthetic significance of large predators as indicators of the vitality of wildlands.

> With the fast-growing appreciation of the true meaning of wilderness, we are beginning to question the idea of the total elimination of predators, realizing that, after all, lions, wolves and coyotes may be an exceedingly vital part of a primitive community, a part of which once removed would disturb the delicate ecological adjustment of dependant types and take from the country a charm and uniqueness which is irreplaceable. To go into a region where the large carnivores are gone, to see hoofed game with its natural alertness lacking, to know above all that the primitive population has been tampered with, is like traveling through a cultivated estate. Wilderness in all its forms is what the true observer wants to see and with this realization dawns a new appreciation of carnivores and the role they play. (Olson 1938b, p. 324)

In the concluding section of his paper in *Scientific Monthly*, Olson indicated just how thoroughly he had rejected his own youthful aversion to wolves, and adopted the scientific language of Shelford's community ecology.

> The presence of the timber wolf in the Superior Area, instead of being a hazard, is a distinct asset to big game types. Long investigation indicates that the great majority of the killings [is] of old, diseased, or crippled animals. Such purely salvage killings are assuredly not detrimental to either deer or moose, for without the constant elimination of the unfit the breeding stock would suffer. Furthermore, the wolf is a natural stimulus to a

herd's alertness and injects the primitive element of danger without which most big game animals lose much of their natural charm. (Olson 1938b, p. 335)

The timber wolf is an integral part of the wilderness community, the destruction of which would destroy the fine balance between related forms. To eliminate as vital a relationship as exists between predatory forms and the animals they prey upon, to destroy a mutual dependence, means that artificiality has entered the wilderness picture. (Olson 1938b, p. 336)

Olson's paper reflected Shelford's influence not only as an ecologist but also as a conservation advocate. Backes (1997, p. 88) notes, "Under Shelford's tutelage … Olson joined the ranks of the wolf advocates. … He concluded with a call – radical for its time – to designate the canoe country's Superior National Forest as a carnivore sanctuary." Olson's study, basic by today's standards, and beholden to the idea of natural "balance" in a way that even his contemporaries had begun to move away from, was nonetheless a milestone. It stands as the first in-depth scientific study of wolves in the region, and perhaps in the world. The findings of that research also provided one of the first widely published calls to conserve the species.

1.5   Green Fire Dying

Errington's pioneering studies of predation, Murie's challenge to predator control orthodoxy, Leopold's reframing of the larger debate over predators, and Olson's original thesis on the wolves of Minnesota – all were indicators of a changing relationship between wildlife science and wildlife policy in the 1930s. Establishment of the Wildlife Society in 1937 symbolized the emergence of wildlife ecology and management within the family of resource management professions. The new field bridged the gap between basic wildlife research and the pragmatic work of the state and federal resource management agencies. The implications for conservation of wolves and other large predators were far-reaching, not only in regions like the Great Lakes where they had grown scarce but also in parts of North America were they still thrived.

The pace of policy debates and research on predators began to accelerate in the late 1930s and early 1940s. In 1937, E. A. Goldman of the BBS published "The wolves of North America" in the *Journal of Mammalogy* (Goldman 1937). Goldman's taxonomic review of the species provided the basis for an expanded book, also entitled *The Wolves of North America*, coauthored with Stanley P. Young and published in 1944 – at that point, the authoritative scientific text on the species (Young and Goldman 1944).

In the West, Olaus Murie's younger brother Adolph completed a ground-breaking ecological study of the coyote in Yellowstone National Park (Murie 1940). Adolph's thorough investigation served to taint his reputation within the National Park Service much as Olaus had been stigmatized within the BBS (Dunlap 1988). Adolph followed up the Yellowstone study by taking his advanced research methods back into the field, this time to examine the ecology of Alaskan wolves at Mt. McKinley (Denali). The results of Murie's work appeared in 1944 as *The*

Wolves of Mt. McKinley. It was the most advanced examination of wolf ecology, life history, behavior, and social dynamics yet undertaken, and its findings influenced wolf protection and restoration efforts nationwide, not the least in the Great Lakes. Commenting on Murie's work, Leopold noted with some understatement that "the publication of authoritative prey-predator studies, like that now given us by Murie, is of great importance to sound conservation" (Leopold 1945b).

For the foresters, wildlife managers, landowners, conservationists, and sportsmen of the upper Great Lakes at the time, greater scientific understanding of wolf ecology was of immediate and pressing importance. Wolves were at the center of a complex, interlocking set of conservation issues (Flader 1974). As the cutover forests of the north began to recover, the deer herd grew quickly, threatening widespread damage to the regenerating forest. Signs of such damage, and of stressed deer populations, were evident in Wisconsin as early as the mid-1930s, becoming even more apparent by the early 1940s. In Wisconsin, wolves were restricted to a few remote northern locations, and in some areas had shown signs of rebounding, but remained subject to state bounties (Thiel 1993). The Great Lakes became the venue for debate over new ideas, new approaches, and new conflicts over conservation policy. The region's wolves were at the epicenter of that debate.

In February 1941, Leopold received a letter from William Hamilton, a zoologist from Cornell University then serving as Chair of the ASMs' Committee on Conservation of Land Mammals (on which Leopold was serving at the time). Hamilton asked Leopold to indicate priorities for committee action. Leopold responded, "I think the most pressing issue in this region is the one of wolf policy. All of the lake states as far as I know continue an official policy of wolf extermination, despite the fact that excess deer are a growing menace to forestry, to conservation of flora, and to their own welfare. I, for one, think the time has come to begin an earnest agitation for reversal of such antiquated policies."

Leopold, with others, had taken steps in Wisconsin to do just that by initiating a research project led by William Feeney ("a full-time deer man who is fully sympathetic with our viewpoint and one who can eventually muster the facts to support it in public debate"). Leopold was also in contact with colleagues in the Michigan and Minnesota conservation departments, and fully understood the regional significance of such research. "The time would seem to be right," he wrote to Hamilton, "for a lake states 'bloc' to advocate reform. I am not sure whether the supposed opposition of sportsmen and of agricultural interests may not be imaginary." Feeney, beginning his work in March 1941 in northern Wisconsin, was not so assured. He wrote to Leopold, "Most of the field workers, wardens, rangers, lumbermen, and settlers are not very receptive to the wolf deer-control idea and do not rate wolves valuable, esthetically or otherwise, except for the bounty they bring." Two weeks later, Feeney wrote again: "We have not yet covered the entire State adequately but it appears that certain tracts in Forest County are, or soon will be, known as the last stand of timber wolves in Wisconsin …. Some of the old-timers state rather convincingly that the timber wolf is doomed to extermination because of logging operations and that they will eventually go out with the timber, regardless of other factors."

A. M. Stebler of the Michigan Department of Conservation concurred with Feeney's observation. "With very few exceptions," Stebler wrote to Hamilton, "we are constantly faced with rather determined opposition toward our ideas. Judging from our experiences it is going to be some time before the public at large really understands predators and their role in the scheme of things." Stebler would make his own key contribution to early wolf conservation literature in 1944, when his article, "Status of the Wolf in Michigan," appeared in the *Journal of Mammalogy* (Stebler 1944). He concluded, soberly, that the available evidence indicated that "the wolf is in real danger of becoming entirely extirpated in Michigan." Widespread clearing and settlement of Michigan's forestland had restricted the wolf to the Upper Peninsula, and even there to "only a few remaining large wilderness areas .... The modification of the primitive habitat by man may have had more of an effect in reducing wolf range and numbers than all the control measures that have been attempted." The earlier studies of Errington and Olaus Murie, Stebler stated, "show plainly that it is possible for both predator and prey species to live together without apparent disadvantage to either .... Considered from a long range viewpoint, predation is not necessarily a harmful influence upon prey species." In a statement that was undoubtedly difficult for a state wildlife biologist to make at the time, Stebler opined that "the loss of so spectacular and notorious a member of the State's native fauna would be unfortunate, to say the least. ... To forestall, or prevent, the passing of the wolf in Michigan, or for that matter, in the Great Lakes' region generally, what measures can be taken?"

That question bedeviled the small community of Great Lakes wolf researchers and advocates through the World War II years. It was intertwined, intimately and inherently, with the emotional issue of deer management. Leopold, serving on the Wisconsin Conservation Commission, dealt with both matters in the public arena, arguing for more liberal deer seasons and for lifting the bounty on wolves (Flader 1974). Through other connections, Leopold supported Feeney's research on Wisconsin's remaining wolves, and used that information to push for reforms in Wisconsin's deer and predator-control policies. Amid the intense political forces swirling around these issues, Leopold agued for lifting the wolf bounty in 1944, then found himself in the awkward position, as a commissioner, of having to vote to reinstate it a year later (Flader 1974).

At the height of Wisconsin's "deer wars," Leopold stepped back from the fray and expressed his mature perspective on predators in more poetic terms. In April 1944, he drafted his famed essay "Thinking Like a Mountain," in which he poignantly expressed the ecological lessons he had garnered since he himself had led the charge against wolves years earlier.

> We reached the old wolf in time to watch a fierce green fire dying in her eyes. I realized then, and have known ever since, that there was something new to me in those eyes—something known only to her and to the mountain. I was young then, and full of trigger-itch; I thought that because fewer wolves meant more deer, that no wolves would mean hunters' paradise. But after seeing the green fire die, I sensed that neither the wolf nor the mountain agreed with such a view. (Leopold 1949, p. 130)

Published posthumously in *A Sand County Almanac*, Leopold's essay would eventually carry the new perspective on predators and prey, people and land, human cause, and ecological effect, to a global readership. Leopold's highly personal account was forged in a crucible defined by the wolves of the Great Lakes region and their tenuous fate.

1.6 "Shall We Save Our Larger Carnivores?"

Soon after drafting "Thinking Like a Mountain," Leopold reviewed Young and Goldman's *The Wolves of North America* (Leopold 1945a). He did not spare ink in his direct criticism.

> Viewed as conservation, *The Wolves of North America* is, to me, intensely disappointing. The next to the last sentence in the book asserts: 'There still remain, even in the United States, some areas of considerable size in which we feel that both the red and gray [wolves] may be allowed to continue their existence with little molestation.' Yes, so also thinks every right-minded ecologist, but has the United States Fish and Wildlife Service no responsibility for implementing this thought before it completes its job of extirpation? Where are these areas? Probably every reasonable ecologist will agree that some of them should lie in the larger national parks and wilderness areas; for instance, the Yellowstone and its adjacent national forests.

Leopold then asked a radical question: "The Yellowstone wolves were extirpated in 1916, and the area has been wolfless ever since. Why, in the necessary process of extirpating wolves from the livestock ranges of Wyoming and Montana, were not some of the uninjured animals used to restock the Yellowstone?"

Leopold's review is regularly cited for its remarkably early recommendation for the restoration of wolves. It was not an isolated proposal. Leopold drafted his book review in August 1944. That same month he wrote to Newton Drury, Director of the National Park Service (NPS), advocating the introduction of wolves to Isle Royale National Park. Drury shied away from the idea, citing "the possibility for adverse public reaction that might do harm to the conservation of an adequate stock of wolves in the lake states region."

Over the next several years, however, Leopold continued to discuss the potential for introducing wolves to Isle Royale with Victor Cahalane, an NPS biologist based in Chicago. Leopold and Cahalane shared ideas and information about the occurrence of large predators in other national parks. In September 1946, Cahalane shared the news that fresh wolfs tracks had been found in Yellowstone. Encouraged perhaps by this event and his correspondence with Leopold, Cahalane published that year in *Living Wilderness* (the magazine of the Wilderness Society) an article entitled "Shall we save the larger carnivores?" (Cahalane 1946). Meanwhile, the prospect of translocating wolves to Isle Royale continued to intrigue Leopold and Cahalane. In 1947, Cahalane made arrangements for Leopold to visit Isle Royale as a consultant, but Leopold had to forego the opportunity for health reasons (Meine 1988).

In his response to Cahalane's news of the Yellowstone wolf, Leopold wrote, "I am letting Bill Feeney and Dan Thompson see your letter. … They will understand that the information is confidential. Both of them share our views about wolves."

Feeney had continued his research on deer–wolf interactions in Wisconsin's north-woods (Thiel 1993). Dan Thompson was a new student of Leopold's, recently returned from the war, who was beginning graduate research on wolf food habits, movements, and population indices.

Leopold died in April 1948, suffering a heart attack while fighting an escaped grass fire at his "sand county" farm. In a real sense, however, Leopold's influence on wolf research and conservation was only beginning to be felt. *A Sand County Almanac* went to press in 1949. Another student of Leopold's, Anton DeVos, also contributed to the literature on wolves of the Great Lakes region, reporting from north of the border in Ontario (DeVos 1949, 1950; DeVos and Allin 1949). Thompson completed his dissertation in 1950, publishing his research in the *Journal of Mammalogy* as "Travel, range, and food habits of timber wolves in Wisconsin" (Thompson 1952). His findings and conclusions echoed those of Stebler's from Michigan, but were distinguished by the strong emphasis he placed on land use and the need to maintain large blocks of road-free forest. "Certain land-use problems and relationships," Thompson wrote, "indicate the precarious state of this species in much of the Lake States area at the present time." He recommended "maintain[ing] areas of at least 150 square miles as wilderness habitat" and surmised that "the timber wolf will eventually be extirpated from Wisconsin" unless such steps were taken.

The most significant development in the story of Great Lakes wolves in the late 1940s occurred far from meeting rooms, agency offices, and academic corridors. Sometime (apparently) in the winter of 1948–1949, an adventurous band of timber wolves set out from Minnesota's north shore, crossed the Lake Superior ice, and arrived on the hard rock shores of Isle Royale. The plans that Leopold and Cahalane explored for bringing wolves to Isle Royale turned out to be unnecessary. In colonizing Isle Royale, the wolves unwittingly opened wide a new chapter in the history of their own ecology and conservation. With their arrival they transformed Isle Royale (and the region in general) into a prime laboratory for the next generation of wolf researchers – Milt Stenlund, Durward Allen, David Mech, and those who would follow in their footsteps (Allen 1979; Peterson 1995; see Vucetich and Peterson, this volume). All, in fact, were following the tracks of the persistent wolves of Minnesota, one small band of which elected to disperse across the frozen water.

1.7   Wolves of the Great Lakes Region and the Extension of Conservation Thinking

At the end of his thesis, Dan Thompson suggested several specific steps that could be taken to maintain suitable wolf habitat in northern Wisconsin: avoid fragmentation, restrict access along fire lanes in the forests, and adhere to rural zoning rules. In 1952, Wisconsin's wolf bounty was still in place. Thompson noted that "some form of legal protection is probably already necessary to perpetuate the timber wolf in Wisconsin; but public opinion today is, of course, unprepared for such an extension of conservation thinking" (Thompson 1952).

Thompson and all his predecessors, colleagues, and contemporaries who first asked new questions about predation and about the wolves of the Great Lakes – Errington, Olaus and Adolph Murie, Leopold, Olson, Cahalane, Feeney, and Stebler – had worked to build a "foundation of scientific facts." They had little reason to expect that their work would prompt the very "extension of conservation thinking" that allowed the region's wolves not only to endure but also to recover.

In 1933, in *Game Management*, Aldo Leopold had posited that there was "social significance" to be found in this new branch of conservation – but that the field itself would need to "[expand] with time into that new social concept toward which conservation is groping" (Leopold 1933, p. 423). Few other species, or places, would contribute so importantly to that new social concept as the wolves in the Great Lakes. In giving wolves the time and space to survive, the people of the region found themselves, too, in a new relationship within the larger land community.

Note

Letters quoted in this chapter are all located in the Aldo Leopold Papers, University of Wisconsin Archives, University of Wisconsin-Madison.

References

Allen, D. L. 1979. Wolves of Minong: Their Vital Role in a Wild Community. Boston, MA: Houghton Mifflin.

Backes, D. 1997. A Wilderness Within: The Life of Sigurd F. Olson. Minneapolis, MN: University of Minnesota Press.

Bailey, V. 1907. Destruction of Deer by the Northern Timber Wolf. U.S. Department of Agriculture, Biological Survey Circular 58: 1–2.

Binkley, D., Moore, M., Romme, W., and Brown, P. 2006. Was Aldo Leopold right about the Kaibab deer herd? Ecosystems 9: 227–241.

Cahalane, V. 1946. Shall we save the larger carnivores? Living Wilderness 17: 17–22.

Callicott, J. B., and Nelson, M. P. 2004. American Indian Environmental Ethics: An Ojibwa Case Study. Upper Saddle River, NJ: Pearson Prentice-Hall.

Cronon, W. 1991. Nature's Metropolis: Chicago and the Great West. New York, NY: Norton.

DeVos, A. 1949. Timber wolves killed by cars on Ontario highways. Journal of Mammalogy 30: 197.

DeVos, A. 1950. Timber wolf movements on Sibley Peninsula, Ontario. Journal of Mammalogy 31: 169.

DeVos, A., and Allin, A. E. 1949. Some notes on moose parasites. Journal of Mammalogy 30: 430–431.

Dunlap, T. 1988. Saving America's Wildlife. Princeton, NJ: Princeton University Press.

Errington, P. L. 1934. Vulnerability of bob-white populations to predation. Ecology 15: 110–127.

Errington, P. L. 1967. Of Predation and Life. Ames, IA: Iowa State University Press.

Flader, S. 1974. Thinking Like a Mountain: Aldo Leopold and the Evolution of an Ecological Attitude Toward Deer, Wolves, and Forests. Columbia: University of Missouri Press.

Goldman, E. A. 1937. The wolves of North America. Journal of Mammalogy 18: 37–45.

Hickerson, H. 1970. The Chippewa and Their Neighbors: A Study in Ethnohistory. New York, NY: Holt, Rinehart and Winston.

Knight, R., and Bates, S., eds. 1995. A New Century for Natural Resources Management. Washington, DC: Island Press.

Leopold, A. 1929, December 2–3. Report of the Committee on American Wild Life Policy. Transactions of the Sixteenth North American Game Conference, 196–210.

Leopold, A. 1933. Game Management. New York, NY: Charles Scribner's Sons.

Leopold, A. 1936. Naturschutz in Germany. Bird-Lore 3: 102–111.

Leopold, A. 1945a. Review of Stanley P. Young and Edward H. Goldman. The Wolves of North America. Journal of Forestry 43: 928–929.

Leopold, A. 1945b. Wolves. Unpublished manuscript. Aldo Leopold Papers, University of Wisconsin Archives, 20 March.

Leopold, A. 1949. A Sand County Almanac and Sketches Here and There. New York, NY: Oxford University Press.

Lopez, B. 1978. Of Wolves and Men. New York, NY: Charles Scribner's Sons.

Mech, L. D., and Boitani, L., eds. 2003. Wolves: Behavior, Ecology, and Conservation. Chicago, IL: University of Chicago Press.

Meine, C. 1988. Aldo Leopold: His Life and Work. Madison, WI: University of Wisconsin Press.

Meine, C. 2004. Correction Lines: Essays on Land, Leopold, and Conservation. Washington, DC: Island Press.

Murie, A. 1940. Ecology of the Coyote in the Yellowstone. Fauna Series No. 4, National Park Service, Department of the Interior. Washington, DC: Government Printing Office.

Murie, A. 1944. The Wolves of Mt. McKinley. Fauna Series No. 5, National Park Service, Department of the Interior. Washington, DC: Government Printing Office.

Murie, O.J. 1935. Food Habits of the Coyote in Jackson Hole, Wyoming. USDA Circular 362. Washington, DC: Government Printing Office.

Newton, J. L. 2006. Aldo Leopold's Odyssey. Washington, DC: Island Press.

Olson, S. F. 1938a. Organization and range of the pack. Ecology 19: 168–170.

Olson, S. F. 1938b. A study in predatory relationship with particular reference to the wolf. Scientific Monthly 46: 323–336.

Peterson, R. O. 1995. The Wolves of Isle Royale: A Broken Balance. Minocqua, WI: Willow Creek Press.

Stebler, A. M. 1944. Status of the wolf in Michigan. Journal of Mammalogy 25: 37–43.

Thiel, R. P. 1993. The Timber Wolf in Wisconsin: The Death and Life of a Majestic Predator. Madison, WI: University of Wisconsin Press.

Thompson, D. Q. 1952. Travel, range, and food habits of timber wolves in Wisconsin. Journal of Mammalogy 33: 420–442.

Young, C. C. 2002. In the Absence of Predators: Conservation and Controversy on the Kaibab Plateau. Lincoln, NE: University of Nebraska Press.

Young, S. P., and Goldman, E. A. 1944. The Wolves of North America. Washington, DC: American Wildlife Institute.

Chapter 2
Long-Term Research on Wolves in the Superior National Forest

L. David Mech

2.1 Background

The seeds for the blossoming of the wolf (*Canis lupus*) population throughout the upper Midwest were embodied in a long line of wolves that had persisted in the central part of the Superior National Forest (SNF) of northeastern Minnesota, probably since the retreat of the last glaciers. This line of wolves had withstood not only the various natural environmental factors that had shaped them through their evolution but also the logging, fires, market hunting of prey animals, and even the bounties, aerial hunting, and poisoning that had exterminated their ancestors and their dispersed offspring only a few wolf pack territories away in more accessible areas. The dense and extensive stretch of wild land that is now labeled the Boundary Waters Canoe Area Wilderness had proven too formidable a barrier even for the foes of the wolf who had strived to eliminate the animal and had succeeded everywhere else in the contiguous 48 states of the United States. The wolves of the SNF became the reservoir for the recolonization of wolves throughout Minnesota and into neighboring Wisconsin and the Upper Peninsula of Michigan.

The only other part of the 48 contiguous United States where wolves still survived in the late 1960s was Isle Royale in Lake Superior, just 32 km off Minnesota's coast (Vucetich and Peterson, this volume). Those wolves had crossed Lake Superior's rare ice bridge to the 540-km^2 island from Ontario or possibly Minnesota in 1949. At that time, Isle Royale was a national park, and the wolves that reached the island were fully protected there from bounties, poisons, and aerial hunting.

The wolves of the central SNF also were those that wildlife biologist, wilderness enthusiast, and writer Sigurd Olson (1938) had trailed in the snow in the late 1930s and that Milt Stenlund (1955) had studied later. Although neither worker realized it, molecular geneticists would eventually debate whether the wolves they studied were an interesting blend of animals descended from the most recent colonization of North America across the Bering land bridge (*Canis lupus*), such as those in northwestern Canada and Alaska, and wolves that evolved in North America (*Canis lycaon*), such as inhabit southeastern Ontario (Wilson et al. 2000). Wolves with both types of genetic markers sometimes live in the same pack, and apparently many wolves in Minnesota are hybrids between the two types (Mech and Federoff 2002; Wilson et al. unpublished data).

When the last remaining 700 or so wolves inhabiting Minnesota, most of them in the SNF, were placed on the federal Endangered Species List in 1967, it was only logical to begin studying them. A few ground-breaking studies had provided some insights into the biology of wolves (e.g., Olson 1938; Murie 1944; Cowan 1947; Stenlund 1955; Mech 1966; Pimlott et al. 1969). However, because wolves were so scarce in the contiguous United States, and lived in such low densities and inaccessible areas where they did survive, much basic information about wolves was unknown. Fortunately, when wolves were declared endangered, wildlife researchers were beginning to apply the revolutionary technology of radio-tracking (Cochran and Lord 1963). G.B. Kolenosky and Johnston (1967) had proved in Ontario that radio-tracking wolves was practical. This technique promised to greatly enhance the ability of researchers to discover many new things about the behavior and ecology of wolves.

In 1968, I began a pilot project in the central SNF using radio-tracking to determine whether wolf packs were territorial (Mech and Frenzel 1971). My preliminary aerial observations during 1966–1967 and 1967–1968 had shown that there were several packs of different sizes and color combinations. However, without reliable identifiers for each pack, and without being able to find packs systematically, I had only a subjective notion that they were territorial. Thus, radio-tracking wolves from aircraft, which allowed both identifying individuals and systematically locating them, was the ideal method to answer this question.

2.2   Study Area

My study area encompassed some 2,060 km^2 immediately east of Ely in the east-central SNF (48°N, 92°W). Although somewhat smaller than the areas I have reported on earlier, this area encompassed the core of that region in which I have been able to monitor the wolf population during the entire 40-year study (Fig. 2.1). The area represents only a small percentage of the total range of wolves in Minnesota.

Topography in the study area varies from large stretches of swamps and uneven upland to rocky ridges, with elevations ranging from 325 to 700 m above sea level. Winter temperatures below −35°C are not unusual, and snow depths (usually from mid-November through mid-April) generally range from 50 to 75 cm on the level. Summer temperatures rarely exceed +35°C. Conifers predominate in the forest overstory, including jack pine (*Pinus banksiana*), white pine (*P. strobus*), red pine (*P. resinosa*), black spruce (*Picea mariana*), white spruce (*P. glauca*), balsam fir (*Abies balsamea*), white cedar (*Thuja occidentalis*), and tamarack (*Larix laricina*). However, as a result of extensive cutting and fires, much of the coniferous cover is interspersed with large stands of white birch (*Betula papyrifera*) and aspen

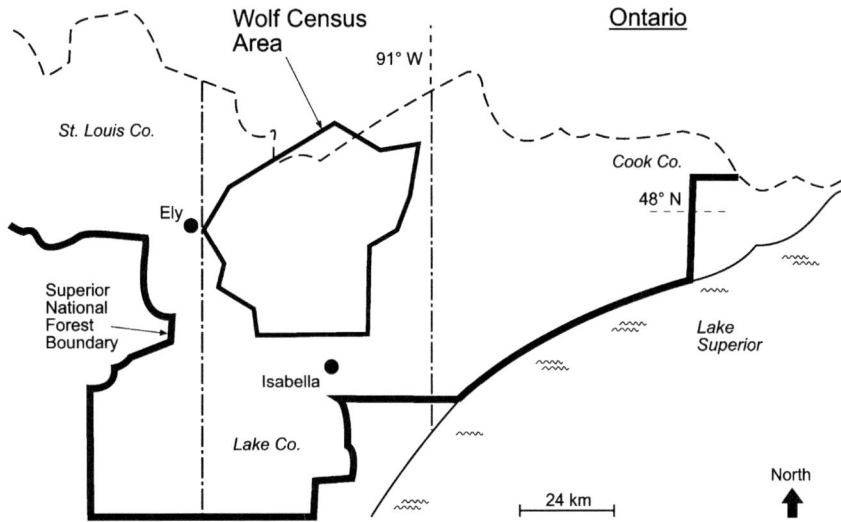

Fig. 2.1 The central Superior National Forest study area

(*Populus tremuloides*). Heinselman (1993) presented a detailed description of the forest vegetation.

In the northeastern half of this area, as well as immediately north and east of it, the overwintering population of white-tailed deer (*Odocoileus virginianus*) was extirpated by about 1975 by a combination of severe winters, maturing vegetation, and high numbers of wolves (Mech and Karns 1977), and the area has remained devoid of wintering deer ever since (Nelson and Mech 2006). Moose (*Alces alces*) inhabit the entire area but occur at a higher density in the northeastern half. In spring, about 32% of the deer inhabiting the southwestern half of the study area migrate into the northeastern half or beyond and return in fall (Hoskinson and Mech 1976; Nelson and Mech 1981). Beavers (*Castor canadensis*) occur throughout the study area, but generally are available as prey only from about April through November. Although all three prey species are consumed by wolves in the region (Van Ballenberghe et al. 1975), since about 1975 the primary prey of wolves inhabiting the northeastern part has been moose, whereas wolves in the southwestern part have consumed primarily deer.

Year-around hunting and trapping of wolves was legal until October 1970 when they were fully protected on federal land within the SNF by the US Forest Service. In August 1974, wolves were protected under the Endangered Species Act of 1973. In 1978, wolves in Minnesota were reclassified to threatened but remained legally protected except for depredation control outside the SNF (Fritts et al. 1992). However, illegal taking of wolves continued, primarily in fall and winter (Mech 1977a, unpublished data). In March 2007, wolves in the upper Midwest, including Minnesota, were removed from the Endangered Species List though this ruling was recently overturned.

2.3   Long-Term Research on Wolves, Wolf Packs, and Population Trends

My main objective at the beginning of the study was to determine spacing in the wolf population, but I also realized that by being able to find and identify each marked pack, I could obtain much other information. For example, during winter I could count pack members, determine how consistently each pack maintained its size, track its movements, find and examine its kills, and locate marked wolves after death. In addition, if the packs were territorial, then by radio-tagging enough packs in the study area, I could determine the total number of wolves there by locating each pack and counting the pack members.

Over the long term, monitoring the population trajectory of wolves in the SNF became my basic objective. The longer this study continued, the more valuable the data on changes in population size. The only other data available on wolf population trends were those from Isle Royale, which began in 1959 (Mech 1966) and continues today (Vucetich and Peterson, this volume). Those data are of great interest. However, they do pertain to an island with no emigration or immigration, and cannot fully represent most populations of wolves. The opportunity to gather long-term data on a population of mainland wolves and determine what drove the changes in that population was highly attractive.

The primary technique used has been live-trapping wolves in modified steel foot-traps, anaesthetizing each of them (except most pups), weighing them, blood sampling them, and outfitting them with a radio-collar (Mech 1974). Since 2000 my assistants, students, associates, and I also estimated the age of each wolf based on tooth wear (Gipson et al. 2000). We aerially radio-tracked the wolves at least weekly during most years, and observed and counted them as often as possible, primarily from December through March (Mech 1973, 1986). The most wolves we saw during winter in each pack was considered the pack size. If a radioed pack territory fell partly outside the census area, the number of wolves I assigned to the census area was proportioned to the proportion of the territory in the area.

2.3.1   Territoriality of Wolf Packs

Each time we located a wolf, we recorded its location. We plotted these locations from October 1 through March 30 and April 1 through September 30 each year, and used minimum convex polygons (MCPs; Mohr 1947) to represent territories (Mech 1973, 1977b, 1986).

Pack territories based on radio locations were delineated for each radioed pack in the study area each year. However, some packs died out, new ones formed, and not all packs were radioed each year. The existence of nonradioed packs in the study area in any year was inferred from voids in the maps of the territorial mosaic. Incidental observations of nonradioed packs and/or their tracks in these voids indicated sizes of these packs. (Some data pertaining to individual packs in some years in this chapter may differ from data presented previously [Mech 1973, 1977a, 1986] because of reinterpretation of the data based on additional experience with these packs.) If data on individual packs were unavailable in any year, pack size estimates were made based on the previous and subsequent year's data for packs occupying those territories. Because an unknown portion of the territories of some of these packs may have fallen outside the census area, these data are not precise. Data in 1966–1967 and 1967–1968 were based solely on observations of nonradioed packs during intensive aerial observations. In the estimates of population trajectory for wolves presented in this chapter, I considered the number of lone wolves inconsequential because their proportion of the population was low and most of these individuals were dispersers accounted for by using the maximum numbers in each pack. During the earlier part of the study, lone wolves were estimated at 7–14% of the population (Mech 1973).

With monitoring the population density of wolves in the study area requiring the maintenance of radio-collars on several adjacent packs, the project became a data-gathering system that allowed several parallel studies. By knowing where wolf packs lived regularly and how many members each contained, Fred Harrington and I could approach on foot and howl to them under various conditions to determine their responses (Harrington and Mech 1979). By tracking known packs in the snow, and examining their scent marks, Roger Peters and I could describe and quantify scent-marking behaviors (Peters and Mech 1975). Russell Rothman and I conducted a similar study on newly formed pairs of wolves (Rothman and Mech 1979).

From 1968 through 2006, we live-trapped 712 wolves (119 female pups, 141 male pups, 239 females ≥1-year old, and 213 males ≥1-year old) in the study area, for a total of 1,044 captures of wolves from 15 or more packs. The number of packs radioed each year varied, and over the 38 years of radio-tracking, some packs disappeared and many new ones were formed. Weights of both males and females peaked at 5 or 6 years of age, with mean peak weights of 40.8 ± 1.5 (SE) kg and 31.2 ± 2.4 (SE) kg, respectively (Mech 2006a). The age structure of the population between 2000 and 2004 was relatively young, with only 12% of animals >1-year-old being >5 years of age (Mech 2006b). Some wolves, however, lived to be 13-years old (Mech 1988). Most females 4–9 years of age had bred based on assessment of nipple sizes; those that had not bred had lower average weights than those that had.

Each radioed pack inhabited a separate territory, the first time that this fact was clearly established (Mech 1973). Pimlott et al. (1969, p. 78) had concluded that "the results are far from conclusive on the question of whether or not pack territoriality is involved." Mech (1970, p. 105) had speculated that wolf packs might even have "spatio-temporal" territories. Radio-tracking wolves in the SNF showed that wolves were territorial and that their territories were spatial (Mech 1973). The wolves advertised and defended their territories by howling (Harrington and Mech 1979), scent-marking (Peters and Mech 1975), and direct aggression (Mech 1994).

Analysis of wolf pack territory size is not in the scope of this chapter. Based on the MCPs of radioed wolf packs, territory sizes through winter 1973 varied from 125 to 310 km² (Mech 1974). However, during 1997–1999, the Farm Lake pack inhabited only 23–33 km², a density of 182–308 wolves per 1,000 km², the highest density ever reported (Mech and Tracy 2004). The overall territorial structure gradually shifted over the years, although some semblance of the early structure is still apparent (Fig. 2.2).

Maximum winter pack sizes during 233 radioed-pack-years (one pack radio-tracked for 1 year = 1 pack-year) varied from 2 to 15 and averaged 5.6 ± 0.20 (SE). Maximum winter pack sizes for 11 packs with at least 11 years of data varied from 2–8 to 2–15 per year with means of 3.7 ± 0.5 (SE) to 7.9 ± 1.1 (SE); the SEs around these means show that individual packs in the study area tended to retain their basic sizes (Appendix). Approximately 67% of the packs included a maximum of two to six members during winter, and 90% included two to nine (Fig. 2.3).

One of the more novel findings of our long-term study was the concept of the buffer zone between wolf pack territories (Mech 1977c). There appears to be an area of 1–2 km around the edge of a wolf-pack territory where neighboring packs travel but spend less time (Mech and Harper 2002), and wolves fight there if an encounter between packs occurs, often to the death (Mech 1994). Thus, prey seems to survive longer in these zones. When deer declined early in the study, most of those remaining inhabited these zones (Hoskinson and Mech 1976; Mech 1977b, c; Nelson and Mech 1981). Even after the deer population increased, we continued to find evidence of this relationship (Kunkel and Mech 1994).

Buffer zones between territories of wolf packs are quite important to territorial maintenance. Besides fighting there, adjacent packs also scent-mark disproportionately there (Peters and Mech 1975). No doubt howling in and near the buffer zone is also important. Harrington and Mech (1979, p. 243) estimated that each pack on

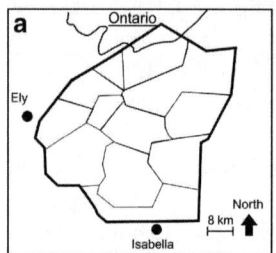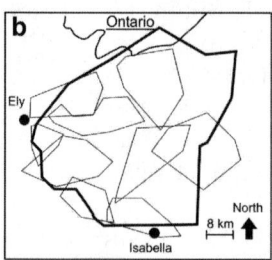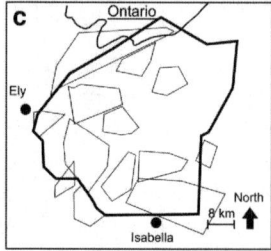

Fig. 2.2 The territorial structure of wolf packs in the central Superior National Forest study area. "A" represents the territorial structure from 1971 to 1973 but arbitrarily extends each pack's minimum convex polygon (*MCP*) to the boundaries of its neighbors (Mech 1973). "B" represents the actual MCPs for radioed packs during winter 1984–1985 (Mech 1986). "C" represents the same for 2006–2007. In 1984–1985, a nonradioed wolf pack with an estimated six wolves occupied an unknown part of the northeastern area, and in 2006–2007, a nonradioed pack of eight wolves occupied the northeastern area. Several aerial surveys over the east-central area indicated no wolves there during winter 2006–2007

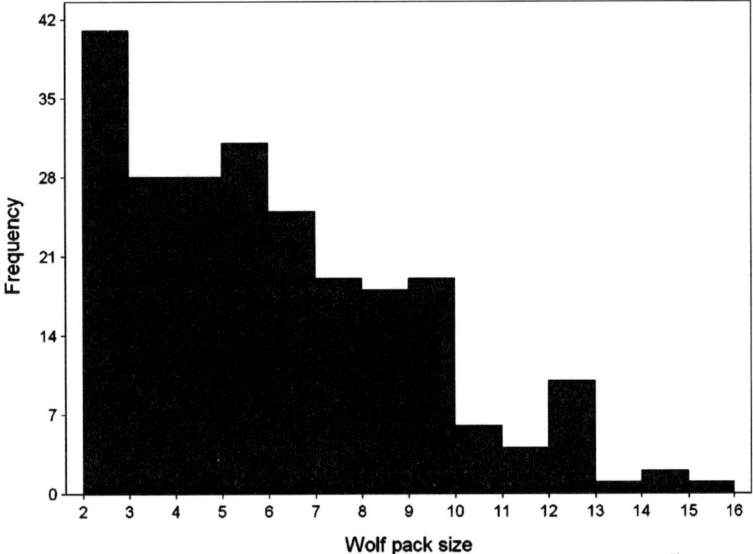

Fig. 2.3 Distribution of maximum winter pack sizes in the central Superior National Forest study area

average is within howling range of at least one neighboring pack about 78% of the time, and "the probability of one pack hearing another, and the probability of encounters both increase when packs approach one another at a common border."

2.3.2   Population Trends

In our 2,060-km^2 study area, numbers of wolves ranged from 35 to 87, with a mean of 59 and a median of 55 (Appendix), a density of 17–42 wolves per 1,000 km^2 with a mean of 28 per 1,000 km^2 and a median of 27 per 1,000 km^2. The population dropped between the winters of 1968–1969 to 1973–1974 and then increased ($r^2 = 0.33$; $P < 0.001$; Fig. 2.4). Mean pack size also increased after winter 1973–1974 ($r^2 = 0.21$; $P < 0.01$). In winter 2006–2007, the population was estimated to be 81 wolves, or 39 wolves per 1,000 km^2. Both the population and average-pack-size trend increased after 1973–1974 at a mean annual rate of 0.01. Annual changes in the estimated size of the wolf population were related to annual changes in mean sizes of radioed packs ($r^2 = 0.35$; $P < 0.001$). Estimates of pack size and population change were accurate because radioed packs were easily located and counted several times each winter.

From the beginning of the study through about the late 1980s, the proportion of wolves on a deer economy in our area dropped, and more wolves had to rely on moose. The decline in wolves through 1982 coincided with the decline in deer (Fig. 2.5), which in turn coincided with maximum cumulative 3-year snow depth

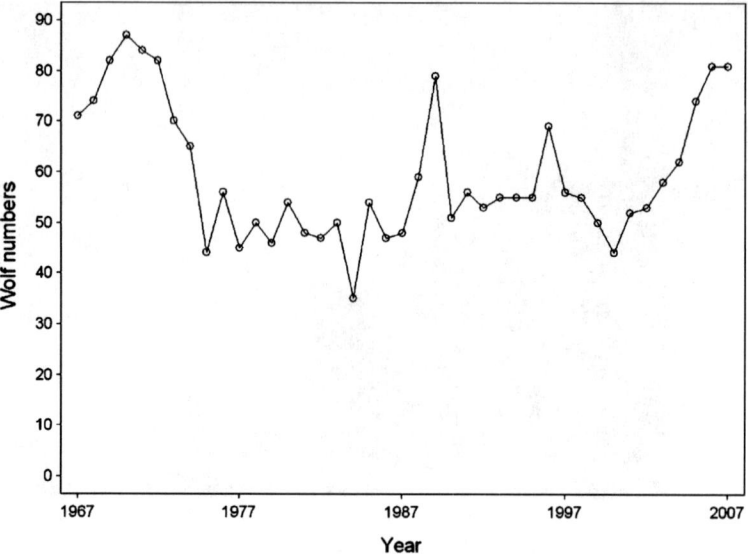

Fig. 2.4 Trend in size of the wolf population in the central Superior National Forest

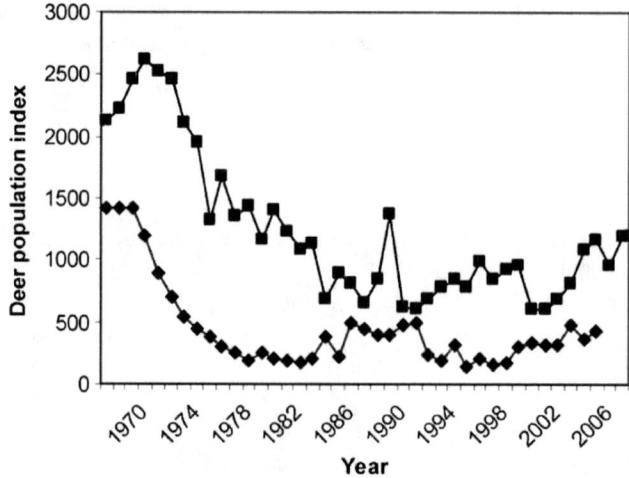

Fig. 2.5 Trend in sizes of the deer (*lower curve*) and wolf (*upper curve*) populations in south-western portions of the central Superior National Forest study area. Wolf trend is actual wolf population times 30. (Updated from Fuller, Mech, and Fitts-Cochrane 2003, Fig. 6.6)

(Mech et al. 1987a). When the snowfall moderated in 1982–1983, deer began increasing again (Fuller et al. 2003). The trend for the wolf population that depended on deer declined curvilinearly, bottoming out about 1991 and gradually increasing through 2007 ($r^2 = 0.86$; $P < 0.00001$). The wolf population in the northern, northeastern, and eastern parts of the area that preyed increasingly on moose showed a reverse-sigmoid increasing trend ($r^2 = 0.80$) from about 1978 through 2007, related ($r^2 = 0.12$; $P = 0.06$) to an increase in abundance of moose from 3,900 individuals in 1978 to 6,460 in 2007 (M. Lennarz, MN DNR, personal communication).

Canine parvovirus (CPV) began affecting the wolf population in the early 1980s and continues to do so (Mech et al. 2008), greatly complicating the relationships among snow depth, wolves, and prey that had been so apparent. From 1984 to 2004, the annual change in the wolf population was negatively related to seroprevalence of CPV ($r^2 = 0.51$; $P = < 0.01$), and the change in the wolf population was related to an index of survival of wolf pups ($r^2 = 0.22$; $P = 0.03$) (Mech et al. 2008).

2.3.3 Dispersal

Our wolf population occurred at a high density, and packs occupied most of the available space. Any excess production of pups, therefore, resulted in their dispersal as 1–3-year olds (Mech 1987; Gese and Mech 1991). Some dispersers floated around their natal population, covering as much as 4,100 km^2 (Mech and Frenzel 1971; Mech 1987). However, others dispersed farther and helped recolonize other parts of Minnesota, as well as Wisconsin and Michigan (Mech et al. 1995; Merrill and Mech 2000).

2.4 Studies of the Ecology of Deer

As I radio-tracked wolves, it became clear that to conduct a thorough study of wolf ecology, I also had to examine the natural history and ecology of its main prey, white-tailed deer. In 1973, I began radio-tagging deer in the same area and traced their movements, survival, and mortality along with those of the radioed wolves. Reed Hoskinson (Hoskinson and Mech 1976), and then Mike Nelson (Nelson and Mech 1981; Nelson 1993), conducted the initial studies of deer. Mike remained with the project as a collaborator in charge of deer research (DelGiudice et al., this volume). Ted Floyd used our radio-tagged deer to pioneer the technique of evaluating observability biases in aerial ungulate censuses and applied an adjustment for observability to our data (Floyd et al. 1979). We used this technique to count deer in winter through 1992 (Nelson and Mech 1986a, unpublished data), until funding constraints forced us to discontinue it. Since 1992, we have used buck harvest in the Isabella part of our area to index deer population trends (Mark Lenarz, MN DNR, personal communication). Deer numbers decreased in our area from the late 1960s and 1970s, bottomed out about 1981, and have slowly and intermittently increased since (Fig. 2.5).

Between 1973 and 2007, we radio-collared 347 deer, mostly females. Besides learning much basic natural history about these deer (e.g., Hoskinson and Mech 1976; Nelson and Mech 1981, 1987, 1990; Nelson 1993; Mech and McRoberts 1990), we also found that during summer wolves only rarely killed adult females (Nelson and Mech 1986c), that wolf predation was greatest during deepest snow (Nelson and Mech 1986b), that daily predation rates during fall migration were 16–107 times that of deer in wintering areas or yards (Nelson and Mech 1991), that survival of adult females was related to the nutritional condition of their mothers, and that survival of yearlings to 2-year olds was related to the nutrition of their grandmothers (Mech et al. 1991).

We learned that condition was an important factor predisposing deer to predation by wolves, and various measures of condition provided evidence. Wolves tended to kill old deer (Mech and Frenzel 1971; Mech and Karns 1977; Nelson and Mech 1986a), deer with abnormalities (Mech et al. 1970; Mech et al. 1971; Mech and Karns 1977), deer with low blood fat (Seal et al. 1978) and low marrow fat (Mech and Frenzel 1971; Mech 2007), and newborn fawns of below-average weight and/ or with low serum urea nitrogen (Kunkel and Mech 1994).

Condition of deer in winter depends on snow depth because the deeper the snow, the harder it is to find food (Verme 1968). Thus, we were not surprised to find that deer numbers and population trend were related to snow conditions (Mech et al. 1971, 1987, 1991; Mech and Karns 1977; McRoberts et al. 1995; c.f. Messier 1995).

2.5   Spin-Offs from, and Adjuncts to, the SNF Wolf Research

2.5.1   Development of a Capture Collar

While trapping wolves in the SNF, I quickly realized that if we could capture them more easily, we could examine them more often and better monitor their weights, blood values, and conditions. Furthermore, the early collars we used often lasted for <1 year, so replacing them was important. The longer data were collected, the more complete a picture we could gain of the natural history of packs and the spatial organization of the population.

To determine if we could use radio signals to remotely dart and recapture a radio-collared wolf, I consulted my former coworker, Bill Cochran, who had pioneered radio-tracking (Cochran and Lord 1963). Cochran suggested using a squib, which was an electrically detonated match-head, like a tiny flashbulb. When a signal sent current through the squib, it flashed. Gunpowder in front of the squib would detonate, drive a dart, and inject a drug. That, however, would require a radio receiver attached to the dart to pick up the signal, and an electrically detonated dart small enough to be attached to a wolf collar. The dart also had to be wolf and water proof, and in a position to inject a drug into a wolf. We designed the mechanism, but needed a talented machinist to produce the experimental prototypes. Lee Simmons, Director of the Henry Doorly Zoo in Topeka, Kansas, came to the rescue. Ulysses (Ulie) Seal, an expert on drugs suitable to use in such a collar (Seal et al. 1970), completed the development team.

The time between conception and a working dart collar was about 10 years. Sometime during the final development, Rick Chapman, a graduate student on the project, was hired by 3M Company, and that company was interested enough in the concept of the collar to invest considerable time and funding to perfect it (Mech et al. 1984). We also tested the capture collar on several deer (Mech et al. 1990) and used it to conduct studies of year-around nutritional condition in deer (DelGiudice et al. 1992) and of capture stress (DelGiudice et al. 1990). We then tested the collar successfully on wild wolves (Mech and Gese 1992) and used it to obtain such elusive types of data as serial weights and blood values of the same wolf over long periods, as well as of field metabolic rates (Nagy 1994). The most important contribution of the capture collars, however, was unexpected. To facilitate recovery of the collar in case it failed, Chapman invented a remote-release mechanism. When that mechanism was applied to Global Positioning System (GPS) collars, then being developed, biologists could retrieve the GPS collars to download the data (Merrill et al. 1998). Unfortunately, commercial companies found it much more lucrative to produce GPS collars than capture collars, so the latter soon became unavailable.

2.5.2   Blood Sampling

During the 1970s, Ulysses Seal began studying blood. I then began a productive collaboration with him, collecting blood from both wolves and deer. Although my main objective was to determine the nutritional condition of my study animals (Seal et al. 1975, 1978), the samples gained more significance in determining seroprevalence of CPV in our wolves (Mech et al. 2008).

2.5.3   Studies of Captive Wolves

As these projects produced new information, they also spawned many questions. Some could be answered with additional field studies, but others required a different approach. Thus, Jane Packard, Ulie Seal, and I set up a colony of captive wolves that could be observed closely and examined frequently, blood-sampled, and otherwise studied intensively (Seal et al. 1987; Seal and Mech 1983; Packard et al. 1983, 1985). As that project grew, Cheri Asa (Asa et al. 1985, 1990), James Raymer (Raymer et al. 1985, 1986), and Terry Kreeger (Kreeger et al. 1990, 1997) became additional collaborators. Glenn DelGiudice made use of both the captive wolf colony (Mech et al. 1987b) and the field studies in the SNF (DelGiudice et al. 1988, 1989) to begin investigations on the nutritional condition of various animals using analyses of urine in the snow.

2.5.4 Beyond the SNF

Several other spin-offs of research in the SNF contributed to increased knowledge of wolves and wolf recovery in the Midwest and elsewhere. Because radio-tracking was so productive in the SNF where the wolf population had been long established and occurred at high density, I wanted to use the same techniques to examine a recently colonized wolf population. For this I recruited Steve Fritts to study a wolf population just getting a toehold 290 km away in northwestern Minnesota (Fritts and Mech 1981).

We also assisted the Minnesota Department of Natural Resources in starting a research project on wolves in north-central Minnesota similar to the SNF study. We taught colleagues, students, and technicians how to live-trap, anesthetize, radio-tag, and radio-track wolves. Many of them continued research on wolves in other areas (Berg and Kuehn 1982; Ream et al. 1991; Boyd et al. 1995; Meier et al. 1995; Fuller et al. 2003; Burch et al. 2005). Furthermore, we conducted an experimental reintroduction of four wolves into northern Michigan that demonstrated that translocated wolves held for only a week before release tended to return homeward (Weise et al. 1979).

Biologists in other areas became interested in doing similar studies, so I was invited to Italy, to Riding Mountain National Park, Canada, and to Alaska to help organize their first radio-tracking studies of wolves (Boitani and Zimen 1979; Carbyn 1980; Peterson et al. 1984). Some of my technicians helped start projects in Portugal and Romania. Furthermore, the SNF project hosted biologists from Sweden, Israel, Portugal, Poland, Spain, Croatia, India, Italy, Mexico, Norway, Turkey, and Austria to receive training in wolf research techniques.

2.5.5 Wolf Depredation Control Program

Responses to complaints about livestock depredation had been conducted by the Animal Damage Control Branch of the US Fish and Wildlife Service, but in 1978 when wolves in Minnesota were reclassified from endangered to threatened, I was asked to design a control program for wolves. This program had to keep within the directives of a court order while still attempting to reduce wolf depredations on livestock, taking minimal wolves, yet satisfying farmers and ranchers. I was appointed to direct the program, and I put Steve Fritts, with his new Ph.D. degree, in charge of it. Bill Paul, a recent technician on the SNF project, was hired as his main assistant. These two workers conducted a well-respected program that continues under the auspices of the USDA Wildlife Services (Fritts et al. 1992).

We tried many alternative nonlethal methods to reduce losses of livestock, such as translocating depredating wolves (Fritts et al. 1985), using "fladry" (flagging), blinking lights, guard dogs, and taste aversion (Fritts et al. 1992), and conceived several others such as radio-controlled shock collars, radio-activated alarm systems, human-applied scent marking, and recorded howling. None was very effective or practical because the law allowed lethal control and the population was not so low (1,250 in 1978) that every last wolf needed to be preserved at all costs. Some of these concepts have since proven useful where lethal control is not allowed or where wolf numbers are so low that extraordinary means are justified (Musiani et al. 2003; Schultz et al. 2005; Shivik 2006).

2.6 Future Directions

To understand the functioning of natural wolf populations, it is important to follow the long-term trend of at least one long-extant population. The value of the information that science has obtained from the wolf population on Isle Royale over 50 years is immeasurable (Vucetich and Peterson, this volume). However, the fact that population is restricted to an island with no regular immigration or emigration is problematic. The central SNF study is the longest-running, nonisland study of a wolf population. As such, it is extremely important to continue this investigation as long as possible. My hope is that this summary will help serve that end.

Acknowledgments This study was supported by the US Department of the Interior during its entirety through the US Fish and Wildlife Service, the National Biological Survey, and the US Geological Survey and Patuxent Wildlife Research Center, Midcontinent Wildlife Research Center, and Northern Prairie Wildlife Research Center, as well as by the US Forest Service through North Central Forest Experiment Station (now North Central Research Station) and the SNF. Private funding sources included the New York Zoological Society, the World Wildlife Fund, the Mardag Foundation, the Special Projects Foundation of the Big Game Club, Wallace Dayton, and Valerie Gates. I especially thank M. E. Nelson, J. Renneberg, and T. Wallace as well as numerous volunteers and graduate students who assisted with the fieldwork. S. Barber-Meyer and D. J. Demma critiqued an early draft of the manuscript and offered helpful suggestions for improvement.

Appendix

Numbers of wolves in each pack in the east-central Superior National Forest study area. See Mech (1986) for 1966–1967 to 1984–1985. Underlines indicate pack was radioed; zeros, that the pack did not exist or was outside the census area; parentheses, that estimate was subjective; hyphens, that information unknown. Nonunderlined numbers not in parentheses are based on observation of nonradioed pack or its tracks. Entries with two numbers (e.g., 3 + 1) indicate different proportions of a pack in and outside the census area.

Winter	Birch Lake	Clear Lake	Ensign Lake	Farm Lake	Little Gabbro	Jack-pine	Mani-waki	Malberg Lake	Nip Creek	Pagami Lake	Perent	Quadga Lake	Saw-bill	Star-light	Wood Lake	Total[a]	Economy Deer[b]	Economy Moose
1985–1986	6	0	0	0	3	7	8	7	0	1	0	5	2	0	8	47	27	20
1986–1987	8	0	(7)	0	3	3	4	(7)	0	(3)	0	8	2	0	3	48	22	26
1987–1988	5	0	(7)	0	4	2	(8)	(9)	0	5	0	(7)	8	0	4	59	28	31
1988–1989	9	0	4	0	8	6	12	11	0	(5)	0	6	12	0	6	79	46	33
1989–1990	4	0	5	0	5	5	5	3	0	5	0	3	14	0	2	51	21	30
1990–1991	4	0	0	0	5	6	4	2	0	3	0	5	5	0	2	56	20	36
1991–1992	4	0	7	0	6	11	0	7	0	9	0	3	4	0	2	53	23	30
1992–1993	3	-	-	0	5	14	0	6	0	11	0	-	2	0	4	55	26	29
1993–1994	4	0	0	0	4	11	0	(8)	0	15	0	(0)	4	0	9	55	28	27
1994–1995	6	0	5	0	3	12	2	6	0	12	0	0	4	0	5	55	26	29
1995–1996	10	0	4	0	5	12	3	9	0	14	0	0	6	0	4	69	33	36
1996–1997	6	0	6	0	3	10	3	12	0	2	0	0	5	0	7	56	28	28
1997–1998	3	0	6	2	6	9	3	8	1	2	0	0	5	0	5	55	31	24
1998–1999	3	0	6	4	4	9	0	4	2	5	0	3	3	0	6	50	32	18

Winter	Birch Lake	Clear Lake	Ensign Lake	Farm Lake	Little Gabbro	Jack-pine	Mani-waki	Malberg Lake	Nip Creek	Pagami Lake	Perent	Quadga Lake	Saw-bill	Star-light	Wood Lake	Total[a]	Economy	
																	Deer[b]	Moose
1999–2000	2	0	6	3 + 1	1 + 1	9 + 1	0	12	0	3	0	0	4 + 1	0	4	44	20	24
2000–2001	3	0	10	4	2	9	0	11	0	6	0	0	5	0	2	52	20	32
2001–2002	2	0	2	6 + 1	2	6	0	12	1	4	0	0	7	0	3	53	23	30
2002–2003	2	5	7	2	2	7	0	6 + 6	1	2	3	0	7	0	5	58	27	31
2003–2004	3	7	6	5 + 1	1	7	0	4 + 4	3	2	4	0	5 + 1	0	7	62	36	26
2004–2005	1	12	6	6	0	7 + 1	0	2 + 2	4 + 4	0	8	2	9	5	9	74	39	35
2005–2006	0	7	8	10	0	7	0	8	0	0	9	8	8 + 3	8	8	81	32	49
2006–2007	2	8	7	12	0	6	0	8	3	0	7	6	4 + 3	3	9	81	40	41
N[c]	28	14[d]	15	25		28	11	12		13		15	28		30			
Range	2–10	2–9	2–10	2–8		2–14	3–13	2–12		2–15		2–8	2–14		2–9			
Mean	4.1	3.7	6.3	3.9		7.3	7.5	7.9		6.6		4.9	6.0		5.4			
SE	0.4	0.5	0.8	0.4		0.6	0.9	1.1		1.1		0.5	0.6		0.5			

[a]The difference between the total and the sum of the pack numbers represent additional wolves in unknown packs straddling the census area border

[b]From 1977–1978 to 1983–1984, excludes the Malberg Pack; from 1984–1985 through 1988–1989 excludes the Malberg, Quadga, Maniwaki, and Ensign Packs; in 1989–1990, excludes packs excluded previously and Sawbill Pack; from 1990–1991 through 1999–2000 excludes packs excluded previously and Pagami Pack; after 2000–2001, excludes packs excluded previously and Perent and Starlight Packs

[c]These summary data are only for packs with at least 11 years of data, from 1966–1967 to 2006–2007. See Mech (1986) for the pack size data from 1966–1967 through 1984–1985

[d]Data in this summary column are for the Harris L. Pack (Mech 1986) which had disappeared by 1985–1986

References

Asa, C. S., Mech, L. D., Seal, U. S., and Plotka, E. D. 1990. The influence of social and endocrine factors on urine-marking by captive wolves (*Canis lupus*). Hormones and Behavior 24:497–509.

Asa, C. S., Peterson, E. K., Seal, U. S., and Mech, L. D. 1985. Deposition of anal sac secretions by captive wolves (*Canis lupus*). Journal of Mammalogy 66:89–93.

Berg, W. E., and Kuehn, D. W. 1982. Ecology of wolves in north-central Minnesota. In Wolves of the World: Perspectives of Behavior, Ecology, and Conservation, ed. F. H. Harrington and P. C. Paquet, pp. 4–11. Park Ridge, NJ: Noyes Publications.

Boitani, L., and Zimen, E. 1979. The role of public opinion in wolf management. In The Behavior and Ecology of Wolves, ed. E. Klinghammer, pp. 471–477. New York, NY: Garland STPM Press.

Boyd, D. K., Paquet, P. C., Donelon, S., Ream, R. R., Pletsher, D. H., and White, C. C. 1995. Transboundary movements of a colonizing wolf population in the Rocky Mountains. In Ecology and Conservation of Wolves in a Changing World, eds. L. N. Carbyn, S. H. Fritts, and D. R. Seip, pp. 135–140. Edmonton, Alberta: Canadian Circumpolar Institute.

Burch, J. W., Adams, L. G., Follmann, E. H., and Rexstad, E. A. 2005. Evaluation of wolf density estimation from radiotelemetry data. Wildlife Society Bulletin 33:1225–1236.

Carbyn, L. N. 1980. Ecology and management of wolves in Riding Mountain National Park, Manitoba. Canadian Wildlife Service, Report No. 10, Edmonton, Alberta. 184 pp.

Cochran, W. W., and Lord, R. D., Jr. 1963. A radio-tracking system for wild animals. Journal of Wildlife Management 27:9–24.

Cowan, I. M. 1947. The timber wolf in the Rocky Mountain national parks of Canada. Canadian Journal of Research 25:139–174.

DelGiudice, G. D., Mech, L. D., and Seal, U. S. 1988. Comparison of chemical analyses of deer bladder urine and urine collected from snow. Wildlife Society Bulletin 16:324–326.

DelGiudice, G. D., Mech, L. D., and Seal, U. S. 1989. Physiological assessment of Minnesota deer populations by chemical analysis of urine in the snow. Journal of Wildlife Management 53:284–291.

DelGiudice, G. D., Mech, L. D., and Seal, U. S. 1990. Effects of winter undernutrition on body composition and physiological profiles of white-tailed deer. Journal of Wildlife Management 54:539–550.

DelGiudice, G. D., Mech, L. D., Kunkel, K. E., Gese, E. M., and Seal, U. S. 1992. Seasonal patterns of weight, hematology and serum characteristics of free-ranging female white-tailed deer in Minnesota. Canadian Journal of Zoology 70:974–983.

Floyd, T. J., Mech, L. D., and Nelson, M. E. 1979. An improved method of censusing deer in deciduous-coniferous forests. Journal of Wildlife Management 43:258–261.

Fritts, S. H., and Mech, L. D. 1981. Dynamics, movements, and feeding ecology of a newly protected wolf population in northwestern Minnesota. Wildlife Monographs No. 80, pp.1–79.

Fritts, S. H., Paul, W. J., and Mech, L. D. 1985. Can relocated wolves survive. Wildlife Society Bulletin 13:459–463.

Fritts, S. H., Paul, W. J., Mech, L. D., and Scott, D. P. 1992. Trends and management of wolf-livestock conflicts in Minnesota. U.S. Fish and Wildlife Service Resource Publications Series No. 181.

Fuller, T. K., Mech, L. D., and Fitts-Cochrane, J. 2003. Population dynamics. In Wolves: Behavior, Ecology, and Conservation, eds. L. D. Mech and L. Boitani, pp. 161–191. Chicago, IL: University of Chicago Press.

Gese, E. M., and Mech, L. D. 1991. Dispersal of wolves (*Canis lupus*) in northeastern Minnesota, 1969–1989. Canadian Journal of Zoology 69:2946–2955.

Gipson, P. S., Ballard, W. B., Nowak, R. M., and Mech, L. D. 2000. Accuracy and precision of estimating age of gray wolves by tooth wear. Journal of Wildlife Management 64:752–758.

Harrington, F. H., and Mech, L. D. 1979. Wolf howling and its role in territory maintenance. Behaviour 68:207–249.

Heinselman, M. 1993. The Boundary Waters wilderness ecosystem. Minneapolis, MN: University of Minnesota Press.

Hoskinson, R. L., and Mech, L. D. 1976. White-tailed deer migration and its role in wolf predation. Journal of Wildlife Management 40:429–441.

Kolenosky, G. B., and Johnston, D. 1967. Radio-tracking timber wolves in Ontario. American Zoologist 7:289–303.

Kreeger, T. J., DelGiudice, G. D., and Mech, L. D. 1997. Effects of fasting and refeeding on body composition of captive gray wolves (*Canis lupus*). Canadian Journal of Zoology 75:1549–1552.

Kreeger, T. J., Kuechle, V. B., Mech, L. D., Tester, J. R., and Seal, U. S. 1990. Physiological monitoring of gray wolves (*Canis lupus*) by radio telemetry. Journal of Mammalogy 71:259–261.

Kunkel, K. E., and Mech, L. D. 1994. Wolf and bear predation on white-tailed deer fawns. Canadian Journal of Zoology 72:1557–1565.

McRoberts, R. E., Mech, L. D., and Peterson, R. O. 1995. The cumulative effect of consecutive winters' snow depth on moose and deer populations: a defense. Journal of Animal Ecology 64:131–135.

Mech, L. D. 1966. The Wolves of Isle Royale. National Parks Fauna Series No. 7. U.S. Government Printing Office, 210 pp.

Mech, L. D. 1970. The Wolf: The Ecology and Behavior of an Endangered Species. New York, NY: Natural History Press, Doubleday Publishing Co. (Reprinted in paperback by University of Minnesota Press, May 1981).

Mech, L. D. 1973. Wolf numbers in the Superior National Forest of Minnesota. USDA Forest Service Research Paper No. NC-97.

Mech, L. D. 1974. Current techniques in the study of elusive wilderness carnivores. In Proceedings of the 11th International Congress of Game Biologists, eds. I. Kjerner and P. Bjurholm, pp. 315–322. Stockholm: Swedish National Environment Protection Board.

Mech, L. D. 1977a. Productivity, mortality and population trends of wolves in northeastern Minnesota. Journal of Mammalogy 58:559–574.

Mech, L. D. 1977b. Population trend and winter deer consumption in a Minnesota wolf pack. In Proceedings of the 1975 Predator Symposium, eds. R. L. Phillips and C. Jonkel, pp. 55–83. Missoula: Montana Forest and Conservation Experimental Station.

Mech, L. D. 1977c. Wolf pack buffer zones as prey reservoirs. Science 198:320–321.

Mech, L. D. 1986. Wolf numbers and population trend in the Superior National Forest, 1967–1985. St. Paul, MN: USDA Forest Service Research Paper No. NC-270, North Central Forest Experiment Station.

Mech, L. D. 1987. Age, season, and social aspects of wolf dispersal from a Minnesota pack. In Mammalian Dispersal Patterns, eds. B. D. Chepko-Sade and Z. Halpin, pp. 55–74. Chicago, IL: University of Chicago Press.

Mech, L. D. 1988. Longevity in wild wolves. Journal of Mammalogy 69:197–198.

Mech, L. D. 1994. Buffer zones of territories of gray wolves as regions of intraspecific strife. Journal of Mammalogy 75:199–202.

Mech, L. D. 2006a. Age-related body mass and reproductive measurements of gray wolves in Minnesota. Journal of Mammalogy 87:80–84.

Mech, L. D. 2006b. Estimated age structure of wolves in northeastern Minnesota. Journal of Wildlife Management 70:1481–1483.

Mech, L. D. 2007. Femur-marrow fat of white-tailed deer fawns killed by wolves. Journal of Wildlife Management 71:920–923.

Mech, L. D., Chapman, R. C., Cochran, W. W., Simmons, L., and Seal, U. S. 1984. A radio-triggered anesthetic-dart collar for recapturing large mammals. Wildlife Society Bulletin 12:69–74.

Mech, L. D., and Federoff, N. E. 2002. Alpha1-antitrypsin polymorphism and systematics of eastern North American wolves. Canadian Journal of Zoology 80:961–963.

Mech, L. D., and Frenzel, L. D., Jr. 1971. Ecological studies of the timber wolf in northeastern Minnesota. St. Paul, MN: USDA Forest Service Research Paper No. NC-52, North Central Forest Experimental Station.

Mech, L. D., Frenzel, L. D., Jr., and Karns, P. D. 1971. The effect of snow conditions on the ability of wolves to capture deer. In Ecological Studies of the Timber Wolf in Northeastern Minnesota, eds. L. D. Mech and L. D. Frenzel, Jr., St. Paul, MN: USDA Forest Service, North Central Forest Experiment Station.

Mech, L. D., Frenzel, L. D., Jr., Karns, P. D., and Kuehn, D. W. 1970. Mandibular dental anomalies in white-tailed deer from Minnesota. Journal of Mammalogy 51:804–806.

Mech, L. D., Fritts, S. H., and Wagner, D. 1995. Minnesota wolf dispersal to Wisconsin and Michigan. American Midland Naturalist 133:368–370.

Mech, L. D., Goyal, S. M., Paul, W. J, and Newton, W. E. 2008. Demographic effects of canine parvovirus on a free-ranging wolf population over 30 years. Journal of Wildlife Diseases 44:824–836.

Mech, L. D., and Gese, E. M. 1992. Field testing the Wildlink Capture Collar on wolves. Wildlife Society Bulletin 20:249–256.

Mech, L. D., and Harper, E. K. 2002. Differential use of a wolf, Canis lupus, pack territory edge and core. Canadian Field Naturalist 116:315–316.

Mech, L. D., and Karns, P. D. 1977. Role of the wolf in a deer decline in the Superior National Forest. St. Paul, MN: USDA Forest Service Research Paper No. NC-148, North Central Forest Experimental Station.

Mech, L. D., Kunkel, K. E., Chapman, R. C., and Kreeger, T. J. 1990. Field testing of commercially manufactured capture collars on wild deer. Journal of Wildlife Management 54:297–299.

Mech, L. D., and McRoberts, R. E. 1990. Survival of white-tailed deer fawns in relation to maternal age. Journal of Mammalogy 71:465–467.

Mech, L. D., McRoberts, R. E., Peterson, R. O., and Page, R. E. 1987a. Relationship of deer and moose populations to previous winters' snow. Journal of Animal Ecology 56:615–628.

Mech, L. D., Nelson, M. E., and McRoberts, R. E. 1991. Maternal and grandmaternal nutrition effects on deer weights and vulnerability to wolf predation. Journal of Mammalogy 72:146–151.

Mech, L. D., Seal, U. S., and DelGiudice, G. D. 1987b. Use of urine in the snow to indicate condition of wolves. Journal of Wildlife Management 51:10–13.

Mech, L. D., and Tracy, S. 2004. Record high wolf, Canis lupus, pack density. Canadian Field Naturalist 118:127–129.

Meier, T. J., Burch, J. W., Mech, L. D., and Adams, L. G. 1995. Pack structure dynamics and genetic relatedness among wolf packs in a naturally regulated population. In Ecology and Conservation of Wolves in a Changing World, eds. L. D. Carbyn, S. H. Fritts, and D.R. Seip, pp. 293–302. Edmonton, Alberta: Canadian Circumpolar Institute, Occasional Publication 35.

Merrill, S. B., Adams, L. G., Nelson, M. E., and Mech, L. D. 1998. Testing releasable GPS collars on wolves and white-tailed deer. Wildlife Society Bulletin 26:830–835.

Merrill, S. B., and Mech, L. D. 2000. Details of extensive movements by Minnesota wolves. American Midland Naturalist 144:428–433.

Messier, F. 1995. Is there evidence for a cumulative effect of snow on moose and deer populations. Journal of Animal Ecology 64:136–140.

Mohr, C. O. 1947. Table of equivalent populations of North American small mammals. American Midland Naturalist 37:223–249.

Murie, A. 1944. The Wolves of Mount McKinley. U.S. National Park Service Fauna Ser. No. 5. Washington, DC: U.S. Government Printing Office.

Musiani, M., Mamo, C., Boitani, L., Callaghan, C., Cormack, G. C., Mattei, L., Visalberghi, E., Breck, S., and Volpi, G. 2003. Wolf depredation trends and the use of fladry barriers to protect livestock in western North America. Conservation Biology 17: 1539–1547.

Nagy, K. A. 1994. Field bioenergetics of mammals: what determines field metabolic rates. Australian Journal of Zoology 42:43–53.

Nelson, M. E. 1993. Natal dispersal and gene flow in white-tailed deer in northeastern Minnesota. Journal of Mammalogy 74:316–322.

Nelson, M. E., and Mech, L. D. 1981. Deer social organization and wolf depredation in northeastern Minnesota. Wildlife Monographs No. 77, pp. 1–53.

Nelson, M. E., and Mech, L. D. 1986a. Deer population in the central Superior National Forest, 1967–1985. St. Paul, MN: USDA Forest Service Research Paper No. NC-271, North Central Forest Experimental Station.

Nelson, M. E., and Mech, L. D. 1986b. Relationship between snow depth and gray wolf predation on white-tailed deer. Journal of Wildlife Management 50:471–474.

Nelson, M. E., and Mech, L. D. 1986c. Mortality of white-tailed deer in northeastern Minnesota. Journal of Wildlife Management 50:691–698.

Nelson, M. E., and Mech, L. D. 1987. Demes within a northeastern Minnesota deer population. In Mammalian Dispersal Patterns, eds. B. D. Chepko-Sade and Z. Halpin, pp. 27–40. Chicago, IL: University of Chicago Press.

Nelson, M. E., and Mech, L. D. 1990. Weights, productivity, and mortality of old white-tailed deer. Journal of Mammalogy 71:689–691.

Nelson, M. E., and Mech, L. D. 1991. White-tailed deer movements and wolf predation risk. Canadian Journal of Zoology 69:2696–2699.

Nelson, M. E., and Mech, L. D. 2006. Causes of a 3-decade dearth of deer in a wolf-dominated ecosystem. American Midland Naturalist 155:361–370.

Olson, S. F. 1938. Organization and range of the pack. Ecology 19:168–170.

Packard, J., Mech, L. D., and Seal, U. S. 1983. Social influences on reproduction in wolves. In Proceedings of the Canadian Wolf Workshop, ed. L. Carbyn, pp. 78–85. Ottawa, ON: Canadian Wildlife Service.

Packard, J. M., Seal, U. S., and Mech, L. D. 1985. Causes of reproductive failure in two family groups of wolves (Canis lupus). Zeitschrift fur Tierpsychology 68:24–40.

Peters, R., and Mech, L. D. 1975. Scent-marking in wolves: A field study. American Scientist 63:628–637.

Peterson, R. O., Woolington, J. D., and Bailey, T. N. 1984. Wolves of the Kenai Peninsula, Alaska. Wildlife Monographs No. 88, pp. 1–52.

Pimlott, D. H., Shannon, J. A., and Kolenosky, G. B. 1969. The ecology of the timber wolf in Algonquin Provincial Park, Ontario. Ottawa, ON: Ontario Department of Lands and Forests Research Report (Wildlife) No. 87.

Raymer, J., Wiesler, D., Novotny, M., Asa, C., Seal, U. S., and Mech, L. D. 1985. Chemical investigations of wolf (Canis lupus) anal sac secretions in relation to the breeding season. Journal of Chemical Ecology 2:593–608.

Raymer, J., Wiesler, D., Novotny, M., Asa, C., Seal, U. S., and Mech, L. D. 1986. Chemical scent constituents in the urine of wolf (Canis lupus) and their dependence on reproductive hormones. Journal of Chemical Ecology 12:297–313.

Ream, R. R., Fairchild, M. W., Boyd, D. K., and Pletscher, D. H. 1991. Population dynamics and home range changes in a colonizing wolf population. In The Greater Yellowstone Ecosystem: Redefining America's Wilderness Heritage, eds. R. K. Keiter and M. S. Boyce, pp. 349–366. New Haven, CT: Yale University Press.

Rothman, R. J., and Mech, L. D. 1979. Scent-marking in lone wolves and newly formed pairs. Animal Behavior 27:750–760.

Schultz, R. N., Jonas, K. W., Skuldt, L. H., and Wydeven, A. P. 2005. Experimental use of a dog training shock collar to deter depredation by gray wolves (Canis lupus). Wildlife Society Bulletin 33:142–148.

Seal, U. S., Erickson, A. W., and Mayo, J. G. 1970. Drug immobilization of the Carnivora. International Zoological Yearbook 10:157–170.

Seal, U. S., and Mech, L. D. 1983. Blood indicators of seasonal metabolic patterns in captive adult wolves. Journal of Wildlife Management 47:704–715.

Seal, U. S., Mech, L. D., and VanBallenberghe, V. 1975. Blood analyses of wolf pups and their ecological and metabolic interpretation. Journal of Mammalogy 56:64–75.

Seal, U. S., Nelson, M. E., Mech, L. D., and Hoskinson, R. L. 1978. Metabolic indicators of habitat differences in four Minnesota deer populations. Journal of Wildlife Management 42:746–754.

Seal, U. S., Plotka, E. D., Mech, L. D., and Packard, J. M. 1987. Seasonal metabolic and reproductive cycles in wolves. In Man and Wolf, ed. H. Frank, pp. 109–125. Boston, MA: Dr. W. Junk Publishers.

Shivik, J. A. 2006. Tools for the edge: What's new for conserving conservation. BioScience 56:253–259.

Stenlund, M. H. 1955. A field study of the timber wolf (*Canis lupus*) on the Superior National Forest, Minnesota. Minnesota Department of Conservation Technical Bulletin 4:1–55.

Van Ballenberghe, V., Erickson, A. W., and Byman, D. 1975. Ecology of the timber wolf in northeastern Minnesota. Wildlife Monographs 43, 43 pp.

Verme, L. J. 1968. An index of winter severity for northern deer. Journal of Wildlife Management 32:566–574.

Weise, T. F., Robinson, W. L., Hook, R. A., and Mech, L. D. 1979. An experimental translocation of the eastern timber wolf. In The Behavior and Ecology of Wolves, ed. E. Klinghammer, pp. 346–419. New York, NY: Garland STPM Press.

Wilson, P. J., Grewal, S., Lawford, D., Heal, J. N. M., Granacki, A. G., Pennock, D., Theberge, J. B., Theberge, M. T., Voigt, D. R., Waddell, W., Chambers, R. E., Paquet, P. C., Goulet, G., Cluff, D., and White, B. N. 2000. DNA profiles of the eastern Canadian wolf and the red wolf provide evidence for a common evolutionary history independent of the gray wolf. Canadian Journal of Zoology 78:2156–2166.

Chapter 3
Wolf and Moose Dynamics on Isle Royale

John A. Vucetich and Rolf O. Peterson

3.1 Background

Moose (*Alces alces*) arrived on Isle Royale in the early 1900s (Mech 1966). For 50 years moose interacted with the forest without being exposed to predation or significant human harvest. By the late-1920s the impact of moose on the forest had become noticeable and the population probably comprised between 2,000 and 3,000 moose (Murie 1934). By the mid-1930s many moose had died of malnutrition and the population declined to probably a few hundred animals (Hickie 1936).

Although there were suggestions and one attempt to introduce gray wolves (*Canis lupus*) to Isle Royale in the 1940s and 1950s, the attempt failed in 1952 (Mech 1966). While humans were trying to reintroduce wolves, they arrived on their own in the late 1940s by crossing an ice bridge connecting Isle Royale and Canada. Analysis of mitochondrial deoxyribonucleic acid (mtDNA) indicated that the population of wolves on Isle Royale was founded by a single female (Wayne et al. 1991). Since the founding event, the wolf population on Isle Royale has, to our knowledge, remained genetically isolated.

Humans do not harvest the wolves, moose, or forest on Isle Royale. Although present on the nearby mainland, white-tailed deer (*Odocoileus virginianus*), coyotes (*Canis latrans*), and black bears (*Ursus americanus*) are absent from Isle Royale. The diet of Isle Royale wolves is ~95% moose during winter, and the diet in summer is >85% moose. Most of the remaining diet consists of beavers (*Castor canadensis*). The only significant causes of death for moose on Isle Royale are wolf predation and malnutrition, both of which are sometimes exacerbated by severe winters and winter ticks (*Dermacentor albipictus*). Between 40% and 60% of the diet of moose in winter is a single species, balsam fir (*Abies balsamea*).

Although the wolf–moose system on Isle Royale is commonly characterized as a single-prey/single-predator system, this characterization may not be entirely justified. The importance of other factors such as canine parvovirus (Wilmers et al. 2006), moose ticks (Peterson and Vucetich 2006), and winter severity (Vucetich and Peterson 2004a) have been clearly documented. Nevertheless, compared with many communities of large vertebrates, the wolf–moose system on Isle Royale seems simple (Smith et al. 2003).

A.P. Wydeven et al. (eds.), *Recovery of Gray Wolves in the Great Lakes Region of the United States*,
DOI: 10.1007/978-0-387-85952-1_3, © Springer Science+Business Media, LLC 2009

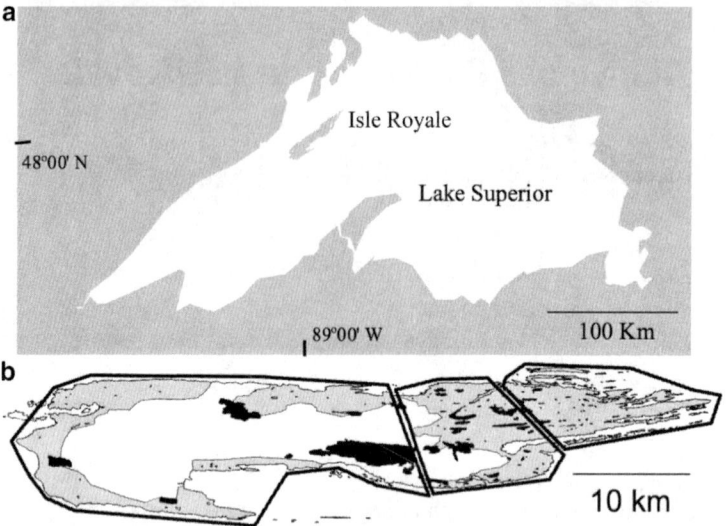

Fig. 3.1 a The location of Isle Royale within Lake Superior, North America. **b** Isle Royale is in most years inhabited by three wolf packs, whose typical territorial boundaries are indicated by the *thick-lined polygons*. The *gray regions* of the island represent area with higher moose density. The *white region* is roughly associated with a forest fire that burned in 1936. *Black areas* are inland lakes

Isle Royale is a long (72 km) and narrow (~7.5 km) archipelago with one main island (544 km²) and ~150 smaller surrounding islands (most <0.1 km²). The island is located in Lake Superior, ~24 km from the Lake's north shore (Fig. 3.1). The topography is rough due to glacial scouring of ridges and valleys running the length of the island. Elevation ranges from 180 m to 238 m above sea level.

The island is almost completely forested. The island's forests are usefully characterized by three distinct regions. The northeast region is transitional boreal forest, dominated by spruce (*Picea glauca*), balsam fir, aspen (*Populus tremuloides*), and paper birch (*Betula papyrifera*). The middle region was burned in 1936 and is currently dominated by 80-year-old stands of birch and spruce. The southwest region is covered with mixed stands of maple (*Acer saccharum*), yellow birch (*Betula alleghaniensis*), cedar (*Thuja occidentalis*), and spruce. Swamps and other wetlands are common in the numerous valleys on the island, but are more numerous in the eastern two-thirds of Isle Royale. The vegetation of Isle Royale, especially as it relates to moose herbivory, is further described in Pastor et al. (1998).

3.1.1 History of Research on Isle Royale

Continuous research on the wolves and moose of Isle Royale began in the summer of 1958 (Fig. 3.2). At that time, the research was primarily based on an annual winter census of wolves and moose. Beginning in the early 1970s, long-term

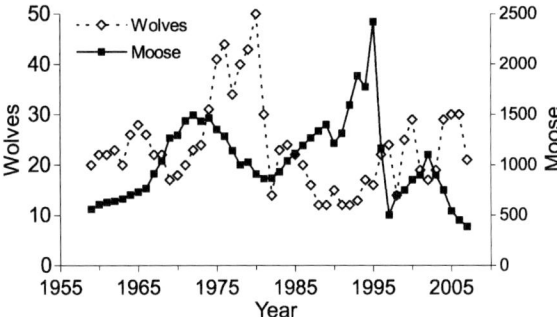

Fig. 3.2 Population trajectories of wolves and moose on Isle Royale, 1959–2007. Each year the entire wolf population is counted from a small aircraft (details in Peterson and Page 1988). The number of moose is estimated from population reconstruction (prior to 1995, see Solberg et al. 1999) and aerial surveys (after 1995, details in Peterson and Page 1993)

monitoring expanded to include per capita kill rate – a key statistic connecting populations of predator and prey – and systematic and more concerted effort to collect specific skeletal remains of dead moose (including skull, mandible, and metatarsus). The skeletal remains of approximately one-third of all moose that have ever lived in the population are eventually sampled; currently we have skeletal remains of more than 4,000 different moose. By the mid-1990s, long-term monitoring had expanded again to include aspects of forest structure and demography (especially tree-ring growth patterns of balsam fir).

3.1.2   Some Basic Demography

The density of moose on Isle Royale varies among the three basic habitat types. Typical densities in winter are 0.6 moose/km² in the island's middle region and 2.5 moose/km² in habitat types at the east and west ends of Isle Royale. For context, typical moose densities at other sites in North America tend to be <1.0 moose/km² and commonly <0.2 moose/km² (Karns 1997). Each January and February, calves represent 15% of the population, on average (coefficient of variation = 0.39). During the 1960s, twinning rates (proportion of cows with calves that had twins) were high (0.25). In the early-1970s, the rate dropped to ~0.10. In recent decades, the twinning rate has been less than ~0.05.

Empirical and analytical assessments suggest that the wolf population on Isle Royale is extremely inbred, has lost ~80% of it neutral genetic diversity since being founded, and continues to lose ~13% of its neutral diversity each generation (i.e., the effective population size is approximately three, and one wolf generation is ~4 years; Peterson et al. 1998). The ultimate impact of inbreeding on the wolves of Isle Royale is unclear. Although wolves on Isle Royale exhibit high rates of skeletal deformities (J. Räikkönen et al., unpublished data), whether fitness is affected by such deformities is unknown. Wolves on Isle Royale have vital rates (survival and recruitment) that are comparable with other healthy wolf populations (mean pack size = 4.9 [Coefficient of variation {CV} = 47] for 1967–2006, mean number of pups in mid-winter = 3.0 [CV = 90] for 1997–2006, mean annual mortality rate = 0.28 [CV = 60] for 1975–2006). However, since 1980 the number of wolves for every old (vulnerable) moose has been substantially less than before 1980 (Vucetich and Peterson 2004b).

3.2   Some Perspectives from Isle Royale

Here we present an annotated list of observations and inferences derived from studies of the wolves and moose of Isle Royale.

1. *The functional response and numerical response, fundamental elements of conventional predator–prey theory, represent inadequate bases for understanding kill rates and wolf–moose dynamics.*

The density of prey populations and the prey:predator ratio each have an important influence on per capita kill rate (Fig. 3.3a, b). This empirical finding conflicts with several models for Isle Royale wolves and moose that have assumed otherwise (e.g., Eberhardt 1997). However, "having an important influence" is critically different than claiming kill rate is adequately understood or predictable (contra Messier 1994). Because neither prey density nor prey:predator ratio predict more than about one-third of the variation in kill rate, neither seems adequate for understanding annual fluctuations in kill rate. Similarly, with respect to the numerical response, the kill rate explains about 22% of the variation in wolf population growth rate (Fig. 3.3c; Vucetich and Peterson 2004c). Clearly, kill rate has an important influence on growth rate of the wolf population. However, to merely identify kill rate as an important influence is far from concluding that one can reliably predict growth rate of the wolf population from kill rate. This ability is still beyond our reach, at least for Isle Royale wolves.

Much of the unexplained variation in kill rate and growth rate of the wolf population is likely attributable to factors such as climatic variation, age structure of the moose population, sampling error (in the measurement of kill rates), demographic stochasticity (Vucetich and Peterson 2004b), and behavioral factors such as pack size, experience, and leadership. The prospects for predicting some of these factors (e.g., annual climate) are severely limited.

The poverty of these relationships is meaningful because the functional response (Fig. 3.3a, b) and the numerical response (Fig. 3.3c) are the fundamental elements of conventional predator–prey theory. An important approach to predicting population dynamics is to assemble mechanisms such as the functional response and numerical response into a population model for the purpose of better understanding or predicting predator–prey dynamics (Messier 1994; Turchin 2003; Varley and Boyce 2006).

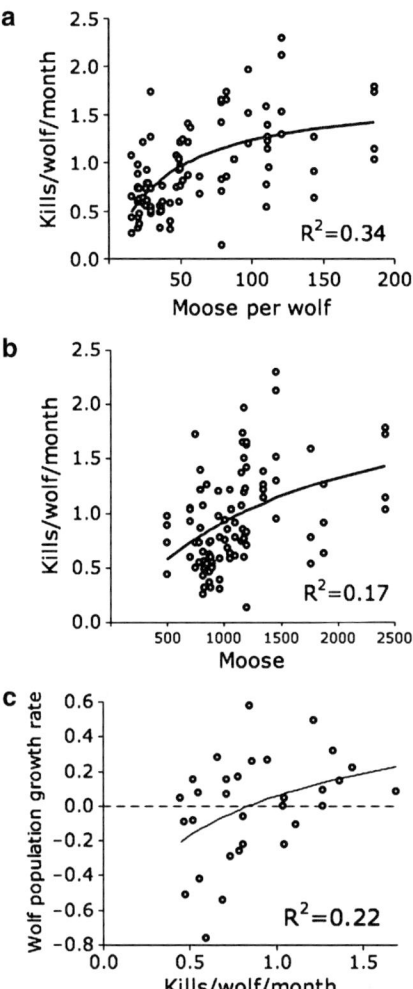

Fig. 3.3 The relationship between moose per wolf and per capita kill rate (**a**), the functional response (**b**), and numerical response (**c**) for wolves and moose on Isle Royale. Panels (**a**) and (**b**) are adapted from Vucetich et al. 2002, depict pack-specific kill rates, and represent data from 1971 to 1998. Panel (**c**) is adapted from Vucetich and Peterson (2004b), depicts population-level kill rates, and represents data from 1971 to 2002

The extent to which such models are valuable is limited if the underlying relationships are inadequate. Below we inspect the inferences and attitudes arising from three efforts to analyze such models in the context of wolf management.

First, Messier (1994) conducted a graphical analysis of deterministic models of the functional and numerical responses. The parameters for these responses were based on data collected from 27 studies from across North America where wolves and their prey have been observed for five or fewer years. Messier (1994, p. 484) concludes, from this graphical analysis that in the presence of a single predator, the wolf, moose would stabilize at ~1.3 moose/km^2, compared to an equilibrium density of 2.0 moose/km^2 with no predators" and that the addition of bear predation would reduce moose abundance to less than ~0.5 moose/km^2. Both [of these] equilibria are *caused* by density dependent food competition. If moose growth rate is reduced by only 5–10%, because of either a less productive habitat or a density invariant predation rate by an alternate predator like grizzly or black bear, a low density [<0.5 moose/km^2] equilibrium is predicted. This low equilibrium is the *result* of density dependent predation by wolves. The most striking feature of the model is the fact that a multiple-equilibrium system is practically impossible to generate [italics added].

We suppose it is the single-factor causal inferences (highlighted by the italicization) and/or the relative precision of the predictions that causes Messier (1994, p. 486) to ultimately conclude: "There is now a good theoretical and empirical understanding of the effect of wolf predation on moose population dynamics."

Varley and Boyce (2006) conducted simulations to predict population dynamics of wolves and elk (*Cervus elaphus*) in Yellowstone National Park. Their simulations were based on stochastic functional and numerical responses derived from empirical considerations. Compared with Messier (1994), the modesty of their conclusions is striking (Varley and Boyce 2006, p. 331).

[Our] models consistently have predicted neither an insignificant effect of wolves on elk numbers as some had once believed, or enormous effects that are tantamount to ecological collapse as has been popularized outside the scientific community. Rather, the predictions are of moderate reductions in elk numbers with a sustainable, moderate hunter harvest.

They predicted that two very extreme cases are unlikely. Although one may be struck by the modesty of the inference, it is significant that the inference seems reasonably justified from the model.

Turchin's (2003) analysis of a similarly structured model led him to be struck by that which most modelers of wolf populations had overlooked. Turchin (2003, p .382) wrote

wherever deer populations are not heavily affected by humans, oscillations with a period of roughly 30–50 years appear to be the rule, rather than the exception. If cervid dynamics indeed turn out to be more prone to oscillation, then … the current discourse about the limiting and regulating factors largely misses an important point… the main question becomes what factors are responsible for the oscillatory nature of dynamics, which factors ensure the oscillations do not get out of hand, and which factors are responsible for stochastic fluctuations in the realized per capita rate of change. (A partial answer to the last question, for moose on Isle Royale, is reflected in Fig. 3.4.)

Turchin may disagree (we are not sure), but his writing seems consistent with an important attitude expressed by Sir D'arcy Thompson, the so-called father of mathematical biology, and analyzed by the philosopher, E. Keller, in her book *Making Sense of Life* (Fox 2003). Thompson wrote, in his classic *On Growth and Form* (1942, p. 643) that "It is the principle involved, and not its ultimate and very complex results, that we can alone attempt to grapple with." Although we may be able to understand and explain past dynamics, it seems overly optimistic to entertain even modestly precise predictions about future states of wolf and prey populations.

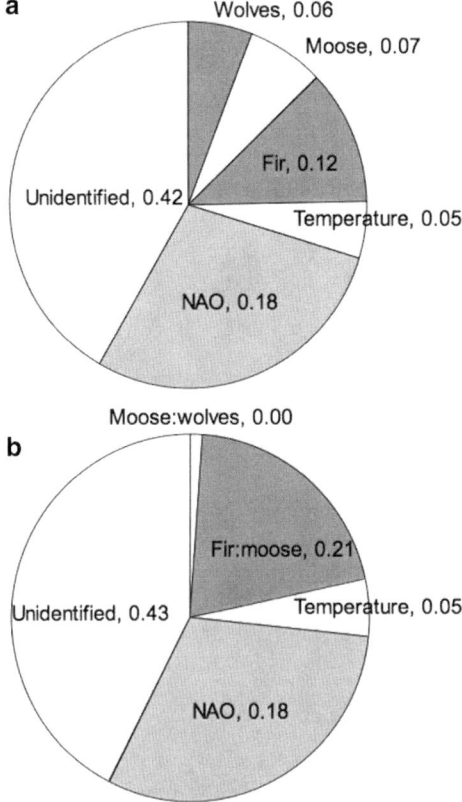

Fig. 3.4 The partitioning of variance for the growth rate of the moose population on Isle Royale (1959–2006) associated with two multiple regressions. Panel (**a**) relates growth rate to abundances, and panel (**b**) relates growth rate to ratios of abundances. Fir refers to balsam fir, the primary winter forage for moose; temperature is summer temperature; and NAO is the North-Atlantic Oscillation, which is an index of winter severity. Adapted from Vucetich and Peterson (2004b)

2. *Recruitment of moose is an important predictor of moose population growth, but it is not well predicted by the abundance of wolves.*

It is well established that wolves have a strong preference to prey upon calves relative to healthy, prime-aged ungulates (Peterson 1977; Wright et al. 2006). From this observation, some argue for wolf control because they believe that increased abundance of wolves reduces ungulate abundance (National Research Council 1997). However, preferring to prey on calves and reducing ungulate abundance are entirely different propositions. Observations from Isle Royale indicate that while calf recruitment *is* an important determinant of moose population growth rate (Fig. 3.5a), the rate of calf recruitment *is not* significantly impacted by the abundance of wolves (Fig. 3.5b). (On Isle Royale, calf recruitment is measured as the proportion

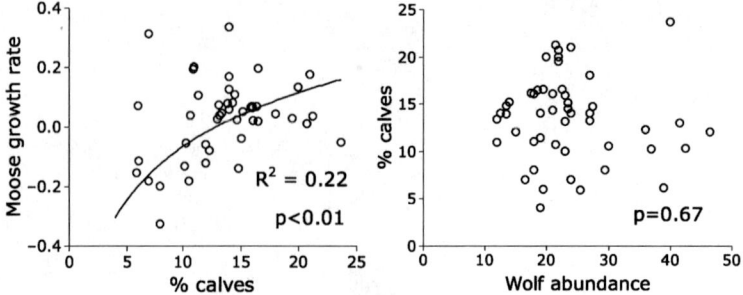

Fig. 3.5 Calf production has an important influence on growth rate of the moose population (*left panel*). However, the number of wolves is not well correlated with annual calf production (*right panel*). Percent calves is the percent of the total moose population in early February that are calves. Calf production depends on the combined effects of many factors (predation, food, and climate). The details are poorly understood. The number of wolves is the average of the current year and previous year. Data are from 1959 to 2007

of moose in the population that are calves during mid-winter.) Other evidence also lends support to the notion that fluctuations in wolf abundance do not necessarily impact the population growth rate of moose (Fig. 3.4), even though wolves prefer calves and calf recruitment has an important influence on moose population dynamics (Gaillard et al. 2000).

Two plausible explanations for the difference between the observations from Isle Royale and those of other studies, which suggest increased wolf abundance does reduce recruitment of ungulate prey (Gasaway et al. 1992; and references in Mech and Peterson 2003), are: (1) these other studies may be too short in duration to obtain an accurate perspective (see below) and (2) these studies also may be biased by a tendency to make observations when low recruitment happens to coincide with high wolf abundance, which does occasionally happen, even on Isle Royale (Fig. 3.5b).

3. *The relative influence of forage, predation, and climate on the population dynamics of moose on Isle Royale depends on the time scale.*

We recently analyzed data from Isle Royale with time series analyses to quantify the proportion of variation in moose population growth rate that could be attributed to annual fluctuations in predation (as indexed by wolf abundance), winter forage abundance (as indexed by dendrochronology of balsam fir and moose density), and climate (as indexed by the North Atlantic Oscillation). Using data observed between 1959 and 1999, that analysis indicated (Vucetich and Peterson 2004a) that (1) most fluctuations in moose population growth were not explained by any of these factors, (2) wolf abundance represented the least important factor, and (3) climate was more important than winter forage for explaining year-to-year fluctuations in the moose population (Fig. 3.4).

By contrast, alternative considerations were used to suggest that wolf predation represented a very strong top-down force on the moose of Isle Royale (McLaren and Peterson 1994). These considerations begin by observing that the accidental introduction of canine parvovirus to wolves on Isle Royale in 1980 represented a perturbation that was exogenous to wolf–moose interactions. The disease triggered a nearly decade-long reduction in the abundance of wolves and was associated with a two- to threefold increase in abundance of moose over the same time period. Moose abundance was eventually reduced in 1996 by the convergence of three extreme events: the most severe winter of the twentieth century, a severe outbreak of winter ticks (*Dermacentor albipictus*), and a severe reduction in the abundance of forage.

These two analyses differ not only in their conclusion but also in the time scale being assessed. Whereas the regression analysis of Vucetich and Peterson (2004b) is primarily focused on short-term, year-to-year fluctuations in moose abundance, the historical analysis of McLaren and Peterson (1994) focused on a longer-term, decadal time scale. That is, McLaren and Peterson (1994) assessed how longer-term, sustained reduction in wolf abundance resulted in a decade long period of moose population growth.

More recently, we repeated the time series analyses described above on two subsets of the Isle Royale data, prior to and after the introduction of canine parvovirus. That analysis indicated that during the first two-decade period wolf abundance was the most important predictor of moose fluctuations, and that during the second two-decade period wolf predation was trivial in its importance, but climate was very important (Wilmers et al. 2006).

The most important lesson to derive from these analyses is that our sense about the relative strength of top–down, bottom–up, and abiotic factors can vary, even within a single system, with the time scale being considered and even the time period under consideration.

4. *Age structure of the moose population is an important predictor of wolf–moose dynamics.*

Relatively few models and predictions of ungulate population dynamics account for the influence of age structure. However, when it has been examined, the effect of age structure on ungulate population dynamics seemed to be important (e.g., Coulson et al. 2001; Festa-Bianchet et al. 2003). Age structure is an important element of population dynamics when several conditions hold: (1) age structure fluctuates over time, (2) such fluctuations are not entirely associated with other predictors of population growth (e.g., population density), and (3) population growth rate varies systematically with age structure. Moose on Isle Royale seem to be characterized by these conditions (Fig. 3.6).

It has long been known that wolves prefer to prey upon calves and senescent moose (Peterson 1977). However, it is not known to what extent varying age structure is caused by wolf predation (relative to, say, climatic variation) or the extent to which population dynamics of wolves are affected by variations in the age structure of their prey population. These underappreciated interactions likely account for important complexities in wolf–prey systems.

5. *The future dynamics of a managed population are as likely to be like past dynamics, as not.*

In the context of informal settings (e.g., public talks and discussions with managers and colleagues), we often characterize the most general conclusion of

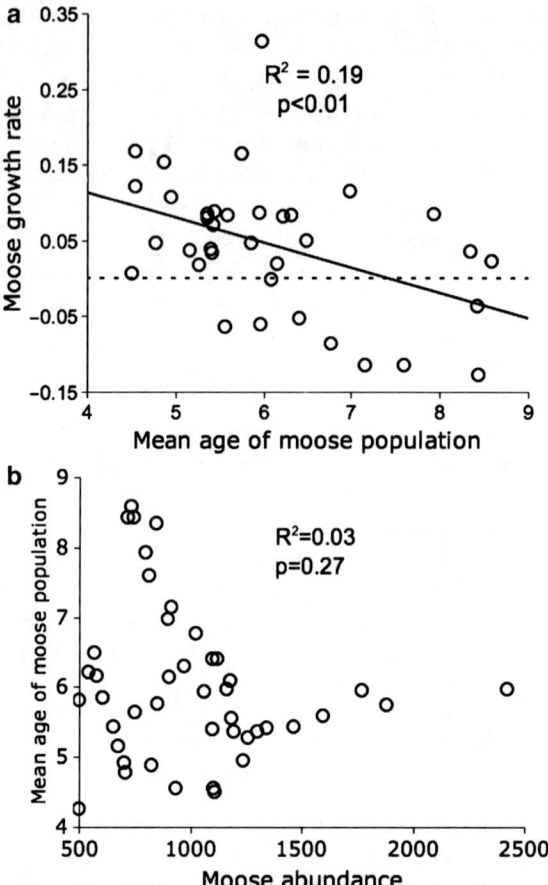

Fig. 3.6 The relationship between mean age of the population and population growth rate for moose on Isle Royale, 1959–1995 (**a**) and between moose abundance and mean age for the same time period (**b**). Mean age is derived from reconstructed population structure, which is based on necropsies of ~4,000 moose collected between 1958 and 2006. From the necropsies we learn the date of death and age at time of death. Data do not extend beyond 1995 because the reconstructed population structure cannot be estimated until most of the moose in a particular cohort are dead

research on wolves and moose on Isle Royale in two ways. First, even after 50 years of observation, each 5-year period of the wolf–moose chronology seems to be importantly different from every other 5-year period. Second, the longer we study, the more we realize how poorly we understand the population dynamics of wolves and moose on Isle Royale. We scrutinize these conclusions in formal analyses presented elsewhere (Vucetich et al. in press) and summarize the results of those analyses here.

– For no other purpose than as a heuristic, suppose that a simple explanation for a positive correlation between predator abundance and prey abundance is that abundance of prey largely determines abundance of predators; a negative correlation may suggest that predators determine prey abundance and a weak correlation may indicate either a more complex interaction or a weak interaction. Between 1959 and 2006, the correlation between wolf abundance and moose abundance has been negative, but not strongly so ($r = -0.26$, $R^2 = 0.07$, $P = 0.08$). However, estimated correlations for shorter time intervals have fluctuated greatly throughout the first 50 years of the study.

– To assess quantitatively how the estimated correlation has fluctuated over time, and how it has depended on the length of observation, we calculated a set of correlations, each depending on a different subset of the data. First, we estimated the correlation (and R^2) for each 5-year, consecutive set of observations (e.g., 1959–1963, 1960–1964, ... 2002–2006). There are 44 such sets of data. Then we estimated the correlation (and R^2) for each 10-year, consecutive set of observations (e.g., 1959–1968, 1960–1969, ... 1997–2006). There are 39 such sets of data. We continued this procedure for sets of data that were 15, 20, 25, 30, 35, 40, 45, and 50 years in length. The result is depicted in Fig. 3.7.

We appreciate that these data sets are not independent. We are careful to limit inferences drawn from this analysis (see below) to those that would be insensitive to this lack of independence. Our inferences are motivated by appreciating that one could have observed the wolves and moose of Isle Royale beginning in any year and continuing for any period.

– Estimated values of r range from nearly –1 to 1, and instances of strong positive and strong negative correlation are common (Fig. 3.7a). The variation in r is substantially reduced for periods of observation that are 15 years and greater. The average R^2 declines with increasing periods of observation (Fig. 3.7b). Keep in mind that R^2 is sometimes taken as a measure of the explanatory power of a model.

Studies of shorter duration (5–10 years) frequently suggest that simple explanations may provide high levels of predictive or explanatory power. However, studies of longer duration (>15 years) make clear that such simple explanations are of less value. Moreover, long-term research is also necessary for developing complex ideas. This is because in the context of multiple regressions, detecting the influence of even moderately important predictor variables requires about ten observations per predictor variable (i.e., a model with five predictor variables may require upwards of 50 observations).

Even at the longest periods for which we can judge, wolf–moose dynamics from one time period differ from those of the previous time period. More precisely, the first 25-year period of wolf–moose dynamics was characterized by significantly stronger top–down influences than the second 25-year period of observation (Wilmers et al. 2006). It seems more likely than not that during the next 50 years wolf–moose dynamics will be different than they have been for the previous 50 years. Models that provided useful explanations of past dynamics were, as it turned out, a poor basis for inferring future dynamics, at least on Isle Royale. It seems reasonable to presume that this pattern would characterize many natural systems.

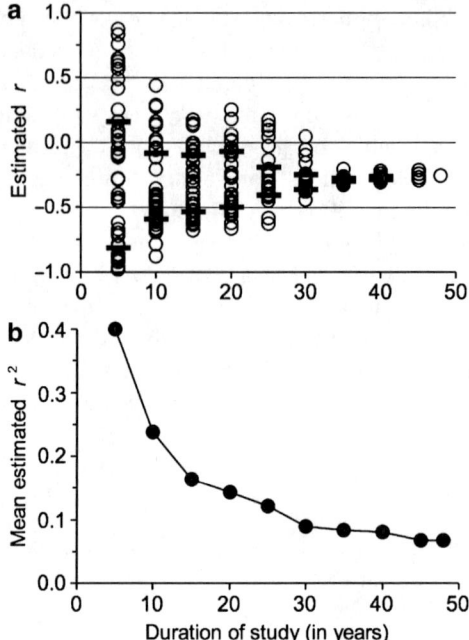

Fig. 3.7 **a** Estimated correlation coefficient (r) between abundance of wolves and abundance of moose across years (see Fig. 3.2). Each estimate is based on a different subset from the time series of abundances of wolves and moose. Each subset of data represents abundances from consequent years and is characterized by the number of years of observation (x-axis). Many of the data subsets are overlapping, and, therefore, not entirely independent. Heavy bars represent the interquartile range for each duration of observation. **b** Mean value of estimates for R^2 for the various subsets of data representing different durations of observation. Insomuch as r [panel (**a**)] represents a simple model of wolf-moose dynamics, R^2 represents the explanatory power of that simple model. Panel (**b**) suggests that with increased duration of observation, the explanatory power of this simple model tends to decline substantially over time. Adapted from Vucetich et al. (in press)

To distinguish good explanations of the past from reliable predictions of the future may have broad implications for conservation. Others have argued (e.g., Holling and Meffe 1996; Ludwig et al. 1993) that many conservation problems arise from our obsession for controlling and managing nature, an obsession fueled by a confident belief about our ability to control nature more reliably and precisely than may in fact be possible. An overconfident sense about one's ability to control nature is liable to arise from an overconfident sense about one's ability to predict natural phenomena, such as population growth rates. The Isle Royale experience suggests how overconfidence about our ability to predict nature may arise from confusing a model's ability to offer good explanations of the past with reliable and precise predictions of the future.

Students, managers, and members of the general public regularly ask us whether the limited ability to predict future dynamics of wolves and moose on Isle Royale is occasion for discouragement about our effort to understand nature. Our reaction to such questions is that it seems we can be proud of the undeniably impressive knowledge we have about nature. However, it also remains true that our knowledge is pale when compared to nature's complexity. This juxtaposition – that we know much in an absolute sense, but very little in a relative sense – is not occasion for discouragement, but an occasion to be filled with wonder about and respect for the natural world. We find this attitude enriching, not discouraging.

References

Coulson, T., Catchpole, E., Albon, S., Morgan, B. J. T., Pemberton, J. M., Clutton-Brock, T. H., Crawley, M. J., and Grenfell, B. T. 2001. Age, sex, density, winter weather, and population crashes in Soay sheep. Science 292:1528–1531.

Eberhardt, L. L. 1997. Is wolf predation ratio-dependent? Canadian Journal of Zoology 75:1940–1944.

Festa-Bianchet, M., Gaillard, J. M., and Cote, S. D. 2003. Variable age structure and apparent density dependence in survival of adult ungulates. Journal of Animal Ecology 72:640–649.

Fox, E. K. 2003. Making Sense of Life: Explaining Biological Development with Models, Metaphors, and Machines. Cambridge, MA: Harvard University Press.

Gaillard, J. M., Festa-Bianchet, M., Yoccoz, N. G., Loison, A., and Toigo, C. 2000. Temporal variation in fitness components and population dynamics of large herbivores. Annual Review of Ecology and Systematics 31:367–393.

Gasaway, W. C., Boertje, R. D., Grangaard, D. V., Kelleyhouse, D. G., R. O. Stephenson, R. O., and Larsen, D. G. 1992. The role of predation in limiting moose at low-densities in Alaska and Yukon and implications for conservation. Wildlife Monographs 120:1–59.

Hickie, P. F. 1936. Isle Royale moose studies. Transactions of the North American Wildlife Conference 1:396–399.

Holling, C. S., and Meffe, G. K. 1996. Command and control and the pathology of natural resource management. Conservation Biology 10:328–337.

Jost, C., G. Devulder, J. A. Vucetich, R. O. Peterson, and R. Arditi. 2005. The wolves of Isle Royale display scale-invariant satiation and density dependent predation on moose. Journal of Animal Ecology 74:809–816.

Karns, P. D. 1997. Population distribution, density and trends. In Ecology and Management of the North American moose, eds. A. W. Franzmann and C. C. Schwartz, pp. 125–140. Washington, DC: Smithsonian Institution Press.

Ludwig, D., Hilborn, R., and Walters, C. 1993. Uncertainty, resource exploitation, and conservation: Lessons from history. Science 260:17–36.

McLaren, B. E., and Peterson, R. O. 1994. Wolves, moose, and tree rings on Isle Royale. Science. 266:1555–1558.

Mech, L. D. 1966. The wolves of Isle Royale. National Park Service Faunal Series Scr. No. 7. Washington, DC: U.S. Government Printing Office.

Mech, L. D., and Peterson, R. O. 2003. Wolf-prey relations. In Wolves: Behavior, Ecology, and Conservation, eds. L. D. Mech and L. Boitani, pp. 131–160. Chicago, IL: University of Chicago Press.

Messier, F. 1994. Ungulate population-models with predation – a case-study with the North-American moose. Ecology 75:478–488.

Murie, A. 1934. The moose of Isle Royale. Miscellaneous Publications of the Museum of Zoology No. 25. Ann Arbor, MI., pp. 1–11.

National Research Council. 1997. Wolves, bears, and their prey in Alaska: Biological and social challenges in wildlife management. Washington, DC: National Academy Press.

Pastor, J., Dewey, B., Moen, R., Mladenoff, D. J., White, M., and Cohen, Y. 1998. Spatial patterns in the moose-forest-soil ecosystem on Isle Royale, Michigan, USA. Ecological Applications 8:411–424.

Peterson, R. O. 1977. Wolf ecology and prey relationships on Isle Royale. Natl. Park Service Sci. Monogr. Ser. No. 11. Washington, DC: U.S. Government Printing Office.

Peterson, R. O., and Page, R. E. 1988. The rise and fall of Isle Royale wolves, 1975–1986. Journal of Mammalogy 69:89–99.

Peterson, R. O., and Page, R. E. 1993. Detection of moose in midwinter from fixed-wing aircraft over dense forest cover. Wildlife Society Bulletin 21:80–86.

Peterson, R. O., Thomas, N. J., Thurber, J. M., Vucetich, J. A., and Waite, T. A. 1998. Population limitation and the wolves of Isle Royale. Journal of Mammalogy 79:828–841.

Peterson, R. O., and Vucetich, J. A. 2006. Ecological studies of wolves on Isle Royale, 2005–2006 Annual Report. Michigan Technological University, Houghton, MI. Available at http://www.isleroyalewolf.org

Smith, D. W., Peterson, R. O., and Houston, D. B. 2003. Yellowstone after wolves. Bioscience 53:330–340.

Solberg, E. J., Saether, B.-E., Strand, O., and Loison, A. 1999. Dynamics of a harvested moose population in a variable environment. Journal of Animal Ecology 68:186–204.

Turchin, P. 2003. Complex Population Dynamics: A Theoretical/Empirical Synthesis. Princeton, NJ: Princeton University Press.

Varley, N., and M. S. Boyce, M. S. 2006. Adaptive management for reintroductions: Updating a wolf recovery model for Yellowstone National Park. Ecological Modeling 193:315–339.

Vucetich, J. A., and Peterson, R. O. 2004a. The influence of prey consumption and demographic stochasticity on population growth rate of Isle Royale wolves (*Canis lupus*). Oikos 107:309–320.

Vucetich, J. A., and Peterson, R. O. 2004b. The influence of top-down, bottom-up, and abiotic factors on the moose (*Alces alces*) population of Isle Royale. Proceedings of the Royal Society of London, B 271:183–189.

Vucetich, J. A., and Peterson, R. O. 2004c. Long-term population and predation dynamics of wolves on Isle Royale. In Biology and Conservation of Wild Canids, eds. D. Macdonald and C. Sillero-Zubiri, pp. 281–292. Oxford, UK: Oxford University Press.

Vucetich, J. A., Peterson, R. O., and Schaefer, C. L. 2002. The effect of prey and predator densities on wolf predation. Ecology 83:3003–3013.

.Vucetich, J. A., Peterson, R. O., and Nelson, M. P. in press. Will the future of Isle Royale wolves and moose always differ from our sense of their past? In The World of Wolves: New Perspectives on Ecology, Behaviour and Policy, eds. M. Musiani, L. Boitani, and P. Paquet. Calgary, AB, Canada: University of Calgary Press

Wayne, R. K., Lehman, N., Girman, D., Gogan, P. J. P., Gilbert, D. A., Hansen, K., Peterson, R. O., Seal, U. S., Eisenhawer, A., Mech, L. D., and Krumenaker, R. J. 1991. Conservation genetics of the endangered Isle Royale gray wolf. Conservation Biology 5:41–51.

Wilmers, C. C., Post, E. S., Peterson, R. O., and Vucetich, J. A. 2006. Disease mediated switch from top-down to bottom-up control exacerbates climatic effects on moose population dynamics. Ecological Letters 9:383–389.

Wright, G. J., Peterson, R. O., Smith, D. W., and Lemke, T. O. 2006. Selection of Northern Yellowstone elk by gray wolves and hunters. Journal of Wildlife Management 70:1070–1078.

Chapter 4
An Overview of the Legal History
and Population Status of Wolves in Minnesota

John Erb and Michael W. DonCarlos

4.1 Introduction

The modern history of wolf populations in Minnesota, like other areas, is as much a story about humans as it is about wolves. While competition among wolves is in large part driven by the availability of resources such as food and space (Fuller et al. 2003; Packard 2003), wolf–human interactions are a function of real "competition" as well as perceived conflict or fear. Cultural attitudes have clearly played a major role in the dynamics between wolves and humans (Boitani 1995; Fritts et al. 2003). Such human–wolf dynamics are not constant. Cultural attitudes change, the number of people (and livestock) living in close proximity to wolves changes, and the availability and degree of dependence on shared resources changes. Our goal is to examine the legal and population history of wolves in Minnesota, but in so doing we provide a manifestation of human–wolf dynamics, and provide context for understanding the changes in these dynamics through time. We summarize changes in the legal status of wolves and changes in wolf distribution and abundance. We also highlight ecological factors associated with a changing wolf population, and provide an overview of the methods used by the Minnesota Department of Natural Resources (DNR) to monitor the statewide wolf population.

4.2 The Legal History of Minnesota's Wolves

The legal history of wolves in Minnesota is a tale of public policy extremes. From statehood (1858) until about 1970, wolves were completely unprotected in all of Minnesota, and for an extensive portion of this time, wolves were actively persecuted by the federal and state government, as well as by private citizens. Following passage of the Endangered Species Preservation Act of 1966, a precursor to the Endangered Species Act of 1973, the Department of Interior classified the eastern timber wolf as endangered in 1967. This law allowed for legal protections

A.P. Wydeven et al. (eds.), *Recovery of Gray Wolves in the Great Lakes Region of the United States*,
DOI: 10.1007/978-0-387-85952-1_4, © Springer Science+Business Media, LLC 2009

to be instituted on federal lands, and in 1970, the majority of the Superior National Forest (SNF) was closed to the taking of wolves. While the SNF is a small portion of Minnesota, this closure probably protected a significant proportion of wolves present in Minnesota at that time. After passage of the Endangered Species Act of 1973, wolves in all of Minnesota were afforded complete federal protection in 1974. In effect, with the signing of a pen, Minnesota's wolves went from no legal protections to complete legal protection. From 1974 to 1978, wolves could only be killed in defense of human life.

4.2.1   Bounty Era (1849–1965)

Soon after the Minnesota territory was organized in 1849, the Minnesota legislature authorized counties to pay a $3 bounty for wolves. Authorization for bounty payments was made biennially by the state legislature, and the bounty system remained in place until 1965. During this period, numerous changes were made to the bounty system, including the payment amounts, funding source (county and/or state), and requirements for payment approval. Bounty payments ranged from $3 per animal in the beginning of the program, to $35 per animal in the latter years. Initially, bounty payments were the responsibility of counties, followed by various cost-sharing arrangements between counties and the state, with the state assuming full responsibility in the latter years (Minnesota DNR, unpublished data).

From 1946 to 1964, it was legal under the bounty program for private citizens to obtain permits to shoot wolves from airplanes. However, aerial shooting over the Boundary Waters Canoe Area (BWCA), a "stronghold" for wolves at the time, was eliminated in 1950 when all flights under 1,200 m were prohibited. During the first year that aerial shooting by private citizens was allowed, Stenlund (1955) noted that one operator took 38 wolves by this method. He also noted that wolves quickly learned to avoid open lakes when they heard airplanes approaching.

Early records of the total bounty take are sparse and complicated by the lack of record-keeping distinction between coyotes and wolves. The most reliable records for wolves are for the period 1952–1964, when an average of 188 wolves were submitted for bounty payment each year, at an average annual cost of $6,144. Stenlund (1955) also reported 290 and 295 wolves submitted for bounty in 1950 and 1951, respectively.

In addition to the bounty program, state personnel were involved in wolf removal from the late 1940s through the mid-1950s, including via aerial shooting. Outside the BWCA, aerial shooting by state personnel continued until 1954, and other forms of wolf control (shooting and trapping) by state employees ended in 1956. From 1949 to 1953, ~140–150 wolves were taken annually by state employees (Minnesota 1980). The take by state employees dropped to ~80 animals per year during the final years of the program (1954–1956).

4.2.2   Postbounty, Pre-ESA Era (1965–1973)

In 1969, the Minnesota legislature authorized a directed predator control program. The program was a stark contrast to the bounty system, which encouraged unrestricted and widespread take to reduce the wolf population. Under the new program, private trappers certified by the state were authorized to remove wolves only from designated areas where losses of livestock had been verified. Hence, the program was focused on ameliorating localized conflicts, not population reduction. As compensation for control work, certified trappers were paid $50 per wolf. From 1969 until wolves were federally protected in 1974, an average of 65 wolves were removed annually as part of this program (Minnesota DNR 1980), substantially fewer than were removed during the era of bounties and government control. After wolves became federally protected in 1974, management of human–wolf conflicts in Minnesota shifted to federal agencies.

4.2.3   Federal Protection Era (1974–2007)

Wolves were neither federally nor state protected (except on the SNF) in Minnesota until 1974 when they were listed under the US Endangered Species Act of 1973. From 1974 to 1978, wolves were federally classified as an endangered species (Refsnider, this volume) with no provisions for lethal control in response to depredations on livestock. In 1978, the federal status of Minnesota's wolves changed from endangered to threatened, thereby allowing lethal control of depredating wolves under federal guidelines. From 1978 to 2007, an average of 91 (range 6–216) wolves were taken annually by federal employees as part of depredation control activities (Ruid et al., this volume).

After passage of a state endangered species act in 1974, wolves were state-classified as an endangered species, down-listed to state threatened in 1984, and removed from the state's list of threatened and endangered species in 1996. However, the federally threatened status of Minnesota's wolves did not change until 2007, when federal protection of wolves in the Great Lakes region was finally removed. Management authority for wolves in Minnesota now resides with the state and Indian tribes, and wolves are classified as a protected mammal.

4.2.4   Current State Management

The state's wolf management plan (Minnesota DNR 2001) went into effect with federal delisting in 2007. Under the plan, control of depredating wolves continues in Minnesota using guidelines specified in state statutes, and in a majority of wolf range is similar to federal guidelines for wolf depredation take from 1978 to 2007.

Two state wolf management zones were established in the plan, differing only in depredation policies. In Zone A (Fig. 4.1), which constitutes ~85% of current wolf range, wolves may be taken by private citizens, under certain conditions, if they pose an immediate threat (as defined in state statutes) to livestock or domestic pets under owner supervision. Furthermore, when losses of livestock or pets have been verified as wolf depredations, the state will provide a governmental or state-certified private trapper to remove wolves in a defined area. These same rules apply in Zone B, but landowners, under certain conditions, are given added flexibility (immediate threat does not apply) to take wolves to protect livestock, and they may individually hire a state-certified trapper to protect livestock in a defined area. A Minnesota state statute prohibits the public harvest of wolves for the first 5 years after federal delisting. After that period, the Minnesota DNR is authorized to prescribe and regulate public harvest of wolves, but must provide an opportunity for public comment.

It is difficult to quantify the specific effects various historic control programs had on Minnesota's wolf population. In northeastern Minnesota, the area in which wolves had largely been restricted by the 1950s, Stenlund (1955) estimated that the combination of take by private citizens (bounty) and state employees in the early 1950s may have removed 41% of the wolf population, annually. He further pointed to evidence that the wolf population during this time was relatively stable, concluding that control efforts neither reduced wolf populations nor allowed the population to potentially increase. Even if Stenlund (1955) was correct, it is likely that statewide

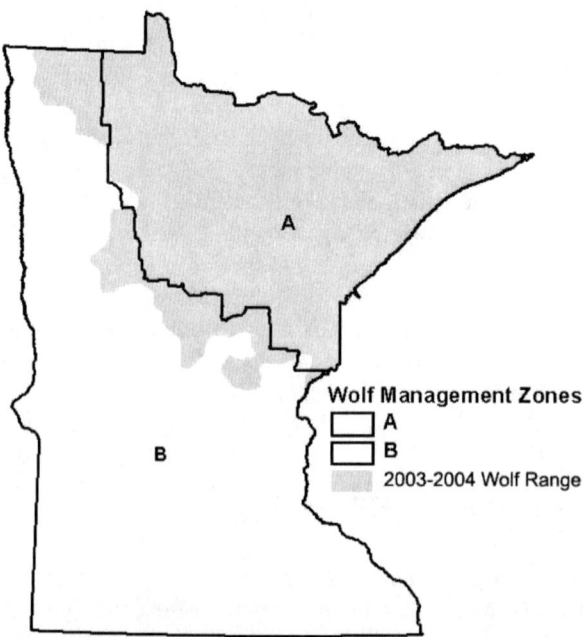

Fig. 4.1 State wolf management zones in Minnesota in relation to the contiguous wolf range delineated in 2003–2004

control efforts had spatially and temporally varying effects on Minnesota's wolf population. Persecution undoubtedly played an important role in once eliminating wolves from all but the remote forests of Minnesota adjacent to Canada. However, population changes are complicated by other factors also affecting wolf ecology, in particular major changes in Minnesota's landscape that affected the distribution and abundance of prey species of wolves (DelGiudice et al., this volume). Hence, the population history of wolves in Minnesota is not only a manifestation of human attitudes and the legal history of wolves, but also the history of their prey.

4.3 The Population History of Minnesota's Wolves

4.3.1 Pre-European Settlement Era

When the Endangered Species Act of 1973 was passed, the only remaining population of wolves in the lower 48, excluding those on Isle Royale, was in northern Minnesota. As early as 1938, Olson (1938), with reasonable accuracy, offered such a prediction of future wolf distribution. As one progresses backward in time, the distributional "picture" of wolves in Minnesota, while still quite coarse, is much clearer than their abundance. At the time Minnesota was settled by Euro-Americans, the distribution of wolves likely encompassed the entire state (Herrick 1892; Surber 1932). Based on reported density estimates for wolves exposed to prey populations similar to those that historically existed in Minnesota, Mech (2000) speculated that the original wolf population may have numbered between 4,000 and 8,000 wolves.

4.3.2 Bounty Era (1849–1965)

By 1900, and perhaps sooner, wolves were rare in the southern and western portions of Minnesota (Herrick 1892; Surber 1932), no doubt a combined result of wolf persecution and extirpation (or near so) of the bison and elk herds that once roamed this region (DelGiudice et al., this volume). By 1930, the range of wolves in Minnesota had further contracted to the north, with the remaining wolves surviving primarily in portions of the forested counties bordering Canada (~31,000 km²; Stenlund 1955). Stenlund (1955) noted that early Forest Service records indicate that 300–600 wolves may have occupied the SNF from 1914 to 1931. However, interpretation is complicated by the changing acreage of the SNF during this time, and different personnel responsible for making the reports. Based on his work in the 1920s and 1930s, Olson (1938) provided an estimate of 250 wolves occupying a 6,500-km² area of the SNF immediately adjacent to Canada, or 1 wolf per 26 km². In the late 1940s to mid-1950s, Stenlund (1955) estimated that between 205 and 273 wolves occupied a 10,600-km² area in northeastern Minnesota. Stenlund's

(1955) larger study area overlapped that of Olson's (1938), suggesting a population decline in this area from 1 wolf per 26 km^2 in around 1930 to 1 wolf per 44 km^2 in around 1950. Stenlund (1955) attributed this wolf decline to a reduction in deer populations. The decline in the deer herd was attributed primarily to maturation of the forest following turn-of-the-century logging (which initially led to deer population increases), and overbrowsing by the abundant deer population in the mid- to late 1930s.

While these studies provided some indication of the density of the wolf population on portions of the SNF, we do not know whether such density estimates applied to the remainder of the wolf range, nor do we know the precise extent of statewide wolf distribution at that time. Stenlund (1955) extrapolated his results to an 18,000-km^2 area of northeastern Minnesota and estimated 300–400 wolves. However, this 18,000-km^2 area represented only 60% of what he delineated as primary wolf range at the time. If the density estimates were applicable to all primary wolf range, we estimate as many as 700 wolves in Minnesota between 1920 and 1960. If we use Olson's (1938) estimate of 1 wolf per 26 km^2 when prey was more abundant, and assume a similar primary wolf range, we estimate ~1,200 wolves in Minnesota at that time. This may be an overestimate because control efforts outside of these study areas (i.e., outside SNF) were likely successful at reducing wolf density. But it is also possible that Stenlund (1955) underestimated *primary* wolf range in the early 1950s, as evidenced by the consistent bounty take of wolves in areas outside, but adjacent to, the primary range. Clearly, any estimate of wolf numbers in Minnesota from 1920 to 1960 remains somewhat speculative. We conclude that the wolf population in Minnesota following European settlement did not likely drop below 300–400 animals, and that the population from 1920 to 1960 may have ranged from 400 to 800 animals, perhaps highest during the late 1930s and early 1940s when deer were abundant and persecution of wolves may have temporarily diminished as a result of World War II.

While wolf control efforts may have increased following World War II, several "protective" changes subsequently occurred from the mid-1950s to the late 1960s. Aerial shooting of wolves was eliminated in the BWCA in the early 1950s, aerial shooting by state personnel outside this area ended in 1954, and all forms of wolf control by state personnel ended in 1956. In addition, the bounty system was terminated in 1965. Mech et al. (1971) and Mech (1973) believed that the wolf population during this period increased due to less wolf control, but there was no indication of major increases in the population. Cahalane (1964) reported between 350 and 700 wolves in Minnesota in the early 1960s, based on a questionnaire survey to game departments and other independent professionals.

4.3.3   Postbounty, Pre-ESA Era (1965–1973)

The Minnesota DNR estimated 750 wolves were present in Minnesota in 1970 based on a survey of field personnel (Leirfallom 1970), but Nelson (1971) believed this represented a minimal estimate. This is near the upper end of the population estimates for the period 1920–1960. While much uncertainty remains in these numbers, the apparent lack of a significant increase in wolf numbers might suggest wolf control was not a major limiting factor at the time. Alternatively, the potential for wolves to increase in the 1960s may have been hampered by declining habitat quality for deer, overharvest of deer by humans, and the severe winters of the late 1960s that negatively affected deer populations (Mooty 1971). In spite of the potentially significant legal changes that occurred from the late 1950s through the 1960s, we believe the best information available suggests the wolf population was relatively stable to slightly increasing.

4.3.4   Federal Protection Era (1974–2007)

The mid-1970s represents the beginning of notable wolf population recovery in Minnesota. As noted above, the Minnesota DNR estimated there were ~750 wolves in the state in 1970 (Leirfallom 1970; Nelson 1971), with primary wolf range (~39,000 km^2) ~25% larger than reported by Stenlund (1955) in the early 1950s. Fuller et al. (1992) later reviewed all available data, and independently concluded that the wolf population numbered at least 736–950 wolves in 1970, supporting the opinion that 750 was a minimal population estimate.

While wolf density appears to have declined in parts of northeastern Minnesota in the early 1970s (Mech 1973, 1986), a radio-telemetry study from 1972 to 1976 in the expanding northwestern portion of the wolf range documented that a rapid increase in the wolf population was occurring there, apparently a result of greater protections provided by the Endangered Species Act (Fritts and Mech 1981). By the mid-1970s, Mech (in Bailey 1978), using the best available information, estimated the statewide population at 1,000–1,200 wolves, an increase from what Mech and Rausch (1975) had tentatively estimated in 1973 (500–1,000). Shortly thereafter, the Minnesota DNR conducted another statewide survey in winter 1979–1980 and derived a population estimate of 1,235 wolves (Berg and Kuehn 1982). While Berg and Kuehn (1982) concluded primary wolf range had changed little since 1970, peripheral wolf range had expanded south and west. In the year that survey report was published (1982), the first wolf pack was confirmed to have recolonized Agassiz National Wildlife Refuge, an area that remains along the northwestern border of Minnesota's wolf range 25 years later.

4.3.4.1   Recent Efforts to Monitor Wolf Populations

The 1970 population estimate represented the first official effort by the state to estimate the statewide wolf population, and for the next 30 years, population estimates were derived at ~10-year intervals (every 5-year starting in 1998). The 1970 estimate was based on a questionnaire to wildlife and enforcement personnel in the DNR in which they were asked to provide their best estimates of wolf numbers and distribution in their work areas. Ten years later, the winter 1979–1980 survey also relied on field knowledge to document areas occupied by wolves, but used density information derived from five study areas where wolves were radio-marked to supplement field observations of wolf density (Berg and Kuehn 1982). While all subsequent surveys followed a similar conceptual approach, numerous changes were made in 1988–1989.

Details of the methodology used since 1988 were first provided by Fuller et al. (1992). While the advent of radio-telemetry, geographic information systems (GIS), and global positioning systems (GPS) has allowed more detailed monitoring and mapping of wolf populations, Minnesota's survey is an ad hoc method that relies on multiple pieces of information. Counting elusive carnivore populations over large areas, particularly in forested habitats, remains a difficult task. Minnesota's 400+ wolf packs occupy nearly 80,000 km^2. Radio-marking all (or most) packs in a given year or using mark-recapture (without harvest as an option for "recapture") would be impractical, if not impossible. Distance sampling methods (e.g., Buckland et al. 2004) are not logistically feasible for secretive carnivores in densely forested landscapes. Other approaches have been employed for predicting or estimating abundance of large carnivores. Approaches based solely on prey or habitat assessments (Fuller 1989; Boyce and Waller 2003) may be useful for estimating potential abundance of large carnivores, but may not always match realized abundance due to other time-varying factors (e.g., disease, weather). Newer aerial sampling methods (Becker et al. 1998; Patterson et al. 2004) show promise. However, they may be difficult to apply during the course of a single winter in broad expanses of dense forest, particularly where abundant deer populations make aerial confirmation of wolf tracks challenging. Recent attempts to use such aerial surveys in Minnesota have not succeeded due to poor snow conditions (2006) and other logistical limitations. Nevertheless, we feel the current survey has served its intended purpose, and adequately documented changes in the distribution and abundance of wolves.

4.3.4.2 Steps in Conducting Recent Surveys

The steps in conducting Minnesota's recent wolf population surveys are listed below (see also Table 4.1):

1. A majority of natural resource field personnel in the state (county, state, tribal, and federal) were provided maps and asked to record all detections of wolf sign during the course of their normal winter (~October through April) work duties. Primary data included the sign location and the estimated number of wolves. This information was supplemented with data obtained from two annual carnivore track surveys (scent station and winter track surveys), USDA-verified depredations, and known territories from radio-marked packs.
2. Using this information, in conjunction with data on forest cover, deer density, human and road density, and professional opinion of field staff, a contiguous *total* wolf range was delineated.

3. To estimate the amount of area within total wolf range that was *occupied*, observations of wolf sign from the current winter survey were entered into a GIS System. Any township (9.7 km × 9.7 km survey block) within total wolf range that included an observation of a pack (>1 wolf traveling together) was deemed occupied.

4. To account for lack of sampling in some areas, townships within total wolf range were also deemed occupied by a pack if the density of humans and roads was below the thresholds reported in Fuller et al. (1992), specifically road density <0.7 km/km^2 and human density <4 per km^2, or road density <0.5 km/km^2 and human density <8 per km^2.

5. Summing (3) and (4) yielded an estimate of the amount of occupied wolf range.

6. The average territory size (minimum convex polygon) obtained from all current radio-marked packs in the state was multiplied by 1.37 (Fuller et al. 1992) to account for vacant spaces between adjacent packs, which may be real or a byproduct of imperfect delineation of territories. This number was divided by the amount of occupied range to estimate the number of packs across the state.

7. The estimated number of packs was multiplied by mean winter pack size obtained from repeat aerial observations of marked packs, yielding an estimate of population size (for pack wolves).

8. This estimate of pack wolves was divided by 0.85, under the assumption that ~15% of the total population is composed of lone wolves (Fuller 1989; Fuller et al. 2003).

9. Confidence intervals (CI) were generated using bootstrap resampling of the data on pack and territory size, and did not incorporate uncertainty in estimates of occupied range, percent lone wolves, or size of interstitial spaces.

4.3.4.3   Recent Population Growth and Range Expansion

We compared key results from surveys among the three most recent surveys (Table 4.1). Population estimates generated from this survey were for mid-winter, near the low point of the annual cycle. Based on these surveys, Minnesota's wolf population appears to have quadrupled in size between 1970 (~750 wolves) and 2004 (~3,000 wolves), while total contiguous wolf range more than doubled to >88,000 km^2. Population increases in Minnesota up until 1998 appear to have largely been through range expansion, though some density increases in previously occupied areas appear to have occurred as well. Results from the winter 2003–2004 survey suggest that range expansion ceased, at least temporarily, around the mid- to late 1990s. The increase in the wolf population estimate for winter 2003–2004 compared to that for winter 1997–1998 was primarily through increased wolf density (Erb and Benson 2004), attributed to a reduction in average size of pack territories. Assuming linear rates of change between the periodic population surveys, growth rate of Minnesota's wolf population ranged from 3% to 6% annually from 1970 to 2004. From 1978 to the present, an additional 2–8% of the population was removed in response to verified depredations (Ruid et al., this volume).

Table 4.1 Summary of methods and key results from Minnesota's wolf surveys, 1988–2004

	Winter 1988–1989 (Fuller et al. 1992)	Winter 1997–1998 (Berg and Benson 1998)	Winter 2003–2004 (Erb and Benson 2004)
1.	Field personnel from natural resource agencies ($n = 154$ work stations) submitted maps with wolf sign observations, including the estimated number of wolves at each location	Field personnel from natural resource agencies ($n = 179$ work stations) submitted maps with wolf sign observations, including the estimated number of wolves at each location	Field personnel from natural resource agencies ($n = 102$ work stations) submitted maps with wolf sign observations, including the estimated number of wolves at each location
2.	Opportunistic field observations, supplemented by carnivore scent station surveys, USDA-verified depredations, and territories delineated for marked packs ($n = 1$ "point" per pack), yielded 1,244 observations...	Opportunistic field observations, supplemented by carnivore scent station and winter track surveys, USDA-verified depredations, and territories delineated for marked packs ($n = 1$ "point" per pack), yielded 3,659 observations	Opportunistic field observations, supplemented by carnivore scent station and winter track surveys, USDA-verified depredations, and territories delineated for marked packs ($n = 1$ "point" per pack), yielded 1,719 observations
3.	Contiguous wolf range was delineated including 93% of townships with pack observations, and covered 60,229 km² of northern Minnesota. To compensate for a lack of systematic sampling, townships with <0.7 km/km² roads and <4 humans/km² or <0.5 km/km²/roads and <8 humans/km² were included even if wolf packs were not detected in this survey	Contiguous wolf range was delineated including 99% of townships with pack observations, and covered 88,325 km² of northern Minnesota. To compensate for a lack of systematic sampling, townships with <0.7 km/km² roads and <4 humans/km² or <0.5 km/km²/roads and <8 humans/km² were included even if no wolf packs were detected in this survey	Contiguous total wolf range was delineated including 99% of townships with packs as determined from all databases, and was the same as 1998 (88,325 km²). To compensate for a lack of systematic sampling, "modeled" townships with <0.7 km/km² roads and <4 humans/km² or <0.5 km/km²/roads and <8 humans/km² were included even if no wolf packs were detected in this survey
4.	Within total wolf range, unoccupied townships (~8,000 km²) with no pack (>1 wolf) detections and not fitting the human/road model, were subtracted from total area to estimate **total occupied wolf range (~53,000 km2)**	Within total wolf range, unoccupied townships (14,405 km²) with no pack (>1 wolf) detections and not fitting the human/road model, were subtracted from total area to estimate **total occupied wolf range (73,920 km2)**	Within total wolf range, unoccupied townships (20,473 km²) with no pack (>1 wolf) detections and not the fitting human/road model, were subtracted from total area to estimate **total occupied wolf range (67,852 km2)**

5.	**Mean MCP territory size (166 km2)** derived from telemetry studies conducted from 1970 to 1989 ($n = 108$ packs) was divided into occupied range (after multiplying by 1.37 for interstitial pack area) to estimate number of packs (~233).	**Mean MCP territory size (140 km2)** derived from current telemetry studies ($n = 36$ packs) was divided into occupied range (after multiplying by 1.37 for interstitial pack area) to estimate number of packs (~385)	**Mean MCP territory size (~102 km2)** derived from current telemetry studies ($n = 24$ packs) was divided into occupied range (after multiplying by 37% for interstitial pack area) to estimate number of packs (~485)
6.	**The mean winter pack size (5.55)** derived from telemetry studies conducted from 1970 to 1989 ($n = 108$ packs) was multiplied by the number of packs to estimate total pack wolves (1,293)	**The mean winter pack size (5.4)** derived from current telemetry studies ($n = 36$ packs) was multiplied by the number of packs to estimate total pack wolves (2,079)	**The mean winter pack size (~5.3)** derived from current telemetry studies ($n = 24$ packs) was multiplied by the number of packs to estimate total pack wolves (2,567)
7.	Lone wolves were assumed to account for ~15% of the population and were incorporated (1,293/0.85) to yield a **total estimate of 1,521 wolves (CI 90% = 1,338–1,762). A separate population estimate (1,750 wolves, CI 90% = 1,020–2,400)** was based on wolf/ungulate biomass ratios	Lone wolves are assumed to account for ~15% of the population and were incorporated (2,079/0.85), to yield a **total estimate of 2,445 wolves (90% CI 1995–2905)**	Lone wolves are assumed to account for ~15% of the total population and were incorporated (2,567/0.85), to yield a **total estimate of 3,020 wolves (90% CI 2301–3708)**

USDA United States Department of Agriculture

While there is little question that elimination of coordinated wolf persecution in the 1950s and 1960s, and subsequent legal protections in the early 1970s, played an important role in wolf recovery, we believe that growth of the deer herd (DelGiudice et al., this volume) played an equally important role. Average size of wolf territories, as summarized during the periodic wolf surveys, has steadily declined (Table 4.1), and the estimate of average territory size (102 km^2) during the winter 2003–2004 survey appears to be the smallest published for any multipack study in Minnesota (Fuller et al. 2003; Gogan et al. 2004), as well as smaller than published estimates from most other areas of North America. Continuing range "saturation" also may have played a role in declining territory size, because colonizing wolf populations exhibit declines in average size of pack territories as the populations become established (Fritts and Mech 1981; Hayes and Harestad 2000; Wydeven et al., this volume). Nevertheless, available prey abundance is arguably the most important ecological factor influencing wolf social and population dynamics. Assuming other factors remain constant, prey abundance is negatively correlated with territory size and positively correlated with population size (Mech and Boitani 2003a; Fuller et al. 2003). The deer population in Minnesota's wolf range is at historic highs in the twenty-first century (DelGiudice et al., this volume), allowing individual wolf packs to survive in smaller areas. Given the low correlation between average pack size and prey biomass (Fuller et al 2003), the lack of major change in average pack size (Table 4.1) is not unexpected.

4.4   The Future of Wolves in Minnesota

Range recolonization by wolves in Minnesota appeared to have ceased in the mid- to late 1990s. Total contiguous wolf range in Minnesota was estimated at ~88,000 km^2 in 1998 and 2004, of which ~70,000 km^2 was deemed occupied in 2004. Mladenoff et al. (1995), based on habitat modeling, predicted there was ~50,000 km^2 of "favorable" wolf habitat in what has been considered primary wolf range in Minnesota (Zone A; Fig. 4.1). Minnesota's currently occupied wolf range, which includes areas (Zone B; Fig. 4.1) not considered by Mladenoff et al. (1995), is ~70,000 km^2. As in Wisconsin (Mladenoff et al. 1999), it is clear that wolves have expanded into areas previously thought to be less favorable. Nevertheless, at a coarse spatial scale, the current distribution of wolves in Minnesota is reasonably similar to the projections of Mladenoff et al. (1995).

Whether wolf range in Minnesota will expand in the future is difficult to predict. Minnesota's wolf management plan imposes no geographic or numeric limit on the wolf population, instead focusing on alleviation of wolf-human conflicts where they occur (Minnesota DNR 2001). The extensive distribution of wolves at the time of Euro-American settlement indicates that wolves are not habitat specialists. Although numerous factors can influence wolf distribution and abundance, two factors – prey abundance and human-caused mortality – are likely the best predictors of possible range expansion in the future. Without large herds of bison and elk roaming the former prairie, now primarily an agricultural landscape, we do not believe that current prey (i.e., deer) density in the majority of southern and western Minnesota is capable of sustaining a viable wolf population, even in the absence of human-caused mortality. Nevertheless, deer populations capable of sustaining wolf packs in this region do exist in some areas of fragmented forest along or adjacent to river valleys.

We believe the area with the greatest potential for future re-colonization, based on prey biomass, is in southeastern Minnesota. Although wolves can disperse substantial distances (Treves et al., this volume), this area is distant from the current population of wolves, and establishment of packs may be hindered by the developed landscape they must travel through to get there. However, prey is not limiting, and wolf packs may eventually establish in this region, perhaps similar to, or even connected with, the "isolated" wolf population now established in central Wisconsin (Thiel et al., this volume). There may be greater potential for wolf mortality in southeastern Minnesota associated with livestock depredation control, vehicle collisions, and increased rates of illegal killing (more people, accessible landscape, and wolves will be more visible). Nevertheless, wolf populations can sustain high annual mortality rates, perhaps in excess of 50% (Fuller et al. 2003), and we doubt that potential human-caused mortality will preclude the possibility of a small and relatively isolated wolf population establishing in southeastern Minnesota. However, if packs do establish in isolated patches of southern Minnesota, we doubt they would account for >5% of the total wolf population in the state. We foresee primary wolf range in Minnesota remaining largely contained within the 88,000-km² area delineated by Berg and Benson (1998).

Based on results from the winter 2003–2004 Minnesota wolf survey, wolf density was ~4.5 wolves per 100 km². Overall, winter density estimates from localized studies in Minnesota have ranged from ~1 wolf per 100 km² to 6 wolves per 100 km² (Olson 1938; Stenlund 1955; Mech 1973, 1986; Van Ballenberghe et al. 1975; Berg and Kuehn 1980; Fritts and Mech 1981; Fuller 1989; Gogan et al. 2004). Pimlott (1967) suggested that intrinsic factors likely limit density of wolves to ≤4 wolves per 100 km², but a few studies have since documented localized winter densities of ≥5 wolves per 100 km² (e.g., Van Ballenberghe et al. 1975; Peterson and Page 1988; Fuller 1989). Most researchers now agree that extrinsic factors (i.e., prey biomass) likely impose the upper limit to wolf density.

Mech (1998), using previous growth rates, projected that the Minnesota population would reach 3,500 wolves by 2005, but assumed the increase would likely be through additional range expansion. While his projection is within our confidence bounds for the population estimate in winter 2003–2004, wolves did not continue to expand their range after 1997–1998. A population of 3,500 wolves, but within the area deemed occupied in 2003–2004 (Table 4.1), would yield a range-wide density estimate of ~5.2 wolves per100 km². Whether future surveys will support this scenario is unknown, but it's clear that Minnesota's wolf population is near the highest densities previously reported in the literature, excluding Isle Royale (Fuller et al. 2003). We also believe that wolf density in northern Minnesota is likely higher today than before European settlement, a result of increases in prey biomass from landscape alterations that transformed the ungulate community from a low-density, moose-caribou prey base into a high-density, white-tailed deer prey base (DelGiudice et al., this volume).

While we do not anticipate significant expansion of wolf range in Minnesota, wolf numbers will undoubtedly fluctuate. Prey availability, disease and parasites, and human-caused mortality will, to varying degrees, play a role in future wolf population dynamics. The future of deer seems secure, and it's possible that deer populations in some areas may further increase as a result of alterations of forest structure and lack of severe winters induced by climate change. While some level of illegal killing of wolves will continue to occur, human attitudes toward wolves are much improved and we do not foresee any dramatic changes that will single-handedly dictate the fate of the wolf population in Minnesota. Human development and population growth in northern Minnesota may pose greater challenges by increasing wolf-human conflicts and vehicle collisions, and creating greater opportunity for illegal killing as a result of more roads, people, and forest fragmentation. Coexistence in increasingly developed areas will not only be dependent on human tolerance of wolf activity, but also on the ability of wolves to tolerate increasingly fragmented forests with more human activity. Finally, while much is known about which diseases and parasites can affect wolves (Kreeger 2003), we know less about their role in limiting or regulating wolf populations and factors that may influence their prevalence and persistence in wolf populations (e.g., wolf density, proximity to domestic animals, and weather). Studying potential population effects of diseases or parasites can be extremely challenging, but monitoring prevalence will improve our understanding and ability to predict outbreaks.

Maintaining or improving human attitudes toward wolves will be a key component of future wolf management, with education and responsive conflict management front and center. While human-wolf conflicts need to be addressed, in so doing we must not forget to acknowledge the many values of wolves. Wolves can play an important role in restoring natural ecosystem dynamics (e.g., Mech and Boitani 2003b; Ripple and Beschta 2004; Rooney and Anderson, this volume), are aesthetically valued by many people, may contribute to tourism (Schaller 1996; Fritts et al. 2003) or recreational opportunity, and are culturally important to Native Americans (David, this volume). The return of wolf populations to the Great Lakes region is indeed a success story, for wolves and for many people.

References

Bailey, R. 1978. Recovery plan for the eastern timber wolf. Washington, DC: US Fish and Wildlife Service.

Becker, E. F., Spindler, M. A., and Osborne, T. O. 1998. A population estimator based on network sampling of tracks in the snow. Journal of Wildlife Management 62: 968–977.

Berg, W. E., and Kuehn, D. W. 1980. A study of the timber wolf population on the Chippewa National Forest, Minnesota. Minnesota Research Quarterly 40: 1–16.

Berg, W. E., and Kuehn, D. W. 1982. Ecology of wolves in north-central Minnesota. In Wolves of the World: Perspectives of Behavior, Ecology and Conservation, eds. F. H. Harrington and P. C. Paquet, pp. 4–11. Park Ridge, NJ: Noyes Publishing.

Berg, W., and Benson, S. 1998. Updated wolf population estimate for Minnesota, 1997–98. In Summaries of Wildlife Research Findings, 1998, ed. B. Joselyn, pp. 85–98. St. Paul, MN: Minnesota Department of Natural Resources.

Boitani, L. 1995. Ecological and cultural diversities in the evolution of wolf-human relationships. In Ecology and Conservation of Wolves in a Changing World, eds. L. N. Carbyn, S. H. Fritts and D. R. Seip, pp. 3–11. Edmonton, Canada: Canadian Circumpolar Institute.

Boyce, M. S., and Waller, J. S. 2003. Grizzly bears for the Bitterroot: predicting potential distribution and abundance. Wildlife Society Bulletin 31: 670–683.

Buckland, S. T., Anderson, D. R., Burnham, K. P., Laake, J. L., Borchers, D. L., and Thomas, L. eds. 2004. Advanced Distance Sampling. Oxford, UK: Oxford University Press.

Cahalane, V. H. 1964. A preliminary study of distribution and numbers of cougar, grizzly, and wolf in North America. New York: New York Zoological Society.

Erb, J., and Benson, S. 2004. Distribution and abundance of wolves in Minnesota, 2003–04. St. Paul, MN: Minnesota Department of Natural Resources.

Fritts, S. H., and Mech, L. D. 1981. Dynamics, movements, and feeding ecology of a newly protected wolf population in northwestern Minnesota. Wildlife Monographs 80: 1–79.

Fritts, S. H., Stephenson, R. O., Hayes, R. D., and Boitani, L. 2003. Wolves and humans. In Wolves: Behavior, Ecology, and Conservation, eds.L. D. Mech and L. Boitani, pp. 289–316. Chicago, IL: University of Chicago Press.

Fuller, T. K. 1989. Population dynamics of wolves in north-central Minnesota. Wildlife Monographs 105: 1–41.

Fuller, T. K., Berg, W. E., Radde, G. L., Lenarz, M. S., and Joselyn, C. B. 1992. A history and current estimate of wolf distribution and numbers in Minnesota. Wildlife Society Bulletin 20: 42–55.

Fuller, T. K., Mech, L. D., and Cochrane, J. F. 2003. Wolf population dynamics. In Wolves: Behavior, Ecology, and Conservation, eds. L. D. Mech and L. Boitani, pp. 161–191. Chicago, IL: University of Chicago Press,

Gogan, P. J. P., Route, W. T., Olexa, E. M., Thomas, N., Kuehn, D., and Podruzny, K. M. 2004. Gray wolves in and adjacent to Voyageurs National Park, Minnesota: research and synthesis 1987–1991. Technical Report NPS/MWR/NRTR/2004–01. Omaha, NE: National Park Service.

Hayes, R. D., and Harestad, A. S. 2000. Demography of a recovering wolf population in the Yukon. Canadian Journal of Zoology 78: 36–48.

Herrick, C. L. 1892. The Mammals of Minnesota. Minneapolis, MN: Minnesota Geological and Natural History Survey, Bulletin No.7.

Kreeger, T. J. 2003. The internal wolf: physiology, pathology, and pharmacology. In Wolves: Behavior, Ecology, and Conservation, eds. L. D. Mech and L. Boitani, pp. 192–217. Chicago, IL: University of Chicago Press.

Leirfallom, J. 1970. Wolf management in Minnesota. In Proceedings of a Symposium of Wolf Management in Selected Areas of North America, eds. S. E. Jorgenson, C. E. Faulkner and L. D. Mech, pp. 9–15. Twin Cities, MN: US Department of the Interior.

Mech, L. D. 1973. Wolf numbers in the Superior National Forest of Minnesota. Research Paper NC-97. St. Paul, MN: USDA Forest Service.

Mech, L. D. 1986. Wolf population in the central Superior National Forest, 1967–1985. Research Paper NC-270. St. Paul, MN: USDA Forest Service.

Mech, L. D. 1998. Estimated costs of maintaining a recovered wolf population in agricultural regions of Minnesota. Wildlife Society Bulletin 26: 817–822.

Mech, L. D. 2000. Historical overview of Minnesota wolf recovery. In The Wolves of Minnesota: Howl in the Heartland, ed. L. D. Mech, pp. 15–27. Stillwater, OK: Voyageur Press.

Mech, L. D., and Boitani, L. 2003a. Wolf social ecology. In Wolves: Behavior, Ecology, and Conservation, eds. L. D. Mech and L. Boitani, pp. 1–34. Chicago, IL: University of Chicago Press.

Mech, L. D., and Boitani, L. 2003b. Ecosystem effects of wolves. In Wolves: Behavior, Ecology, and Conservation, eds. L. D. Mech and L. Boitani, pp. 158–160. Chicago, IL: University of Chicago Press.

Mech, L. D., and Rausch, R. A. 1975. The status of the wolf in the United States, 1973. In Proceedings of the First Working Meeting of Wolf Specialists and First International Conference on the Conservation of the Wolf, ed. D. H. Pimlott, pp.83–88. Gland, Switzerland: International Union for the Conservation Nature and Natural Resources.

Mech, L. D., Frenzel, L. D., Ream, R. R., and Winship, J. W. 1971. Movements, behavior, and ecology of timber wolves in northeastern Minnesota. In Ecological Studies of the Timber Wolf in Northeastern Minnesota, eds. L. D. Mech and L. D. Frenzel, Jr., pp. 1–34.Research Paper NC-52. St. Paul, MN: USDA Forest Service, North Central Forest Experiment Station.

Minnesota DNR. 1980. Minnesota Timber Wolf Management Plan. St. Paul, MN: Minnesota Department of Natural Resources.

Minnesota DNR. 2001. Minnesota Wolf Management Plan. St. Paul, MN: Minnesota Department of Natural Resources.

Mladenoff, D. J., Sickley, T. A., Haight, R. G., and Wydeven, A. P. 1995. A regional landscape analysis and prediction of favorable gray wolf habitat in the northern Great Lakes region. Conservation Biology 9:279–294.

Mladenoff, D. J., Sickley, T. A., and Wydeven, A. P. 1999. Predicting gray wolf landscape recolonization: logistic regression models vs. new field data. Ecological Applications 9:37–44.

Mooty, J. J. 1971. The changing habitat scene. In Proceedings of a Symposium on the White-tailed Deer in Minnesota, ed. M. M. Nelson, pp. 27–37. St. Paul, MN: Minnesota Department of Natural Resources.

Nelson, M. M. 1971. Predator management with emphasis on the timber wolf. In Proceeding of a Symposium on the White-tailed Deer in Minnesota, ed. M. M. Nelson, pp. 68–77. St. Paul, MN: Minnesota Department of Natural Resources.

Olson, S. F. 1938. A study of the predatory relationship with particular reference to the wolf. Scientific Monthly 66: 323–336.

Packard, J. M. 2003. Wolf behavior: reproductive, social, and intelligent. In Wolves: Behavior, Ecology, and Conservation, eds. L. D. Mech and L. Boitani, pp. 35–65. Chicago, IL: University of Chicago Press.

Patterson, B. R., Quinn, N. W. S., and Becker, E. F. 2004. Estimating wolf densities in forested areas using network sampling of tracks in snow. Wildlife Society Bulletin 32: 938–947.

Peterson, R. O., and Page, R. E. 1988. The rise and fall of the Isle Royale wolves. Journal of Mammalogy 69: 89–99.

Pimlott, D. H. 1967. Wolf predation and ungulate populations. American Zoologist 7: 267–278.

Ripple, W. J., and Beschta, R. L. 2004. Wolves, elk, willows, and trophic cascades in the upper Gallatin Range of southwestern Montana, USA. Forest Ecology and Management 200: 161–181.

Schaller, D. T. 1996. The Eco-center as a Tourist Attraction: Ely and the International Wolf Center. MS Thesis. Minneapolis, MN: University of Minnesota.

Stenlund, M. H. 1955. A field study of the timber wolf (*Canis lupus*) on the Superior National Forest, Minnesota. Technical Bulletin No. 4. Minneapolis, MN: Minnesota Department of Conservation.

Surber, T. 1932. The Mammals of Minnesota. Minneapolis, MN: Minnesota Department of Conservation.

Van Ballenberghe, V., Erickson, A. W., and Byman, D. 1975. Ecology of the timber wolf in northeastern Minnesota. Wildlife Monographs 43: 1–43.

Chapter 5
Wolf Population Changes in Michigan

Dean E. Beyer, Jr., Rolf O. Peterson, John A. Vucetich, and James H. Hammill

5.1   Introduction

This chapter chronicles changes in wolf abundance and identifies the significant events in gray wolf (*Canis lupus*) management in Michigan from the early 1800s to present (Table 5.1). We recognize three important time periods. Initially, populations declined (1817–1959) due to public policy that sought to eliminate wolves. During the second period (1960–1988), wolves struggled to maintain their existence in the state. Public policy changed and wolves were granted legal protection. Despite this protection and an increasing shift in public attitudes that favored wolves (and the environment in general), a minority of Michigan residents evidently prevented wolves from reestablishing a population. During the third period (1989–present), wolves staged a remarkable comeback. The speed of their recovery surprised even those charged with aiding it. Although many credit a shift in public attitudes as the primary reason for this recovery, perhaps not enough credit has been given to the resiliency of wolves.

This chapter focuses on wolf population changes on the mainland of Michigan. Information on the wolves occupying Isle Royale can be found in Vucetich and Peterson (this volume). Also, information on trends in wolf depredation of livestock during the period of population recovery may be found in Ruid et al. (this volume).

5.2   The Decline of Wolves in Michigan

Wolves have been part of Michigan's wildlife community since retreat of the last glacier some 10,000 years ago (Holman 1975; Hughes and Merry, unpublished). The region that became the state of Michigan consists of two peninsulas (Upper and Lower), which are bordered by the Great Lakes and connecting rivers. Before European settlement, wolves were found in both Upper and Lower Peninsulas, and based on museum specimens and pioneer accounts, were likely present in all counties of the state (Stebler 1951). Large cloven-hoofed mammals, collectively known as ungulates, are the primary prey of wolves (Mech and Peterson 2003; DelGiudice et al., this volume) and five species were present in Michigan before European settlement. Bison (*Bison bison*) and woodland caribou (*Rangifer tarandus caribou*),

A.P. Wydeven et al. (eds.), *Recovery of Gray Wolves in the Great Lakes*
Region of the United States,
DOI: 10.1007/978-0-387-85952-1_5, © Springer Science+Business Media, LLC 2009

Table 5.1 Significant events in the history of wolf management in Michigan, 1817–2007

Year	Event
1817	United States Congress established a wolf bounty for the Northwest Territories, which included what is now Michigan.
1837	Michigan becomes the 26th state
1838	Michigan legislature establishes a wolf bounty; ninth law passed by first legislature
1910	Wolves probably extirpated from the Lower Peninsula
1922	Wolf bounty repealed because of fraudulent activities
1922	State-paid trapper system put in place to eradicate predators
1935	State-paid trapper system ended and replaced by a new bounty
1940	Last unverified bounty record for the Lower Peninsula
1954	Last record of wolf reproduction in the Upper Peninsula
1956	Wolf population status assessment: probably fewer than 100 wolves remain in the Upper Peninsula
1959	Only one wolf submitted for bounty; down from an average of 31 animals, 1935–1956 suggesting the wolf population had crashed
1960	Bounty repealed
1965	Wolves afforded full legal protection by the state
1967	Wolves protected on federal lands by the Endangered Species Preservation Act of 1966
1973	Wolf population status assessment: probably only six wolves remaining in the Upper Peninsula
1974	Wolf listed as an endangered species under the Endangered Species Act of 1973
1974	Four wolves translocated from Minnesota were released in the Upper Peninsula; within 8 months all four wolves were dead from human-related causes
1976	Wolf listed as an endangered species under Michigan's Endangered Species Protection Act
1978	Federal recovery plan for the eastern timber wolf completed
1989	First evidence of wolves establishing a territory since population crash in the late 1950s
1990	Study completed on public attitudes and beliefs about wolves and wolf restoration in Michigan
1990	First documentation of wolf reproduction since 1954
1992	First wolf captured and radio-collared in Michigan
1992	Federal recovery plan for the eastern timber wolf revised
1994	Coyote hunting banned during firearm deer season in the Upper Peninsula
1997	Michigan Gray Wolf Recovery and Management Plan approved
1997	MDNR[a] and Michigan Technological University begin a program of wolf research
1999	Combined wolf population of Wisconsin and Michigan surpasses the population level recovery criterion of the federal recovery plan for a second population outside of Minnesota. This same criterion allows Michigan to reclassify wolves to state threatened status.
2002	Michigan reclassifies wolf to state threatened status
2003	Federal government reclassifies wolves to threatened status in the eastern distinct population segment which includes Michigan; a special rule allows lethal control as an option for managing wolf depredation of domestic animals.
2004	First confirmed wolf in the Northern Lower Peninsula in over 90 years
2004	Upper Peninsula wolf population meets the state delisting goal of greater than or equal to 200 wolves for five consecutive winters
2005	Federal court enjoins and vacates federal reclassification rule; wolves returned to federally endangered status
2005	Survey of Northern Lower Peninsula found no evidence of wolf presence
2005	Michigan receives federal 10(a)(1)(A) subpermit authorizing use of lethal control for depredation management

(continued)

Table 5.1 (continued)

Year	Event
2005	Michigan begins process of revising 1997 wolf plan; ten public meetings held to gather input
2005	Federal court enjoined federal subpermit; ability to use lethal control for depredation management lost
2005	Coyote hunting banned during firearm deer season in Northern Lower Peninsula
2005	Michigan State University and MDNR begin a study of public attitudes and beliefs toward wolves and their management
2006	MDNR convenes Wolf Management Roundtable, an advisory group of diverse stakeholders charged with developing guiding principles for wolf management
2006	Michigan, Wisconsin, and Minnesota jointly request the designation and delisting of a Western Great Lakes Distinct population segment
2006	Federal government rescinds lethal control provisions of Michigan's 10(a)(1)(A)
2006	Wolf Management Roundtable completes their report
2007	Federal government delists wolves in the Western Great Lakes distinct population segment; management authority reverts to the state
2008	Michigan wolf management plan revised

[a]Michigan Department of Natural Resources

although present in Michigan, were not distributed widely (Baker 1983; Evers 1994; Cochrane 1996) and probably were not important prey items. More important prey species included elk (*Cervus elaphus*), white-tailed deer (*Odocoileus virginianus*), and moose (*Alces alces*). Elk were abundant in the Lower Peninsula (LP) but were likely absent from the Upper Peninsula (UP; O.J. Murie 1951; Baker 1983). White-tailed deer occurred throughout the state, although greater numbers were found in the southern part of the LP (Bartlett 1938; Baker 1983). Moose occurred throughout Michigan except the southwestern portion of the LP (Baker 1983; Verme 1984). Thus, it appears there was adequate prey to support wolves, although no estimates of wolf abundance were recorded. An estimate of the maximum number of wolves that could have been present before European settlement can be made by applying the maximum winter wolf density reported in more recent times. Winter wolf densities (outside of Isle Royale) generally have not exceeded 40 wolves/1,000 km^2 (Fuller et al. 2003). Applying this density across all of Michigan (147,155 km^2) suggests the wolf population would have been fewer than 6,000 animals. Wydeven (1993) estimated presettlement wolf numbers in Wisconsin by assuming wolf density ranged from 19 to 39 wolves/1,000km^2. Appling these densities to Michigan suggests that 3,000–6,000 animals may have been possible.

Wolves were (and still are) important to the tribal culture and beliefs of many aboriginal peoples of Michigan. For example, the Anishinabe people (Ojibwe), a tribe found in Michigan, consider the wolf a sacred clan animal (Benton-Banai 1988). By and large, European settlers viewed wolves much differently than Native Americans. Settlers believed wolves were incompatible with civilization (Lopez 1978). Their hatred and desire to kill wolves was a result of their desire for territorial conquest and agricultural settlement. Their actions were further supported by

European folklore that portrayed wolves negatively (Coleman 2004). Persecution
of wolves became the governing rule of all institutional responses to the presence
of wolves, including that of the federal government. In 1817, the United States
Congress enacted a wolf bounty in the Northwest Territories, which included what
is now Michigan [Michigan Department of Natural Resources (MDNR) 1997]. The
timing of the bounty coincided with a surge of settlers entering the Michigan territory. The number of nonaboriginal people increased from about 9,000 in 1820 to
about 32,000 just 10 years later (Michigan Nonprofit Association and Council of
Michigan Foundations 2002). Three years later, this population again almost doubled
(60,000) meeting a requirement for statehood. Michigan became the 26th state in
1837 and the ninth law passed by the first Michigan Legislature was a wolf bounty.
There is no doubt about the intention of the lawmakers because the legislation was
entitled "An act for the destruction of wolves." Wolves were likely already gone
from the southern LP by the time the state enacted a bounty (MDNR 1997). Two
hundred seventy-nine wolves were bountied in 1840 (MDNR, unpublished data).
The number of wolves reported killed and turned in for bounty payments declined
generally until the late 1880s when few if any wolves were turned in (Fig. 5.1).

By 1850, Michigan's increasing human population was taking advantage of
abundant natural resources, especially virgin timber. Habitat changes resulting from
timber harvesting and subsequent large-scale fires, combined with increased hunting
pressure, soon eliminated moose and elk from the LP (Burt 1946). Deer initially

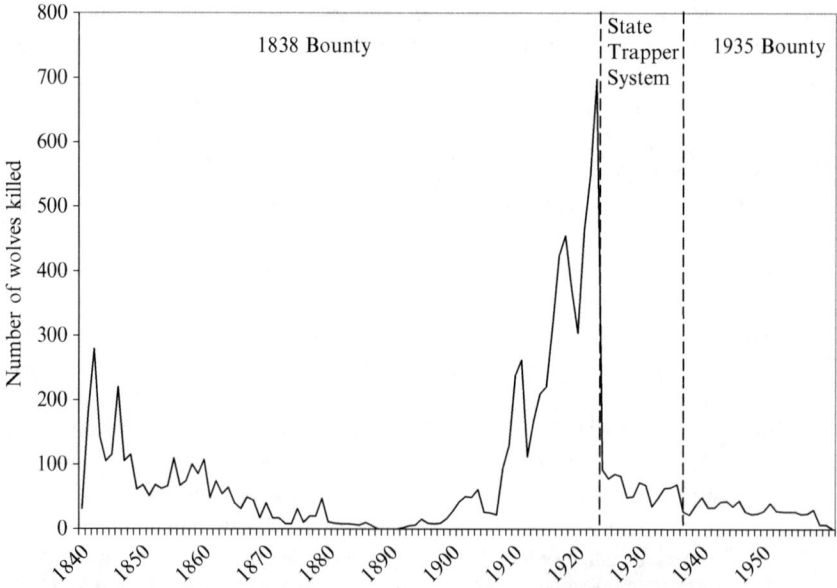

Fig. 5.1 Reported numbers of wolves killed for bounty or by state-paid trappers in Michigan,
1838–1959. Bounty records must be viewed cautiously since these systems were plagued with
fraudulent activities and misidentification of species (*see text*)

increased in response to increased young forest being created by timber cutting (Bartlett 1938), and without persecution, wolves should have increased as well in response to increasing deer herds. However, market hunting and habitat destruction caused by uncontrolled forest fires and settlement reduced deer numbers. The reduced prey base in combination with the excessive killing of wolves continued driving the wolf population in the LP toward extirpation. Wolves probably disappeared from the LP by 1910 (Stewart and Negus 1961), although a few unverified records of bountied wolves from 1935 to 1940 exist (Stebler 1944; Hendrickson et al. 1975). Settlement of the UP occurred later and with less intensity allowing wolves to persist. Still, persecution by humans caused the wolf population to decline in this region as well.

From 1838 to 1921, Michigan amended its wolf bounty eight times. Each amendment made collection of payments easier and many amendments increased payments. During this period, about $154,000 was paid with 70% of the total being paid during 1910–1921. Interestingly, bounty records show a remarkable surge in numbers of wolves killed during 1910–1921, including 698 killed in 1921 alone (Fig. 5.1). Given that wolves were present only in the UP at this time and deer numbers were low (Langenau 1994), these numbers of wolves bountied seem unlikely. Indeed, bounties undoubtedly were subject to widespread fraud and abuse (MDNR, unpublished data; Thiel 1993). In fact, legislators felt the abuse was so great that they repealed the wolf bounty and replaced it with a state-paid trapper system in 1922. However, the goal of the state trapper system was essentially that of the bounty – to eliminate predators from Michigan's landscape. The trapper system, administered by the United States Bureau of Biological Surveys, was supported with funds derived from sales of deer hunting licenses. State trappers never killed more than 92 wolves in any year. The state trapper program continued through 1934 and 855 wolves ($\bar{x} = 66$/year) and more than 7,600 coyotes (*Canis latrans*) were killed at a total cost of just over $530,000 (MDNR, unpublished data).

A new bounty on wolves and coyotes began in 1935 and paid $20 for female wolves and $15 for males (Thiel 1993). An average of 31 wolves were bountied each year during 1935–1956 (Fig. 5.1). The last report of breeding by wolves in the UP was made in 1954 (Gardella et al. 1996). Arnold and Schofield (1956) reviewed bounty records, wolf sightings by full-time Conservation Department employees, and probable pup production and concluded there were probably about 100 animals in the UP distributed among seven areas during 1956. However, by 1957 signs of a population crash were evident because only seven wolves were turned in for payment, suggesting the wolf population may have been smaller than Arnold and Schofield's (1956) estimate (perhaps only 40–50 animals, Hendrickson et al. 1975). Two years later, only a single wolf was bountied. At this time, densities of deer were more than adequate (~9/km²; Eberhardt 1957) to support wolves, suggesting the bounty rather than food supply was the most likely cause of decline in wolf abundance (Hendrickson et al. 1975). Stebler's analysis (1951) also suggested that the bounty was the primary influence on wolf abundance. However, at the time, most analysts believed that bounties were ineffective and that the decline in wolves was due to a decline in the amount of wilderness (MDNR, unpublished data). Apparently, persistence of wolves in remote wilderness areas elsewhere became evidence that wolves needed the unique ecological conditions that wilderness provides rather than a simpler view of wilderness areas as refuges where wolves could avoid persecution by humans. The current recovery of wolves in the Western Great Lakes region demonstrates that wolves do not need wilderness to prosper.

5.3   The Struggle to Maintain the Presence of Wolves in Michigan

Calls for eliminating the bounty on wolves or setting aside areas where wolves would be protected began surfacing during the mid-1940s (Thiel 1993), but the bounty was not repealed until 1960 (Douglass 1970). Wolves received legal protection under state law in 1965 [United States Fish and Wildlife Service (USFWS) 1992], although illegal killing and incidental losses to trapping of coyotes for bounty were thought to be preventing the wolf population from increasing (Hendrickson et al. 1975). Wolves maintained a limited presence in the UP, perhaps through sporadic reproduction but more likely through immigration of animals from Ontario and Minnesota. By 1973, there may have been only six wolves in the UP (Hendrickson et al. 1975).

Wolves were first listed in 1967 as an endangered species under the Federal Endangered Species Preservation Act of 1966, which primarily protected wolves on Federal lands. The Federal Endangered Species Act of 1973 considerably strengthened protections for species at risk of extinction and the wolf was officially listed under this law in 1974. Wolves became protected under Michigan's Endangered Species Protection law in 1976 (MDNR 1976).

The MDNR endorsed restoration of wolves in the UP in 1970 (Weise et al. 1975) and 4 years later (1974) researchers from Northern Michigan University in cooperation with the USFWS and the Huron Mountain Club conducted an experiment to determine if wild wolves translocated from Minnesota could survive in the UP. Four wolves (two males and two females) were livetrapped in Minnesota and transported to Marquette County in the north central part of the UP. Once in Michigan, researchers held the wolves in a pen for a 7-day acclimation period before they were released during mid-March. After release, three of the wolves formed a group and moved in a westerly direction, perhaps trying to return to their Minnesota territory. The fourth animal remained near the release site. By November 1974, all four wolves had died: two were shot, one was trapped and shot, and one was killed by a vehicle (Weise et al. 1975). Despite an observed mating among two of these wolves while being held in Minnesota, no pups were born and this reintroduction did not contribute, at least biologically, to the eventual recovery of wolves in the UP. However, these wolves survived long enough to demonstrate that wild wolves could be translocated successfully and habitat (food and cover) conditions in the UP were favorable

(Weise et al. 1975). It appeared that human persecution was still preventing wolves from gaining a foothold in Michigan.

During the 1970s, while wolves were still struggling to maintain a presence in the UP, wolves in Minnesota were increasing and expanding their range southward. By 1975, a pack established a territory that spanned the Minnesota–Wisconsin border (Thiel 1993). Still, evidence of wolves in the UP suggested that wolves might have never been extirpated. Eight dead wolves were recovered in the UP during 1970–1986. In addition, a live wolf illegally captured by a citizen was confiscated and sent to a wolf colony in Minnesota. All nine of these wolves were yearlings or adults (no pups) and were recovered in close proximity to the Wisconsin or Ontario borders suggesting immigration rather than reproduction was their source (Thiel and Hammill 1988). Six of the eight wolf deaths were attributed to humans. These included shooting (4), trapping (1), and a vehicle strike (1), again suggesting that human persecution continued to be the main factor keeping the wolf population from rebounding (Thiel and Hammill 1988; Robinson and Smith 1977). Cause of death for the remaining two wolves was unknown (Thiel and Hammill 1988). Immigration of wolves into the UP can be inferred from locations of recovered dead wolves and observations of wolf tracks crossing the St. Mary's River (Jensen et al. 1986). However, Thiel (1988) provided the first direct evidence of immigration by documenting the dispersal of a radio-collared wolf from Wisconsin to the central UP in 1986. During the late 1970s and early 1980s, Wisconsin's wolf population was growing, albeit slowly. However, no signs of recovery (i.e., reproduction) were being observed in the UP.

5.4   The Recovery of Wolves in Michigan

5.4.1   Estimating Wolf Abundance

Without question, estimates of wolf abundance have provided critical information about wolf recovery in Michigan as well as the rest of the western Great Lakes region. Wolf abundance in Michigan was estimated by surveying suitable wolf habitat during winter, when snow cover made wolves and their tracks easier to see. Surveys during winter provide estimates of a minimum number of wolves in the UP during a given year. Winter surveys consist of intensive and extensive searches of roads and trails by truck and snowmobile for wolf tracks and other sign (Potvin et al. 2005). The survey is extensive because much of the UP, which encompasses about 43,000 km², is suitable habitat. An average of 12,257 km of roads and trails (about 25% of available roads and trails) was searched at least once each year from 2000 to 2006. The survey is intensive because many roads and trails must be searched multiple times before an accurate count can be made. Searching for wolf sign is systematic and is guided by several sources of information. In the beginning, analysts used observations of wolves or wolf sign made by citizens to help identify areas where wolves might be established. As survey data accumulated, prior results guided subsequent surveys. Perhaps, the most important source of information is the movement pattern of wolves determined from tracking radio-collared animals. Wolves are highly territorial and radio-collared animals provided measurements of the sizes and locations of numerous territories. Counts of packs with radio-collared members are also made from airplane observation of pack members associated with radio-collared wolves; however, heavy forest cover in the UP makes it difficult to observe wolves. In addition, packs often split into smaller hunting groups (Mech and Boitani 2003) so it may take many flights before an entire pack is together and located in open cover where they can be observed.

Wolves are also very mobile, averaging movements of 14–28 km per day (Mech 1966; Ciucci et al. 1997; Jedrzejewski et al. 2001). In winter, this movement behavior results in wolves leaving lots of tracks to find, but the abundance of tracks also presents a challenge to survey crews. Surveyors must use care when they find tracks in nearby areas to ensure that individual wolves are not being counted more than once. Searchers avoid double counting of wolves in adjacent areas by using the territorial boundaries of radio-collared animals to distinguish between discrete groups of wolves. In areas without radio-collared wolves, differentiation of packs depends on finding fresh tracks in adjacent areas with no sign of movement between areas (Potvin et al. 2005). Again, these estimates are minimum counts. Lone wolves are relatively difficult to include in the annual counts because it is difficult to know whether a lone set of tracks represents a lone animal or a pack member that happens to be traveling by itself. During 2001–2006, the percentage of lone wolves in abundance estimates has remained 2–3% (Table 5.2).

Table 5.2 Annual summary statistics for the wolf population in Michigan's Upper Peninsula and survey efforts, 1999–2006

Parameters	1999	2000	2001	2002	2003	2004	2005	2006
Population estimate	169	216	249	278	321	360	405	434
No. of packs[a]	52	63	70	63	68	77	87	91
No. of lone wolves	12	14	5	8	11	6	6	11
Mean pack size (standard error)	3.0 (0.2)	3.2 (0.2)	3.5 (0.3)	4.3 (0.3)	4.6 (0.3)	4.6 (0.3)	4.6 (0.0)	4.6 (0.0)
Kilometers surveyed[b]	8,555	10,161	9,986	11,790	13,023	13,354	13,612	13,876
Field hours		2,550	2,120	2,447	2,385	2,005	2,086	2,122

[a]Packs are defined as groups of wolves with two or more animals
[b]Kilometers of roads and trails searched at least once for wolf sign

Estimates of the proportion of lone wolves in various populations range from 7% to 20% but are generally between 10% and 15% (Fuller et al. 2003). Assuming that lone wolves represent 15% of the population, Michigan's minimum population estimates underestimate actual population size by about 12%.

A more rigorous assessment of the accuracy of wolf population estimates was a comparison of independent wolf surveys conducted for 4 years in a 1,940-km^2 study area in the UP (Huntzinger et al., unpublished data). An independent survey conducted by Michigan Technological University (MTU) was assumed to be more accurate because MTU researchers spent an entire winter in the study area tracking and counting wolves. Overall, the MTU and MDNR counts were similar with an average difference of 4%. The MDNR counts were lower in 3 of 4 years.

5.4.2 Population Trend: 1989–1998

The sporadic occurrence of wolf sign throughout the late 1970s and early 1980s prompted the MDNR to intensify efforts to verify the establishment of a wolf population in the UP. Initial efforts involved following up reports of wolves or wolf sign made by department employees and the public and searching areas where the last wolf packs were known to occur in the 1940s and 1950s. Specifically, the MDNR was looking for evidence wolves had established territories and produced pups.

The beginning of wolf recovery in Michigan can be marked by two intertwined events: documentation of three wolves traveling together and making territorial marks (raised leg urinations) in the central UP during the fall of 1988, and the subsequent birth of pups in this territory during spring 1989 (note: these dates are corrections to those reported in Hammill 1992). This celebrated group of wolves was named the Nordic pack (a contraction of "north" and "Dickinson," the county where the wolves were located) and a wolf from this pack was captured and radio-collared in 1992 (Gardella, et al. 1996). This was a significant event because radio telemetry would become an important tool for documenting wolf recovery in later years.

Counts of wolves increased annually and by 1996, more than 100 animals were estimated in the population (Fig. 5.2). The 1997 count suggested that population growth had stopped. Managers suspected that sarcoptic mange probably reduced survival of pups, but this was never confirmed. However, in 1998, the population count increased by 23% and recovery seemed to be back on track.

Estimates of annual population growth made before 1995 should be viewed cautiously (e.g., 90% growth in 1993; Fuller et al. 2003) because new areas were being searched each year. Thus, some of the annual increases in wolf numbers before 1995 likely included new discoveries of wolves that may have established territories earlier but occurred outside of areas searched in previous years.

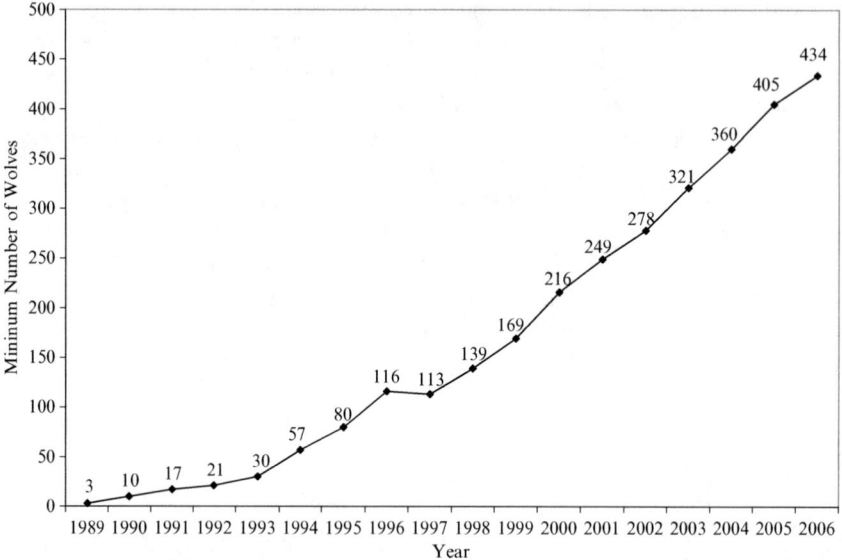

Fig. 5.2 Minimum late-winter estimates of wolf abundance in Michigan's Upper Peninsula, 1989–2006

5.4.3   Population Trend: 1999–2006

Beginning in 1999, managers improved documentation of their wolf survey efforts across the entire UP. All roads and trails searched were highlighted on maps in the field and later entered in a geographic information system (Table 5.2). Because of this documentation, changes in minimum numbers of wolves since 1999 are more reasonably interpreted as annual growth rates (Fig. 5.3). Michigan's wolf population more than doubled from 1999 to 2006, increasing from a minimum of 169 to 434 animals (Fig. 5.2). The annual growth rate of the population has slowed as the wolf population has increased (Fig. 5.3; Huntzinger et al., unpublished data; Van Deelen, this volume). Average pack size during winter grew from about 3.0 wolves/pack to 4.6 during 1999–2003 (Table 5.2), then remained stable in 2003–2006. Average pack size in Michigan is slightly larger than that reported for Wisconsin (Wydeven et al. 2006) but slightly smaller than that reported for Minnesota (Erb and Benson 2004), although survey techniques vary among the three states.

    Recovery of the wolf population begs the question of how many wolves the UP can support. Estimates of the biological carrying capacity for wolves are imprecise. Wolf numbers appear related to availability of food (Mech and Peterson 2003). Mladenoff et al. (1997) and Potvin (2003) estimated carrying capacity of wolves for the UP based on a published relationship of wolf density and prey biomass (Fuller 1989). Although their estimates of prey (deer) density were derived at different spatial scales, the results of both studies suggest the long-term abundance could

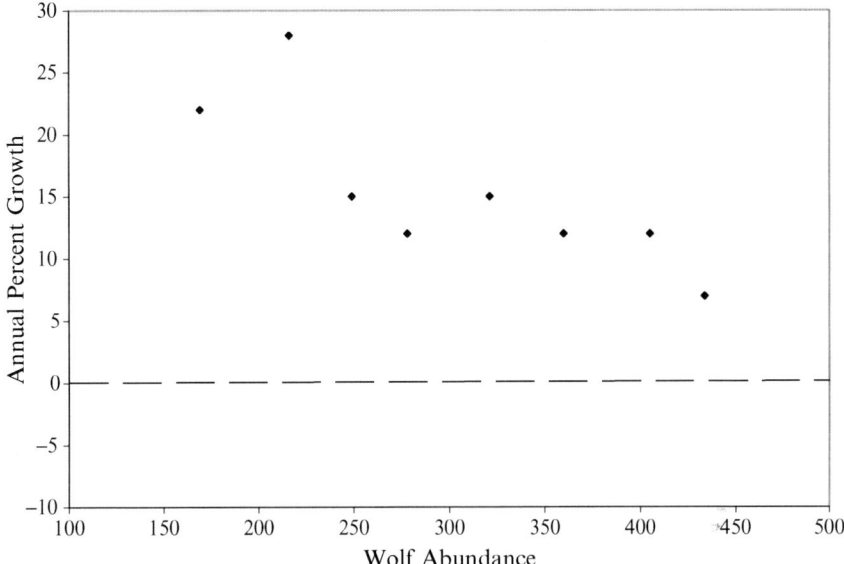

Fig. 5.3 Abundance and annual percent growth of wolves in Michigan's Upper Peninsula, 1999–2006. The dashed line represents zero growth

range from about 600 to 1,350 wolves. A similar estimate of carrying capacity, 620–1,150 wolves, was derived with a density-dependent Leslie matrix model (Miller et al. 2002). Construction of a log-linear regression model of the relationship between wolf abundance and per capita population growth rate suggests an upper limit of about 820 wolves (J. A. Vucetich, MTU, unpublished data). Van Deelen (this volume) estimated an equilibrium number of 1,321 wolves for the combined UP and Wisconsin population.

In early November 2003, managers captured and radio-collared a 1.5-year-old female wolf in the eastern UP. Wolf 4918 was located 11 times over the next 4 months before its signal was lost. Eight months later in October 2004, wolf 4918 was captured and mistakenly killed by a coyote trapper in the northern Lower Peninsula (NLP; Beyer et al. 2006). This was the first verified record of a wolf in the LP since 1910. Just over a month later, tracks from two wolves were observed in the same area. Because of these observations, managers surveyed portions of the NLP during the winters of 2005, 2006, and 2007, although no additional wolf sign was found.

5.5 Judging Wolf Recovery

Species whose populations are very low or declining rapidly are identified for legal protection at two levels of government, federal and state. At the federal level, species at risk are protected by the Endangered Species Act of 1973. In Michigan, species at risk are also protected by Part 365, Endangered Species Protection, of the Natural Resources and Environmental Protection Act (Public Act 451 of 1994). Both statutes identify two levels of risk: endangered species are at risk of extinction and threatened species are at risk of becoming endangered. A species is considered recovered when it is no longer at risk and long-term survival is reasonably assured. For wolves in Michigan, the progress of recovery was judged based on the achievement of goals set in federal and state recovery plans. However, the chronology detailed below illustrates that wolf recovery is not simply a biological construct but has administrative and legal aspects as well.

The Federal Eastern Timber Wolf Recovery Plan identified two delisting criteria: (1) survival of wolves in Minnesota must be assured; and (2) a second (at least one) viable wolf population be reestablished within the species' historic range outside of Minnesota and Isle Royale (USFWS 2007). If the second population was within 160 km of the Minnesota population, it had to maintain a minimum of 100 wolves for 5 consecutive years (interpreted as 6 consecutive winter counts ≥100) to be considered viable (USFWS 2007). The contiguous wolf population in Michigan and Wisconsin qualified as the second population. Wolf numbers in Michigan and Wisconsin combined first exceeded 100 animals in 1994 and delisting criteria were met in 1999 when the combined population totaled 379 animals. Four years later, the federal government designated an Eastern Distinct Population Segment (which included Michigan) and chose to reclassify wolves in this area to threatened status (USFWS 2003) rather than delisting, even though it appeared that the requirements identified in the recovery plan for delisting were met. In 2005, two federal courts enjoined and vacated the rule reclassifying wolves and the species was returned to endangered status. The courts did not agree with the federal government's policy on establishing distinct population segments or their interpretation of what constitutes recovery in a significant portion of the range. In early 2006, the federal government proposed designating a Western Great Lakes Distinct Population Segment and delisting wolves in this area. On March 12, 2007, wolves were removed from the federal list of threatened and endangered species and management authority was transferred to the states (USFWS 2007).

Michigan's Gray Wolf Recovery and Management Plan identified one recovery goal: wolves will be recovered when the population is equal to or greater than 200 wolves for 5 consecutive years (MDNR 1997). This criterion was conservative because it established a minimum sustainable population goal appropriate for an isolated population (USFWS 1992) even though the wolf population in Michigan is not isolated. Regular exchange of wolves occurs among Minnesota, Wisconsin, Michigan, and Ontario (Jensen et al. 1986; Mech et al. 1995; Thiel and Hammill 1988). Results of a population viability analysis suggested that this recovery goal reasonably approximated a viable population size (Hearne et al., unpublished data), although population viability analyses must be viewed cautiously because of model assumptions and uncertainty associated with many model inputs (e.g., frequency of catastrophic events; Fritts and Carbyn 1995). Michigan's wolf population first exceeded 200 animals in 2000 and the recovery goal was reached in 2004 with the fifth consecutive count exceeding 200 animals (Fig. 5.2).

Wolves subsequently were reclassified into threatened status under state law in 2002 and the process to remove wolves from the state list of threatened and endangered species began in 2007.

5.6   Recovery: How Did It Start?

In 1982, Hook and Robinson (1982) noted that wolves had failed to recover in the UP despite state legal protection since 1965, expanded federal protection since 1974, natural immigration of wolves of both sexes, and a reintroduction effort. Persecution by humans was likely preventing wolf recovery (Robinson and Smith 1977; Hook and Robinson 1982; Thiel and Hammill 1988). Using a random survey of residents in six Michigan counties (3 in the UP), Hook and Robinson (1982) concluded that most Michigan residents supported wolf recovery and predators in general in the early 1980s. However, a minority of survey respondents had strong negative attitudes toward predators. Within this minority, those whom researchers reasoned were most likely to kill wolves (hunters with antipredator attitudes) had a greater than average fear of wolves. Hook and Robinson (1982) argued that restoration of wolves would depend primarily on changes in human attitudes.

Coincidentally, while recovery of wolves was beginning in the UP, a second survey of Michigan residents' attitudes toward wolves was conducted (S.R. Kellert, unpublished data). This survey suggested that support for wolf restoration may have increased slightly during the 1980s (Table 5.3), although results of Hook and Robinson (1982) and Kellert's (unpublished data) surveys may not be strictly

Table 5.3 Percent of Michigan residents supporting and opposing wolf restoration in the Upper Peninsula based on three mail survey studies

	Hook and Robinson[a] (1982)	Kellert[b] (unpublished)		Mertig[c] (unpublished)		
Attitude	Statewide	UP	LP	UP	NLP	SLP
Support	54	64	57	46	57	64
Oppose	12	15	9	25	8	5

Comparisons among studies must be done cautiously because of differences in question design (see footnotes), sample populations (*UP* Upper Peninsula, *LP* Lower Peninsula, *NLP* Northern Lower Peninsula, *SLP* Southern Lower Peninsula) and sample size. Columns do not total to 100% because some respondents were uncertain or had no opinion about wolf recovery
[a]Survey question: "Wolves should be restored in the Upper Peninsula"; support includes strongly agree and agree response and opposed includes disagree and strongly disagree responses
[b]Survey question: "In general, how much do you support or oppose reestablishing Timber wolves in the Upper Peninsula." Support includes strongly and moderately support responses and oppose includes strongly and moderately opposed responses
[c]Survey question: "In general, how much do you support or oppose efforts to help wolves recover (come back on their own) in the Upper Peninsula." Support includes strongly and moderately support responses and oppose includes strongly and moderately opposed responses

comparable because of differences in survey questions, populations sampled, and sample size. Significantly, while support for wolf restoration may have increased, opposition to restoration appeared to remain roughly equivalent – the increase in support for recovery may have come from citizens who were previously unsure or uninterested. This is important because supporters of wolf recovery were probably unlikely to kill wolves, whereas some opponents to wolf recovery may have intentionally killed wolves. If a small proportion of citizens opposed to wolves was responsible for preventing wolf recovery in the 1970s, then a decline in opposition to wolf recovery would probably be more important for recovery than an increase in supporters. The 1990 survey suggested that opposition to wolf recovery persisted when wolves began their comeback (S.R. Kellert, unpublished data; Table 5.3). Unfortunately, survey data provide no understanding of whether opponents to wolf recovery were less likely to kill wolves in the 1980s. The only evidence for reduced persecution by humans in the UP during early recovery is an inference from wolf dynamics in neighboring Wisconsin. In Wisconsin, annual survival of adult radio-collared wolves increased from 65% during 1979–1985 to 76% during 1986–1992 (Wydeven 1993). Human-caused mortality dropped from 71% to 17% of known mortalities between these two time periods. These changes along with a suspected drop in prevalence of canine parvovirus and an increasing deer herd facilitated growth of Wisconsin's wolf population (Wydeven 1993). Since most wolves disperse from their natal territories (Mech and Boitani 2003) increases in the Wisconsin population may have increased the number of dispersing wolves that reached the UP. Although the relative roles of reduced persecution by humans, increased immigration, and other factors are unknown, conditions in the 1990s were apparently favorable because wolf numbers in the UP increased steadily. Human-caused mortality still occurred in the UP, but growth of the wolf population was robust and losses did not seem to hinder, what in hindsight was, a relatively rapid recovery.

5.7   Prospects for Wolf Colonization of the Northern Lower Peninsula

The rate of recovery of wolves in the UP was unexpected when the state's recovery and management plan was finalized. Although the plan recognized that wolves were likely to reach the NLP at some point, it established no population recovery goals for this region (MDNR 1997). It took about 13 years after wolf recovery began for the first documented evidence of a wolf dispersing into the NLP (Beyer et al. 2006), although ice conditions would have allowed wolves to cross the Straits of Mackinac during each winter since 1990 (Mark Gill, Supervisor of Vessel Traffic Service St. Mary's River, Soo Coast Guard Station, personal communication). It is important to note that wolf immigration into the NLP does not guarantee colonization. Indeed, three consecutive winter surveys (2005–2007) conducted after wolf tracks were observed in the NLP failed to document the continued presence of wolves. Although obvious, to establish a population at least two wolves of opposite sex would need to disperse to the NLP, find each other, and survive long enough to establish a territory and produce young. A seemingly less likely scenario would be dispersal of a pregnant female. Unfortunately, science-based predictions of these probabilities are unavailable and possibly unachievable.

Although a science-based prediction of when wolves might become established in the NLP is not available, two studies have evaluated the suitability of the region to support wolves (Potvin 2003; Gehring and Potter 2005). Gehring and Potter (2005) applied a model of habitat suitability developed by Mladenoff et al. (1995) which estimates the probability of wolf occupancy based on road density. Potvin (2003) modified the Mladenoff et al. (1995) model and also included deer density as a predictor variable. These studies suggested 4,000–8,000 km² of suitable habitat exists, although the habitat is more fragmented than habitat in the UP. Gehring and Potter (2005) reported that the suitable habitat identified in their study (4,231 km²) could support 46–89 wolves but suggested this estimate may be conservative.

There is lower tolerance for wolves in the NLP than in the UP among interested Michigan citizens (Beyer et al. 2006). In addition, density of livestock farms is much greater in the NLP (1 farm/13 km²) than the UP (1 farm/49 km²; Beyer et al. 2006). Thus, the social and physical environments wolves will encounter if they recolonize the NLP are much different than the UP.

5.8 Overview of Research on Wolves in Michigan

Scientific investigation is the foundation of sound wildlife management. The first investigations of wolves in Michigan were simple compilations of occurrence records to assess wolf distribution within the state (e.g., Wood and Dice 1923). In 1935, the Michigan Department of Conservation (MDC, precursor to the MDNR) began to study food habits of large predators, including wolves. In addition to collecting and examining scats, they paid citizens $0.50 for stomachs (with intestines attached) of predators turned in for bounty. During the winters of 1938 and 1950, Stebler (1951) studied wolf ecology by following wolf tracks in the snow. The status of wolves in the UP received some attention in the mid-1950s when the MDC used bounty records and observations of wolf sign collected by field biologists to estimate the number and distribution of wolves (Arnold 1955; Arnold and Schofield 1956). The number and distribution of wolves was assessed again in the early 1970s using MDNR records and field surveys (Hendrickson et al. 1975). Weise et al. (1975) conducted the experimental translocation project described above in 1974. Biological work during the remainder of the 1970s and 1980s was limited to documenting occurrence records of wolves killed in the UP (Robinson and Smith 1977; Thiel 1988; Thiel and Hammill 1988). Two studies examined public attitudes toward wolves and their potential recovery (Hook and Robinson 1982; S. Kellert, unpublished data).

Efforts to study the ecology of wolves started again shortly after wolves began to recolonize the UP. Two years after reproduction was documented in the Nordic pack (above), a wolf from this pack was captured and radio-collared. The rather humble objective of this effort was to simply learn what area this pack was using. The initial research efforts, although modest, laid the foundation for a more formal research program that began in 1997 with collaboration between MDNR, MTU, and Pictured Rocks National Lakeshore. The focus of this program was to improve population monitoring while the wolf population was recovering (Potvin et al., unpublished data; Johnson 2000; Drummer et al., unpublished data; Huntzinger et al., unpublished data). Broad goals for wolf monitoring in Michigan included determining wolf distribution and abundance, assessing select demographic characteristics (e.g., mortality rates, population growth, and pack sizes), and evaluating some of the ecological interactions between wolves and white-tailed deer.

Radio telemetry has been a fundamental tool for achieving/evaluating most of the goals for monitoring Michigan wolves. Since 1992, 226 wolves have been captured and radio-collared and through 2006, these animals have been relocated over 18,500 times. These marked animals provide information on wolf survival, dispersal, and reproductive status. The radio telemetry information is used to identify wolf territories, determine wolf movements within and among pack territories, and support the winter population survey. The telemetry data were also useful for evaluating and modifying a model predicting wolf distribution that was used for projecting wolf population size and expansion in the UP and evaluating habitat suitability in the NLP (Potvin 2003; Potvin et al. 2005). The marked animals also facilitated an in-depth study of winter predation rates on white-tailed deer (Huntzinger 2006).

As the wolf population increased, it became more difficult and more costly to estimate abundance by surveying the entire UP each year. Researchers have evaluated two alternative survey methods: (1) application of the ad hoc approach used in Minnesota that extrapolates wolf density from small study areas to an estimate of occupied range (e.g., Fuller et al. 1992; Erb and Benson 2004); and (2) extrapolation of sample counts (Potvin et al. 2005; T. Drummer, MTU, unpublished data). The methods used to estimate wolf abundance in Minnesota were deemed unacceptable because they produced an estimate 75% larger than the traditional census (D. E. Beyer, unpublished data). The same conclusion was reached in Wisconsin (J.E. Wiedenhoeft, unpublished data). Evaluation of several sampling plans suggested probability sampling of land areas may be useful for estimating wolf abundance (Potvin et al. 2005). Further work indicated a geographically based, stratified, sampling plan combined with a panel design that takes advantage of the correlation between counts in the same area in successive years would produce unbiased and precise estimates of wolf abundance (T. Drummer, unpublished data). This sampling approach was implemented for the first time in winter 2007. These ecological studies provided science-based documentation of wolf recovery and a better understanding of wolf ecology in the UP (J. A. Vucetich, unpublished data).

Once it became clear the wolf population was recovering, managers recognized that a shift in focus from recovery to management would soon be required as biological and social issues changed. To facilitate this shift, an understanding of current public attitudes toward wolves and various wolf management options was needed. The MDNR funded Michigan State University (MSU) to conduct a comprehensive study of public attitudes. Researchers at MSU surveyed the general public ($n = 8,500$; 53% response rate), livestock producers ($n = 1,000$; 69% response rate), and fur trappers ($n = 1,000$; 69% response rate, Beyer et al. 2006). The survey results refined MDNR's understanding of the range of preferences and tolerances for wolf abundance and wolf–human interactions. The survey results show that citizens interested in wolves vary greatly in their preferences and tolerances for the minimum and maximum levels of wolf abundance and wolf–human interactions. This diversity of values will make management of the recovered wolf population challenging for the MDNR. The survey also provided important information on preferences for various management options. For example, there is broad support for the selective removal of problem wolves (Beyer et al. 2006).

Additional on-going research efforts regarding wolves in Michigan include an analysis of wolf diet via quantitative fatty acid analysis (E. Berkley, University of Wisconsin—Madison); an evaluation of species hybridization in *Canis* (T. Wheeldon, Trent University); and an assessment of nonlethal control options for wolf depredation management (T.M. Gehring, Central Michigan University). Research needs include an evaluation of wolf population responses to selected management options; public responses to selected wolf management practices; an evaluation of information and education efforts; and an assessment of the effect of wolf predation on survival rates of deer.

5.9 Planning for Management After Recovery

The increase in wolf abundance and distribution, in combination with anticipated increased management flexibility after federal delisting, intensified some biological and social concerns and created a few new ones. For example, public support for wolves may be declining in the UP (A.G. Mertig, unpublished data; Table 5.3). Once the federal and state recovery goals were met and the process to remove wolves from federal protection was initiated (USFWS 2004), the MDNR began a process to revise the state's 1997 wolf plan. Revision of the plan was necessary because the original plan was developed when there were relatively few wolves in the state (planning began in 1992) and focused on the biological needs of a recovering population. The revision "addresses the challenges associated with the current biological, social and regulatory context of wolf management in Michigan" (MDNR 2008).

The first step in revision of the plan was to identify important issues associated with wolves and their management. Identification of issues was accomplished through intra- and interagency scoping, ten public meetings held throughout the state, and public comment periods (MDNR 2008). These efforts identified the following eight broad areas of concern and accompanying need for strategic direction: (1) wolf abundance and distribution, (2) wolves and human safety, (3) wolf depredation of domestic animals, (4) wolf–prey relationships, (5) recreational wolf harvest, (6) habitat linkages to neighboring wolf populations, (7) information and education,and (8) funding for wolf conservation (Beyer et al. 2006). Understanding of these issues was refined through focus group meetings with representatives of various stakeholder groups (Bull and Peyton 2005). Managers applied this understanding to the development of public attitude surveys (see general description above) used to characterize public attitudes on: "(1) reasons for having wolves in Michigan, (2) the number of wolves and frequency of wolf-related interactions in different regions of the state, (3) options to address depredation of livestock, hunting dogs and other pets, (4) options to address public concerns regarding human safety, (5) options to address impacts to deer, and (6) a public harvest of wolves"(MDNR 2008).

The results of attitude surveys, along with a summary of pertinent biological science, were shared with an advisory group (Michigan Wolf Management Roundtable) convened by the MDNR. The Roundtable, made up of 20 stakeholder groups representing the diversity of interests in Michigan's wolves, was charged with developing guiding principles for wolf management once the species was removed from state and federal endangered species protection (Michigan Wolf Management Roundtable, unpublished data). The management plan was revised consistent with the spirit of the Roundtable's guidance. A draft of the revised plan was released for public comment, finalized and signed by the DNR Director in 2008.

5.10 Summary

The initial decline in wolf abundance in Michigan was a part of the overall decline of the species in the lower 48 states. The timing of the decline reflected the pattern of settlement by European immigrants. Near elimination of wolves was fostered by public policy and a culture that viewed wolves as incompatible with civilization and a threat to the settlers' interests. In Michigan, concerns for the plight of wolves began surfacing in the mid-1940s, but wolves were not given legal protection (by the state) until after they were nearly eliminated from the state. Eventually, greater environmental awareness by the public resulted in changes in public policy that provided better protection for species like wolves that were at risk of extinction throughout the country. Despite these protections, it took almost two decades before wolves began to recover in the UP. However, once recovery began the population grew quickly and biologists began studying their ecology and documenting the population's rate of growth. During 1988–2006, the population grew from a few lone wolves to a population of over 400 animals. This population growth resulted in issues associated with wolf depredation of domestic animals and nuisance wolf behavior that demanded the attention of managers.

Recent history has shown that the recovery of wolves in the Great Lakes region is not simply judged on the biological viability of the populations. Instead, the recovery of Great Lakes wolves is imbedded in a public debate on the interpretation and administration of the Federal Endangered Species Act of 1973. This public debate is being carried out within our judicial system. Management of wolves in this context will be challenging because the tools available to managers depend in part on whether wolves are listed as a federally threatened or endangered species or under state management authority. Management of wolves is challenging because public opinions on wolves and their management are highly polarized. Divergent public opinion can disrupt agency management. MDNR has addressed this problem by working through the contentious issues in the planning phase of management. The Michigan Wolf Management Roundtable, an advisory group representing a microcosm of society, worked through the difficult issues to develop guiding principles for wolf management in Michigan. These guiding principles were then used to develop a strategic wolf management plan.

Acknowledgments   This work was supported by the Michigan Department of Natural Resources, Michigan Technological University, and Federal Aid in Wildlife Restoration (grant W-147-R). Tim Van Deelen and two anonymous reviewers provided valuable comments and suggestions on an earlier draft of the manuscript. We thank the many MDNR and USDA Wildlife Services staff and MTU students that collected field data that contributed to our understanding of wolves in Michigan.

References

Arnold, D. A. 1955. Status of Michigan timber wolves, 1955. Michigan Department of Conservation Game Division Report No. 2062.

Arnold, D. A., and Schofield, R. D. 1956. Status of Michigan timber wolves, 1954–1956. Michigan Department of Conservation Game Division Report No. 2079.

Baker, R. H. 1983. Michigan mammals. East Lansing: Michigan State University Press.

Bartlett, I. H. 1938. Whitetails, presenting Michigan's deer problem. Lansing: Michigan Department of Conservation Game Division.

Benton-Banai, E. 1988. The Mishomis book-the voice of the Ojibway. St. Paul: The Red Schoolhouse.

Beyer, D., Hogrefe, T., Peyton, R. B., Bull, P., Burroughs, J. P., and Lederle, P. eds. 2006. Review of social and biological science relevant to wolf management in Michigan. Michigan Department of Natural Resources Wildlife Division Report No. 3457.

Bull, P., and Peyton, R. B. 2005. Michigan wolf management focus group meeting results. In Review of social and biological science relevant to wolf management in Michigan, eds. D. Beyer, T. Hogrefe, R. B. Peyton, P. Bull, J. P. Burroughs, and P. Lederle, Appendix IX Michigan Department of Natural Resources Wildlife Division Report No. 3457.

Burt, W. H. 1946. The mammals of Michigan. Ann Arbor: University of Michigan Press.

Ciucci, P., Boitani, L., Francisci, F., and Andreoli, G. 1997. Home range, activity, and movements of a wolf pack in central Italy. Journal of Zoology 243:803–819.

Cochrane, J. F. 1996. Woodland caribou restoration at Isle Royale National Park. United States National Park Service Technical Report No. 96–03.

Coleman, J. T. 2004. Vicious, wolves and men in America. New Haven: Yale University Press.

Douglass, D. W. 1970. History and status of the wolf in Michigan. In Proceedings of the symposium on wolf management in selected areas of North America, eds. S. E. Jorgensen, C. E. Faulkner, and L. D. Mech, pp. 6–8. U.S. Department of Interior Bureau of Sport Fish and Wildlife, Region 3.

Eberhardt, L. L. 1957. The 1956 and 1957 pellet group surveys. Michigan Department of Conservation Game Division Report No. 2133.

Erb, J., and Benson, S. 2004. Distribution and abundance of wolves in Minnesota, 2003–04. Minnesota Department of Natural Resources report, Grand Rapids, Minnesota (http://files.dnr. state.mn.us/natural_resources/animals/mammals/wolves/2004_wolfsurvey_report.pdf).

Evers, D. C. 1994. Elk. In Endangered and threatened wildlife of Michigan, ed. D. C. Evers, pp. 37–81. Ann Arbor: The University of Michigan Press.

Fritts, S. H., and Carbyn, L. N. 1995. Population viability, nature reserves, and the outlook for gray wolf conservation in North American. Restoration Ecology 3:26–38.

Fuller, T. K. 1989. Population dynamics of wolves in north-central Minnesota. Wildlife Monographs 105:1–41.

Fuller, T. K., Berg, W. E., Radde, G. L., Lenarz, M. S., and Joselyn, G. B. 1992. A history and current estimate of wolf distribution and numbers in Minnesota. Wildlife Society Bulletin 20:42–55.

Fuller, T. K., Mech, L. D., and Cochrane, J. F. 2003. Wolf population dynamics. In Wolves; behavior, ecology and conservation, eds. L. D. Mech, and L. Boitani, pp. 161–191. Chicago: The University of Chicago Press.

Gardella Schadler, C. L., and Hammill, J. 1996. Status of the gray wolf in Michigan: a naturally recovering population. In Wolves of America Conference Proceedings. pp. 104–110. Albany, NY.

Gehring, T. M., and Potter, B. A. 2005. Wolf habitat analysis in Michigan: an example of the need for proactive land management for carnivore species. Wildlife Society Bulletin 33:1237–1244.

Hammill, J. H. 1992. Wolf reproduction confirmed on mainland Michigan! International Wolf 2:14–15.

Hendrickson, J., Robinson, W. L., and Mech, L. D. 1975. Status of the wolf in Michigan, 1973. American Midland Naturalist 94:226–232.

Holman, J. A. 1975. Michigan's fossil vertebrates. Michigan State University Publication, Museum Education Bulletin No. 2.

Hook, R. A., and Robinson, W. L. 1982. Attitudes of Michigan citizens toward predators. In Wolves of the world, eds. F. L. Harrington and P. C. Paquet, pp. 382–394. Park Ridge: Noyes Publishing.

Huntzinger, B. A. 2006. Sources of variation in wolf kill rates of white-tailed deer during winter in the U.P. Michigan. MS Thesis, Michigan Technological University, Houghton, Michigan.

Jedrzejewski, W., Schmidt, K., Theuerhauf, J., Jedrzejewska, B., and Okarma, H. 2001. Daily movements and territory use by radio-collared wolves, Canis lupus, in Bialowieza Primeval Forest in Poland. Canadian Journal of Zoology 79:1993–2004.

Jensen, W. F., Fuller, T. K., and Robinson, W. L. 1986. Wolf, Canis lupus, distribution on the Ontario-Michigan border near Sault Ste. Marie, Ontario. Canadian Field Naturalist 100:363–366.

Johnson, J. S. 2000. The return of the gray wolf (Canis lupus) to Upper Michigan. MS Thesis, Michigan Technological University, Houghton, Michigan.

Langenau, E. E. Jr. 1994. 100 years of deer management in Michigan. Wildlife Division Report No. 3213. Michigan Department of Natural Resources, Lansing, Michigan.

Lopez, B. H. 1978. Of wolves and men. New York: Charles Scribner's and Sons.

Mech, L. D. 1966. The wolves of Isle Royale. U. S. Department of Interior, Fauna of the National Parks of the U. S. Fauna Series number 7.

Mech, L. D., Fritts, S. H., and Wagner, D. 1995. Minnesota wolf dispersal to Wisconsin and Michigan. American Midland Naturalist 133:368–370.

Mech, L. D., and Boitani, L. 2003. Wolf social ecology. In Wolves; behavior, ecology and conservation, eds. L. D. Mech, and L. Boitani, pp. 1–34. Chicago: The University of Chicago Press.

Mech, L. D., and Peterson, R. O. 2003. Wolf-prey relations. In Wolves; behavior, ecology and conservation, eds. L. D. Mech, and L. Boitani, pp. 131–160. Chicago: The University of Chicago Press.

Michigan Department of Natural Resources. 1976. Rule 299.1027. Endangered and Threatened Species – Mammals. Effective December 4, 1976. Lansing, Michigan, USA.

Michigan Department of Natural Resources. 1997. Michigan gray wolf recovery and management plan. Lansing: Michigan Department of Natural Resources, Wildlife Division.

Michigan Department of Natural Resources. 2008. Michigan gray wolf management plan. Michigan Department of Natural Resources, Wildlife Division Report No. 3484.

Michigan Nonprofit Association and Council of Michigan Foundations. 2002. Michigan in brief. < http://www.michiganinbrief.org/edition07/Chapter1/Chapter1.htm>. Accessed 6 Feb 2007.

Miller, D. H., Jensen, A. L., and Hammill, J. H. 2002. Density dependent matrix model for gray wolf population projection. Ecological Modelling 151:271–278.

Mladenoff, D. J., Haight, R. G., Sickley, T. A., and Wydeven, A. P. 1997. Causes and implications of species restoration in altered ecosystems: a spatial landscape projection of wolf population recovery. BioScience 47:21–31.

Mladenoff, D. J., Sickley, T. A., Haight, R. G., and Wydeven, A. P. 1995. A regional landscape analysis and prediction of favorable gray wolf habitat in the northern Great Lakes region. Conservation Biology 9:279–294.

Murie, O. J. 1951. The elk of North America. Harrisburg: Stackpole Company.

Potvin, M. J. 2003. A habitat analysis for wolves in Michigan. MS Thesis, Michigan Technological University, Houghton, Michigan.

Potvin, M. J., Drummer, T. D., Vucetich, J. A., Beyer, Jr., D. E., Peterson, R. O., and Hammill, J. H. 2005. Monitoring and habitat analysis for wolves in Upper Michigan. Journal of Wildlife Management 69:1660–1669.

Robinson, W. L., and Smith, G. J. 1977. Observations on recently killed wolves in Upper Michigan. Wildlife Society Bulletin 5:25–26.

Stebler, A. M. 1944. The status of the wolf in Michigan. Journal of Mammalogy 25:37–43.

Stebler, A. M. 1951. The ecology of Michigan coyotes and wolves. PhD Dissertation, University of Michigan, Ann Arbor, Michigan.

Stewart, P. A., and Negus, N. C. 1961. Recent record of wolf in Ohio. Journal of Mammalogy 42:420–421.

Thiel, R. P. 1988. Dispersal of a Wisconsin wolf into Upper Michigan. Jack-Pine Warbler 66:143–147.

Thiel, R. P. 1993. The timber wolf in Wisconsin: the death and life of a majestic predator. Madison: University of Wisconsin Press.

Thiel, R. P., and Hammill, J. H. 1988. Wolf specimen records in Upper Michigan, 1960–1986. Jack-Pine Warbler 66:149–153.

U.S. Fish and Wildlife Service. 1992. Recovery plan for the eastern timber wolf. U.S. Fish and Wildlife Service, Twin Cities, Minnesota, USA.

U.S. Fish and Wildlife Service. 2003. Endangered and threatened wildlife and plants: final rule to reclassify and remove the gray wolf from the list of endangered and threatened wildlife in portions of the conterminous United States; establishment of two special regulations for threatened gray wolves. Federal Register 68(620): 15804–15875.

U.S. Fish and Wildlife Service. 2004. Endangered and threatened wildlife and plants; removing the Eastern distinct population segment of the gray wolf from the list of endangered and threatened wildlife; proposed rule. Federal Register 69(139): 43664–43692.

U.S. Fish and Wildlife Service. 2007. Endangered and threatened wildlife and plants; final rule designating the Western Great Lakes populations of gray wolves as a distinct population segment; removing the Western Great Lakes distinct population segment of the gray wolf from the list of endangered and threatened wildlife; final rule. Federal Register 72(26): 6051–6103.

Verme, L. J. 1984. Some background on moose in Upper Michigan. Michigan Department of Natural Resources Wildlife Division Report No. 2973.

Weise, T. F., Robinson, W. L., Hook, R. A., and Mech, L. D. 1975. An experimental translocation of the eastern timber wolf. Audubon Conservation Report 5. National Audubon Society, New York.

Wood, N. A., and Dice, L. R. 1923. Records of the distribution of Michigan mammals. Michigan Academy of Science, Arts, and Letters 3:425–469.

Wydeven, A. P. 1993. Wolves in Wisconsin: recolonization underway. International Wolf 3:18–19.

Wydeven, A. P., Wiedenhoeft, J. E., Schultz, R. N., Thiel, R. P., Boles, S. R., and Heilhecker, E. 2006. Progress report for wolf population monitoring in Wisconsin for the period October 2005–March 2006. Wisconsin Department of Natural Resources report, Park Falls, Wisconsin (http://dnr.wi.gov/org/land/er/publications/wolfreports/report_oct05-mar06.htm).

Chapter 6
History, Population Growth, and Management of Wolves in Wisconsin

Adrian P. Wydeven, Jane E. Wiedenhoeft, Ronald N. Schultz, Richard P. Thiel, Randy L. Jurewicz, Bruce E. Kohn, and Timothy R. Van Deelen

Preface While we were growing up in Wisconsin during the 1950s and 1960s, gray wolves (we always called them timber wolves, *Canis lupus*) were making their last stand in northern Wisconsin. Wolves were considered a wilderness-dependant relic of Wisconsin's frontier past that no longer belonged in our state. We did not expect wolves to ever again return to the state, at least not in any sizeable numbers. Among us, Dick Thiel was the most tenacious about trying to find evidence of wolves in Wisconsin, even as a student in the 1960s and 1970s. When wolves began returning during the mid-1970s, we dared not hope for any more than a token population of wolves to reestablish. The recovery of wolves in Wisconsin has succeeded beyond our wildest dreams. We have had the pleasure to document and track the amazing return of this powerful predator to our state.

6.1 Introduction

The gray wolf has exhibited a remarkable recovery in Wisconsin during the late twentieth and early twenty-first centuries, despite a common belief during the mid-1900s that the state was no longer wild enough to support populations of large predators such as gray wolves. In some ways, Wisconsin seems like an unlikely place for wolves to have recovered. The state's nickname, "America's Dairyland," reflects the abundance of livestock farming. Wisconsin has over 3.3 million cattle and over 5.5 million people in a land area of 140,663 km². Roughly half the state is forest, and in 2002, 46% was classified as farmland (Wisconsin Legislative Reference Bureau 2003). Public lands include 16.4% of the state, with major land ownership in county forests, national forests, national wildlife refuges, state forests, and state wildlife areas (Wisconsin Legislative Reference Bureau 2003). Wisconsin's largest federal or state designated wilderness area covers 73 km².

Despite few large wild areas, wolves were able to recolonize and again become important elements of forest ecosystems in northern and central Wisconsin. Legal protection, public education and outreach, and sound scientific management of public forest lands enabled wolves to recover and demonstrated that wolves can

A.P. Wydeven et al. (eds.), *Recovery of Gray Wolves in the Great Lakes Region of the United States,*
DOI: 10.1007/978-0-387-85952-1_6, © Springer Science+Business Media, LLC 2009

recover without extensive wilderness, provided there is adequate habitat, prey, legal protection, and public acceptance.

In this chapter, we review the history and management of wolves in Wisconsin, examine the growth and expansion of the wolf population, and speculate on the future of the wolf population with elimination of federal protection and reduction of favorable habitat caused by human landscape developments.

6.2 Early History and Initial Recolonization of Wolves in Wisconsin

Gray wolves probably have occupied Wisconsin since the last glacier receded about 10,000 years ago, and perhaps earlier in portions of southwestern Wisconsin that were not glaciated. Populations of wolves probably fluctuated with the size of ungulate populations. When the first European exploration began in 1634, wolves coexisted with herds of bison (*Bison bison*), elk (*Cervus elaphus* , and white-tailed deer (*Odocoileus virginianus*) in prairies, savannas, and oak (*Quercus*) and maple (*Acer*) forests of southern Wisconsin, and with moose (*Alces alces*), white-tailed deer, and small numbers of caribou (*Rangifer tarandus*) in the hemlock-maple (*Tsuga-Acer*), pine (*Pinus*), swamp conifers, and boreal forests and bogs of northern Wisconsin. Beavers (*Castor canadensis*) also were abundant throughout the state, but probably more so in the streams and glacial lakes of northern Wisconsin. When European settlement started in earnest during the 1830s, beavers were nearly eliminated due to unregulated trapping during the fur trade, and bison were extirpated by Native Americans after acquiring horses and firearms (Thiel 1993). Other prey such as deer, elk, and moose were probably still relatively abundant.

Jackson (1961) speculated that there were 20,000–25,000 wolves in Wisconsin at the beginning of European settlement. This would have represented an unlikely density of 142–177 wolves per 1,000 km². Wolf densities this high have not been documented in modern research on wolves in North America (Fuller et al. 2003). Wydeven (1993) speculated that perhaps 3,000–5,000 wolves existed at the beginning of European settlement, or about 20–35 wolves per 1,000 km². This estimate appears more compatible with likely prey abundance and agrees with recent research on wolf densities.

A bounty for the killing of wolves was offered by the Wisconsin Territory from 1839 through 1847, and following statehood (1848), a state bounty ran nearly continuously from 1865 to 1957 (Thiel 1993). Bounties were paid to private trappers and hunters for killing wolves and coyotes (*Canis latrans*), and both species were listed as wolves in bounty records. After 1947, when wolves had declined to very low numbers, wolves were distinguished from coyotes in the bounty records (Thiel 1993). Unlike western states, federal and state governments made no concerted effort to eliminate wolves in Wisconsin. Rangeland grazing of livestock was not practiced across northern Wisconsin, and livestock were normally kept in small fenced pastures near farmsteads. Nonetheless, unregulated hunting and trapping, as well as the incentive of bounty payments, caused the eventual collapse of the wolf population in Wisconsin.

Thiel (1993) documented the decline of wolves in Wisconsin that occurred from the 1800s to the1950s. The wolf population declined from about 200 in the early 1920s, to a scattered remnant of lone wolves spread across the north in the late 1950s. By 1960, wolves were considered extirpated from the state (Thiel 1993). Despite compiling scattered reports of wolf observations during the 1960s and early 1970s, Thiel (1978) found no evidence of functioning packs in the state.

Recolonization of Wisconsin by wolves began by 1975, and by 1979, five wolf packs were established in two Wisconsin counties. A wolf pack was detected in Minnesota along the Wisconsin border during winter 1974–1975, and between 1975 and 1979, five wolves were found dead in Douglas County, Wisconsin, just east of the Minnesota border (Mech and Nowak 1981; Thiel 1993). Thiel and Welch (1981) documented breeding packs of wolves in the state by 1977 and 1978. In 1979, two wolves were also found dead in Lincoln County, about 200 km southeast of the Douglas County packs (Thiel 1993). The source of colonizing wolves was likely the large Minnesota population to the west, although the appearance of a pack in Lincoln County in north-central Wisconsin in 1979 may indicate that some wolves had persisted in parts of Wisconsin. The Lincoln County pack already consisted of 12 wolves in 1979, indicating that the pack had probably been in the area for ≥ 2 years.

6.3 Federal and State Endangered and Threatened Listing of Wolves

Because of the decline of gray wolves across the USA, the eastern timber wolf (*Canis lupus lycaon*), defined at the time to include wolves in the western Great Lakes region, was listed as endangered in 1967 on the first list of endangered species promulgated by the US Fish and Wildlife Service (USFWS 1992). Following passage of the federal Endangered Species Act in 1973, the eastern timber wolf was again listed in 1974, and in 1978 all forms of gray wolves were listed as endangered in the contiguous USA, except in Minnesota where wolves were listed as threatened (USFWS 1992).

The Wisconsin Department of Natural Resources (WDNR) also maintained a separate list of state endangered and threatened species, and with their return, gray wolves were listed as endangered species under state law in 1975. In 1979, the WDNR began a program of formal monitoring of the wolf population (Wydeven et al. 1995).

The WDNR developed a state recovery plan in 1989. The plan mandated that wolves would be down-listed to threatened status if the population remained above 80 for ≥ 3 years consecutively (WDNR 1989). These criteria were also adopted by the USFWS for federal reclassification to threatened status (USFWS 1992). The USFWS also decided that wolves could be removed from the federal list of endangered and threatened species when the population exceeded 100 wolves for ≥5 years in Wisconsin and Michigan, along with a population of 1,251–1,400 wolves in Minnesota (USFWS 1992). These goals were based on late-winter counts when wolves were at the lowest level in their annual population cycle, and were most easily counted from tracks in the snow and observations from the air.

The WDNR developed a state management plan for wolves in 1999. This plan set a state delisting goal of 250 wolves outside of Indian reservations, and a long-term management goal of 350 wolves outside of Indian reservations (WDNR 1999). WDNR goals excluded wolves living on Indian reservations because the state had no management authority for wildlife on Indian reservations. Normally ≤6% of Wisconsin's wolf population occurs on Indian reservations. Under state law, wolves were down-listed to threatened status in 1999 when the statewide count was 205 wolves. Wolves were delisted from state threatened status in 2004 when 335 wolves occurred in the state. Wolves have been classified in Wisconsin as protected wild animals since August 1, 2004. This classification is given to non-game mammals that are not endangered or threatened.

Federal delisting and reclassification has been a more complex and difficult process (Refsnider, this volume). Wolves in Minnesota were down-listed to federal threatened status in 1978, but wolves in Wisconsin and Michigan were still designated as endangered until 2003 when they were classified as threatened as part of the Eastern Distinct Population Segment (Refsnider, this volume). In 2005, wolves in Wisconsin and other states in the Eastern Distinct Population Segment, except Minnesota, reverted back to federal endangered status as a result of lawsuits by environmental and animal welfare groups (Refsnider, this volume). Wolves were removed from the federal list of endangered species in Wisconsin on March 12, 2007, and all management authority for the species reverted to the state.

6.4   Methods for Monitoring Wolves in Wisconsin

6.4.1   Wolf Population Monitoring

Since 1979, we (as WDNR employees) have used a combination of snow-track surveys, aerial radiotracking, summer howling surveys, and collection of observations of wolves to estimate the size of wolf populations annually (Wydeven et al. 1995). We used territory mapping (Fuller et al. 2002) to determine the location of all wolf territories and determine the number of wolves in each territory. Territories were mapped for packs and lone wolves that appeared to occupy regular home range areas, but not for lone wolves that seemed to be dispersing. This survey system likely underestimates lone wolves that occur outside of established territories.

We have live-trapped and radiocollared wolves since 1979, usually during May and June using modified foot-hold traps (Kuehn et al. 1986). Only limited late-summer trapping was done to avoid capture of bear hounds which are trained or used for bear hunting during that time. Trapping was avoided during fall and winter because of risks of freezing of toes and capturing of hunting dogs. Recently, a few wolves were captured with cable restraints outside of May and June (Olson and Tischaefer 2004). Wolves >14 kg usually were tranquilized with a 5:1 mixture of Ketamine at 0.1 ml/kg and Xylazine at 0.02 ml/kg, and were reversed with Yohimbine at 0.15 mg/kg (Kreeger 2003). Wolf trapping and handling occurs under oversight by the WDNR Animal Care and Use Committee.

Captured wolves were generally fitted with standard VHF radiocollars (Telonics, Mesa Arizona), although a limited number were also fitted with satellite and Global Positioning System (GPS) collars, and some pups were fitted with ear-tag transmitters (Heilhecker et al. in press). Transmitter-equipped wolves were generally located once per week from the air using fixed-wing aircraft, although flights were sometimes more frequent during intense research or less frequent during periods of budget shortfalls. Ground-based telemetry was used for some intense research, and to recover wolves that died. Most transmitters emitted mortality signals after 5.5 or 6 h of inactivity. Year-round radiotracking enabled us to determine annual pack territories. We made special efforts during December–March to observe and count radiocollared wolves and other members of their packs. Radiocollared wolves facilitated aerial observations of packs roughly 30% of the time they were relocated during winter; packs without radiocollared individuals (hereafter non-collared packs) were rarely observed.

We conducted snow-track surveys every winter since 1979–1980 to supplement radio tracking and search new areas for wolf sign. Since 1995, we have used ≥133 survey blocks to provide more systematic coverage of potential wolf range (Wydeven et al. 1996). Each survey block averaged about 500 km^2, and was bordered by highways, public roads, waterways, and state boundaries. Track surveys were focused on areas with historical wolf presence, recent observations of wolves, or areas of highly suitable habitat (Mladenoff et al. 1995, 1999). Northern and central Wisconsin has an extensive network of roads, and all areas used by packs seemed to contain some roads useable by four-wheel drive vehicles. Initially, trained biologists and technicians conducted surveys, but since 1995, volunteer trackers have supplemented and enhanced survey coverage. Volunteers were trained in wolf ecology and animal tracking by agency trackers, and agency and volunteer trackers received special training by animal tracker, James Halfpenny (Halfpenny 1986).

Numbers of tracks observed within survey blocks were used to estimate numbers of wolves in non-collared packs. We conducted surveys 1–3 days after new snowfalls, and attempted to cover most snow-covered roads in survey blocks. Trackers located wolf tracks while slowly driving snow-covered roads and trails, or on foot. Observed wolf tracks were followed to determine where they entered and left roads. Discrete packs were determined by distances between track and sign observations, directions of movements, timing of observations, presence of radiocollared packs, historical pack use of an area, and knowledge of focal points such as den sites and rendezvous sites.

6.4.2   Home Range and Territory Mapping

We mapped territories and distribution on non-collared packs by creating polygons that contained all locations of sign and tracks, and reports of observations of wolves within assumed packs. If a pack was collared in the past, we used the previous year's territory area of that pack for current-year area, unless field sign indicated the territory areas had shifted.

The presence of raised-leg urinations (RLUs), especially double RLUs (urinations by both breeding male and female), was used to infer territory marking and likeliness of breeding activity (Peters and Mech 1975). Proestrus and estrus discharges in urine in the snow associated with RLUs of alpha females provided further evidence of breeding activity (Rothman and Mech 1979; Harrington and Asa 2003). Breeding was also determined from observations of heavily trampled areas at copulation sites where copulation ties had occurred (Mech 1970) and observations of excavated den sites. We assumed breeding occurred during most winters in large packs with histories of regular breeding activity.

We used minimum convex polygons to estimate home ranges of radiocollared wolves using ≥ 20 radio locations (Mohr 1947), and this area was assumed to represent the territory of these wolves. Outlier locations >5 km from other locations were considered extra-territorial movements (Fuller 1989), but small clusters (>2) of radio locations greater than 5 km from other locations were assumed to be connected to the main territory area if there were regular movements between the clusters. The annual monitoring period used for wolves was 15 April to 14 April of the following year, and we defined summer as 15 April through 14 September, and winter as 15 September through 14 April.

We estimated the total area of occupied wolf range by summing the current winter territory area for collared packs, the most recent territory area for packs collared within previous three years, and statewide average territory area for non-collared packs. Lone wolves occupying regular territories were also mapped. The total occupied territories were multiplied by 1.37 to include interstitial areas of 37% between pack territories (Fuller et al. 1992), and this total area was assumed to be the occupied range of territorial wolves.

6.4.3   Productivity and Survival

We estimated numbers of pups present during winter from changes in wolf numbers from previous surveys, knowledge of presence of pups from summer howling surveys (Harrington and Mech 1982), reports of observations, and knowledge of pack composition from previously captured wolves. This estimate of pup production might be biased somewhat by sub-adults dispersing into packs, but from our experience in Wisconsin, most such dispersers became members of the breeding pair and would not have been included in the pup count. These methods gave us a

range of estimates of pups present, and we used the midpoint of that range to estimate pup survival. Midpoint estimates of pups present during late winter, numbers
of breeding females the previous winter, and a fetal rate of 5.2 fetuses/breeding
female (based on placental counts of five adult female wolves found dead in
Wisconsin in the 1980s and early 1990s) were used to estimate pup survival from
birth to the end of their first winter. Numbers of breeding females the previous
winter were determined by assuming one breeding female per pack with evidence
of breeding activity. Pup survival was estimated as follows:

$$\hat{S}_{\text{pups}} = \frac{N}{(N_{\text{bf}} \, 5.2)}$$

where $\hat{S}_{\text{pups}}$ = pup survival through their first winter; N = pups alive during the late
winter; N_{bf} = estimated number of breeding females the previous winter.

We analyzed the survival of wolves that were radiocollared from 1979 to 2003
using a staggered entry Kaplan-Meier approach (Pollock et al. 1989). We compared
the annual survival functions by age (pup, yearling, adult), sex, and by early
(pre-1995) and late (post-1994) periods in wolf recovery using log-rank tests
(Pollock et al. 1989). Annual survival was estimated using a biological year defined
as 1 May through 30 April.

6.5 Population Trends and Ecology of Wolves in Wisconsin

6.5.1 Growth and Expansion of the Wolf Population

We monitored growth of the wolf population in Wisconsin during the winters of
1979–2007 (Table 6.1). Monitoring was facilitated by 2–63 radiocollared wolves
(8–37% of the estimated minimum population) that were tracked each winter. The
fewest wolves were radio monitored in 1979–1980, the first year of the surveys, and in
1990–1991, when a change of personnel occurred in the wolf-monitoring program.
Excluding these anomalies, an average of 27% (±6.6 SD) of the winter wolf population
was collared and monitored from 1980 to 1990. This declined to 16% (±4.7 SD) for
winters 1991–2007. Overall, a mean of 46% (±14.8 SD) of the packs monitored during
1979–2007 contained at least one radiocollared wolf. The percentage of packs with at
least one member radiocollared declined from a mean of 56% (±16.8 SD) during
1980–1990 to a mean of 43% (±6.4 SD) during 1991–2007. In general, the number of
radiocollared wolves we tracked each year increased, but the percentage of the wolf
population and percentage of packs collared declined as the population increased.

WDNR trackers conducted 760–6,571 km of snow-track surveys annually to
estimate number of wolves in non-collared packs, and to supplement wolf counts on
collared packs. Volunteer trackers started in 1995, and conducted 526–7,952 km of
snow-track surveys annually during the late 1990s and early 2000s. Overall, track
surveys increased from 760–1,622 km in the early 1980s when mainly 2 counties

Table 6.1 Efforts associated with Wisconsin's winter survey to estimate the state wolf population sizes (1979–2007)

Winter period	No. of wolves collared	Wolf population collared (%)	Packs collared (%)	DNR snow track surveys (km)	Volunteer track surveys (km)
1979–1980	2	8	20	760	
1980–1981	5	24	40	1,541	
1981–1982	6	22	50	1,622	
1982–1983	7	37	80	1,342	
1983–1984	6	35	75	1,129	
1984–1985	4	27	50	N/A*	
1985–1986	2	13	25	N/A	
1986–1987	5	28	60	N/A	
1987–1988	8	31	66	N/A	
1988–1989	8	26	71	N/A	
1989–1990	8	24	40	N/A	
1990–1991	2	5	17	4,178	
1991–1992	8	18	38	3,957	
1992–1993	10	25	50	6,208	
1993–1994	12	21	50	6,143	
1994–1995	18	22	55	6,253	526
1995–1996	24	24	52	3,447	4,540
1996–1997	22	15	43	3,802	5,341
1997–1998	24	13	43	2,606	4,887
1998–1999	27	13	40	4,457	2,533
1999–2000	32	13	46	3,731	6,347
2000–2001	39	15	43	6,571	5,732
2001–2002	42	13	42	5,428	5,883
2002–2003	63	19	46	4,620	6,094
2003–2004	49	13	35	5,885	7,839
2004–2005	46	11	32	4,466	7,952
2005–2006	43	9	33	4,579	7,884
2006–2007	63	12	40	5,843	6,701

*N/A = data not available.

were surveyed, to 10,000–13,000 km in the 2000s when surveys were conducted in ≥30 Wisconsin counties.

Wolves recolonized extensive areas of northern Wisconsin during 1979–2006 (Fig. 6.1). During the first winter of surveys, we located four packs in Douglas County and one pack in Lincoln County in heavily forested areas of northern Wisconsin (Fig. 6.1a). By winter 1989–1990 we detected 10 pack territories in 8 counties (Fig. 6.1b), and by winter 1994–1995 a total of 22 territories (2 were occupied by lone wolves) were found across 12 Wisconsin counties (Fig. 6.1c).

The first pack of wolves to colonize the Central Forest region (Thiel et al., this volume) was found during winter 1994–1995 about 109 km south of the nearest pack

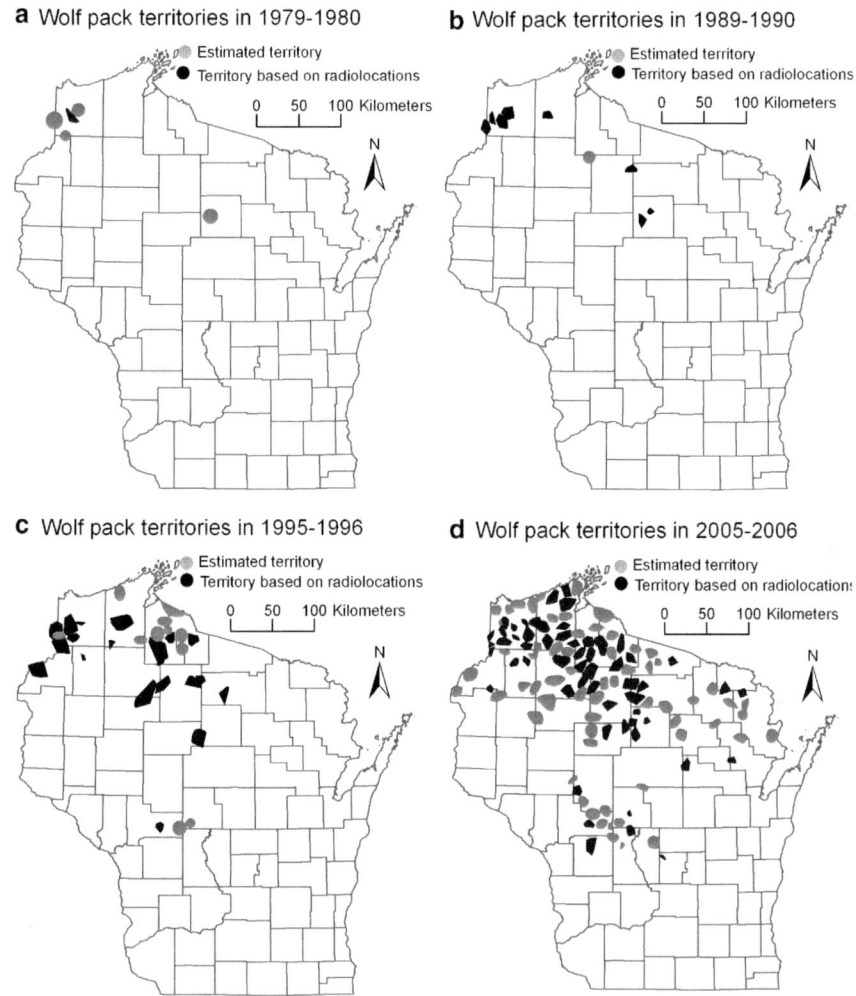

Fig. 6.1 Growth and expansion of the wolf population in Wisconsin for (**a**) 1979–1980, (**b**) 1989–1990, (**c**) 1994–1995, and (**d**) 2005–2006

in the Northern Forest. Wisconsin's Central Forest is an island of extensive forest in the middle of the state and is separated from the Northern Forest region by mostly open farmland. Wolves probably recolonized the Central Forest about 1992 or 1993.

By the early 2000s, wolves had occupied most of the large blocks of public forest land in northwestern and north-central Wisconsin, and wolf packs were beginning to occupy areas of mixed forest and farmland at the southern edge of the northern forests, as well as pockets of agricultural land east of Superior and west of Ashland. By 2005–2006, wolf territories were spread across 31 counties in the Northern and Central Forests (Fig. 6.1d). These territories included at least 116 packs and 5 lone wolves.

6.5.2 Wolf Population Increase and Growth Rates

Our minimum estimate of Wisconsin's wolf population grew from 25–28 wolves in 1979–1980 to 540–577 wolves in winter 2006–2007 (Table 6.2). The wolf population declined to 14–16 wolves in 1984–1985, apparently due to high mortality associated with canine parvovirus (Wydeven et al. 1995). After 1985, the population grew to 34 by 1990 (annual growth [λ] = 1.18). Between 1990 and 2000, the wolf population grew at a rapid annual rate (λ = 1.22), but annual growth rate declined to λ = 1.12

Table 6.2 Estimated characteristics of the wolf population during winter in Wisconsin (1980–2007)

Year	Estimated wolf population	Packs	Mean pack size ± SD	Largest pack	Loners detected	Loners (%)	Area occupied by wolf territories (km²)	Wolf density per 1,000 km²
1980	25–28	5	5.0 ± 4.0	12	0?	0?	1,469	17.0
1981	20–24	5	4.0 ± 2.4	7	0?	0?	1,752	12.0
1982	23–27	4	5.2 ± 2.5	9	2	9	1,310	20.6
1983	19–20	5	3.4 ± 1.3	5	2+	11	1,752	10.8
1984	18–19	4	4.0 ± 2.8	8	2+	11	1,352	12.6
1985	14–16	4	3.3 ± 2.5	7	1	7	963	15.5
1986	15	5	2.6 ± 0.9	4	2	13	1,504	10.6
1987	18–20	5	3.2 ± 1.8	6	2	11	1,188	15.2
1988	26–27	6	3.8 ± 1.2	6	3+	12	1,243	22.5
1989	31	7	4.0 ± 1.8	6	3	10	1,756	17.7
1990	34	10	3.1 ± 1.4	5	3	9	2,799	12.1
1991	39–41	12	3.1 ± 1.0	5	2	5	2,874	13.9
1992	45–52	13	3.0 ± 1.4	5	6	13	2,235	20.1
1993	40–42	12	2.8 ± 0.8	4	6	15	1,909	21.0
1994	54–61	16	3.1 ± 1.3	6	5	9	3,367	16.9
1995	83–86	21	3.6 ± 1.7	8	9+	11	4,299	19.3
1996	99–105	31	3.1 ± 1.3	7	3	3	6,255	15.8
1997	148–151	35	4.1 ± 2.1	10	5	3	5,698	26.0
1998	178–184	47	3.7 ± 1.5	8	6	3	8,547	20.8
1999	205–211	57	3.5 ± 1.6	8	7	3	8,856	23.1
2000	248–259	66	3.6 ± 1.9	11	13+	5	9,301	26.6
2001	257–259	70	3.6 ± 1.5	9	7	3	9,013	28.5
2002	327–343	83	3.8 ± 1.9	10	8+	2	12,986	24.8
2003	335–353	94	3.4 ± 1.5	8	12	4	15,644	21.0
2004	373–410	108	3.2 ± 1.4	9	14	4	13,367	29.7
2005	435–465	113	3.7 ± 1.8	9	14+	3	16,506	27.2
2006	467–504	116	3.9 ± 1.8	12	13	3	14,116	34.8
2007	540–577	138	3.8 ± 1.7	9	17	3	15,869	35.5

between 2000 and 2007, suggesting that habitat was becoming saturated (Van Deelen, this volume). A minor decline occurred in 1993, 2 years after sarcoptic mange was first identified. However, the decline was also linked to two small packs shifting their territories into adjacent states. Minor levels of mange persisted in Wisconsin wolves during the later 1990s and early 2000s without impacting population growth.

6.5.3   Pack Size and Territory Size

We detected 4–138 wolf packs across Wisconsin during winter surveys (Table 6.2). Mean pack sizes ranged from 2.6 to 5.2 wolves annually. Packs were relatively larger during early years (a bias produced by a few large packs), and lowest during population declines. Recently, mean pack size was 3.2–4.1 wolves per pack. The largest packs observed in the state each year declined during the mid-1980s and early 1990s, but increased in the late 1990s and 2000s. During the 28 years of surveys, packs of ≥10 wolves were detected in only 5 years, and only occurred during 1 year when <148 wolves were found in the state.

Mean size of wolf pack territories evidently declined as wolves increased in Wisconsin (Fig. 6.2). The annual mean territory size was determined for 2–36 pack territories for which ≥20 radio locations were obtained. Prior to 1993, the annual mean territory size was based on less than 7 packs annually, and often only 2–3 packs. Since 1999, annual territory size was based on ≥21 pack areas. Mean pack

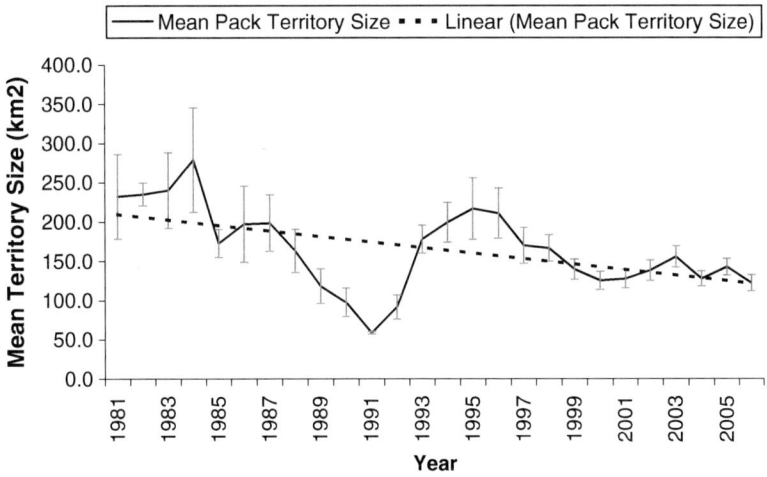

Fig. 6.2 Annual mean size of wolf territories in Wisconsin from 1981 through 2006, and standard errors of mean pack areas

territory size seemingly declined from 180 km² (±85 SD) in 1981–1990 to 165 km² (±94 SD) in 1991–2000, and was down to 136 km² (±67 SD) in 2001–2006. The apparent decline in the early 1990s in Fig. 6.2 may be an artifact of the small number of packs (2 or 3) with relatively small territories being sampled those years.

6.5.4   Lone Wolves

The percentage of the wolf population we detected as lone wolves ranged from 0% to 15% (mean 6.5% ± 4.3 SD). The percentage of the population detected as loners was higher between 1981 and 1995, with ≥83 wolves detected in the state (mean 9.7% ± 3.5 SD), than during the period 1996–2007, with ≥99 wolves in the state (mean 3.2% ± 0.7 SD). We probably routinely underestimated lone wolves because wolf surveys were focused on wolves living in territories.

6.5.5   Production of Pups and Survival

Our estimated numbers of wolf pups ranged from a low of 3 pups during winter 1985–1986 to 190 during winter 2006–2007 (Table 6.3). Estimated survival of pups ranged from 14% to 58%, with a mean of 29.4% (±8.6 SD). Lowest survival of pups occurred during the mid-1980s, coincident with an outbreak of parvovirus (Wydeven et al. 1995), and in 1993, when sarcoptic mange seemed to be having some impacts on survival. Highest survival of pups occurred during the early stages of wolf recolonization, when a few packs had very high pup survival. Although mean survival of pups was similar during 1979–1990 (29.7% ± 12.4 SD) and 1990–2007 (29.1% ± 4.9 SD), survival of pups was more variable during early colonization.

An average of 32.2% of packs (±15.8 SD) had no surviving pups detected during late winter (range: 0–75%). During the first 11 years of surveys, a mean of 36.5% (±22.9 SD) of packs had no surviving pups, but during the last 17 years a mean of 29.4% (±7.3 SD) of packs had no surviving pups detected and annual fluctuations were less variable.

Radiotracking between 1979 and 2003 resulted in 445, 163, and 84 wolf-years of telemetry records for adults, yearlings, and pups, respectively. The survival of radio-collarred wolves was remarkably consistent across sex and age classes and between age class estimates for early and late periods of wolf recovery. Survival functions did not differ by sex for adults ($X^2_1 = 0.51$, $P = 0.48$), yearlings ($X^2_1 = 0.13$, $P = 0.71$), or pups ($X^2_1 = 1.15$, $P = 0.28$). With sexes pooled, survival functions did not differ in pairwise comparisons of age class (adult vs yearling: $X^2_1 = 0.06$, $P = 0.80$; adult vs pup: $X^2_1 = 0.12$, $P = 0.73$; yearling vs pup: $X^2_1 = 0.18$, $P = 0.67$). In addition, survival did not differ between early and late recovery for adults ($X^2_1 = 0.66$, $P = 0.41$), yearlings ($X^2_1 = 0.03$, $P = 0.86$), or pups ($X^2_1 = 0.93$, $P = 0.33$). Survival rates were 0.75 (95% confidence interval [CI]: 0.69–0.79) for adults, 0.75 (CI: 0.59–0.89) for yearlings,

Table 6.3 Estimated numbers and survival of wolf pups in Wisconsin (1979–2007)

Winter period	Estimated number of pups in winter	Midpoint of pup estimates	Estimate of pup survival (%)	Packs with no surviving pups (%)
1979–1980	10–15	12	58	0
1980–1981	6–8	7	34	25
1981–1982	7–11	9	43	0
1982–1983	3–7	5	24	25
1983–1984	5–7	6	38	33
1984–1985	3–5	4	19	75
1985–1986	3	3	14	50
1986–1987	5–8	6	19	67
1987–1988	8–10	9	29	33
1988–1989	11	11	30	43
1989–1990	6–10	8	19	50
1990–1991	11–15	13	23	27
1991–1992	10–16	13	25	50
1992–1993	10	10	19	30
1993–1994	12–20	16	24	38
1994–1995	24–28	26	33	27
1995–1996	29–34	31	30	25
1996–1997	56–66	61	40	24
1997–1998	60–72	66	33	24
1998–1999	58–78	68	28	29
1999–2000	77–98	88	31	37
2000–2001	74–101	88	28	30
2001–2002	89–151	120	34	19
2002–2003	92–129	110	26	30
2003–2004	105–150	128	26	33
2004–2005	118–192	155	31	25
2005–2006	151–222	186	32	19
2006–2007	148–232	190	32	34

and 0.72 (CI: 0.51–0.94) for pups. Survival rates of pups represent survival to the end of a wolf-year for pups captured in late summer or early fall at 3–6 months of age, and thus are much higher than the indirect method used above. Survival functions indicated relatively steady mortality rates over time (Fig. 6.3).

6.6 The Wisconsin Wolf Management Plan

Primary authority for wolf management in Wisconsin returned to the WDNR on March 12, 2007 when wolves were removed from the federal list of endangered and threatened species. The 1999 Wisconsin wolf management plan (WDNR 1999) and

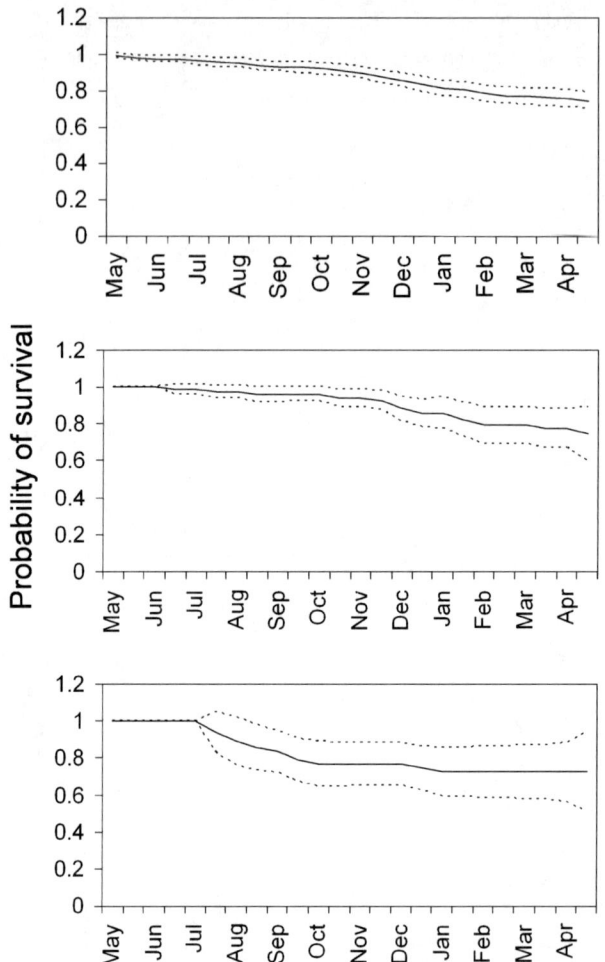

Fig. 6.3 Kaplan-Meier survival functions for wolves radiocollared in Wisconsin 1979–2003, showing adults (top), yearlings (middle), and pups (bottom)

its updates subsequently direct wolf management in the state. Between 1980 and 2007, the wolf population in Wisconsin grew beyond all the listing and management goals set for the population (Fig. 6.4).

A goal of Wisconsin's wolf management plan is maintenance of a viable and healthy population of wolves while attempting to minimize wolf depredation problems. The plan allowed more progressive control as the wolf population was down-listed from endangered to threatened to a delisted, protected wild animal under state law. When wolves attained threatened status (>80 wolves for ≥3 years), reactive lethal control by government trappers was allowed for wolves verified as depredators on domestic animals on private lands. When wolves met the criteria for state delisting (>250 outside Indian reservations), landowner control of problem

Fig. 6.4 Growth of the Wisconsin wolf population as represented by minimum counts in late winter statewide, and outside of Indian reservations. The state management designations for the wolf population at different sizes for areas outside of Indian reservations are listed

wolves could be authorized. When wolf numbers exceeded the population goal (>350 outside Indian reservations), proactive control by government trappers could occur, and a public hunting or trapping season could be considered. Wisconsin's current wolf management goal of 350 wolves was set at a time when there were <205 wolves in the state, and it was assumed to roughly represent the level of social acceptance of wolves. The population goal will be reexamined periodically to accommodate changing understanding of the interaction between wolf life history and human acceptance.

Although Wisconsin's wolf management plan allowed progressive levels of lethal control, many of these controls were not possible until federal delisting occurred. Limited lethal control was authorized for Wisconsin by the federal government in 2003–2006 to control wolves depredating domestic animals on private land, but it was not until federal delisting occurred in 2007 that the state was able to fully implement its management plan.

Mladenoff et al. (1997) estimated the potential equilibrium wolf population for Wisconsin using habitat area and prey-based models. Their estimate of potential wolf numbers based on habitat analysis was 380 (90% confidence interval [CI]: 324–461), and their estimate by the prey-based model was 462 (90% CI: 262–662). Consequently, WDNR used a population of 500 wolves as the estimated potential biological carrying capacity of the state. Although the wolf population in Wisconsin exceeded this number in 2007, recent declines in rate of growth suggest the wolf population may be approaching an equilibrium level (Van Deelen, this volume).

The WDNR management goal was set lower than the biological carrying capacity because managers assumed that acceptance by humans (social carrying capacity) would be less than the biological potential. Numerous livestock depredations during 2004–2006 suggested that a population of 373–467 wolves already exceeded the social carrying capacity for some stakeholders (Wydeven et al., in press; Ruid et al., this volume).

Wisconsin's wolf management plan includes four wolf management zones (Fig. 6.5, Wisconsin DNR 1999). Zones allow for maximum levels of wolf protection in areas with most suitable habitat, but allow more freedom to control problem wolves in areas of marginal or poor habitat. In 2007, Zone 1 contained 81% of the wolves in the state, Zone 2 had 13%, Zone 3 had 6%, and Zone 4 had <1%. Zones 1 and 2 represent large, forested, and wildland areas in large blocks of public land where

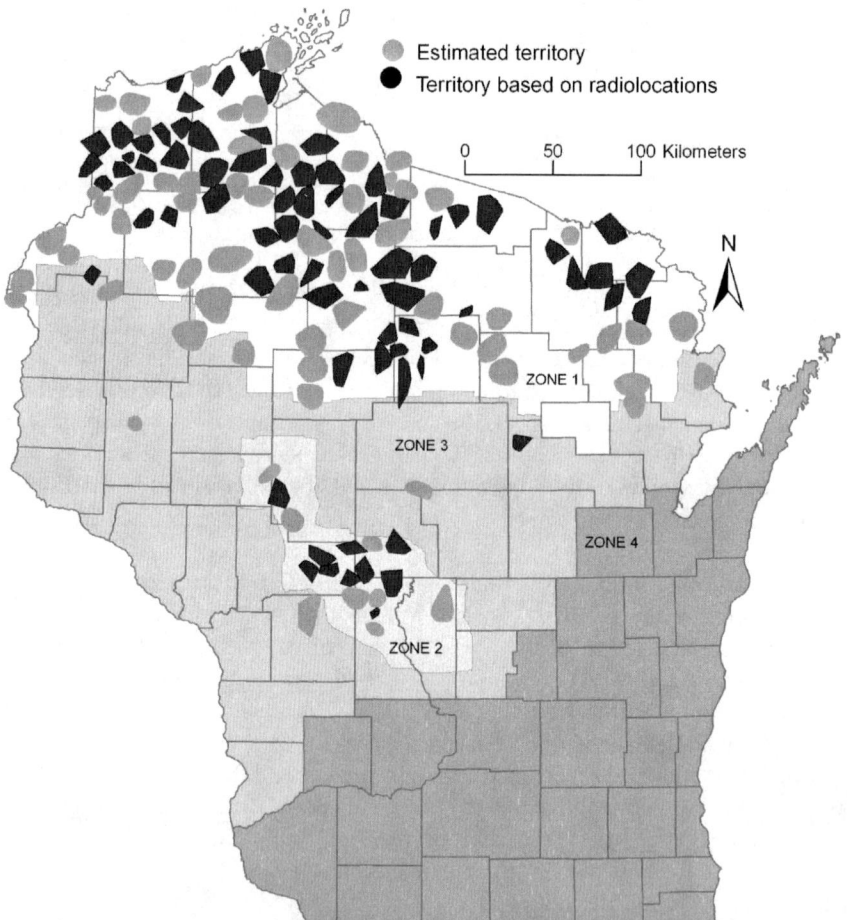

Fig. 6.5 Wolf management zones in Wisconsin with the distribution of wolf territories in 2007 illustrated. Wolf conservation activities are concentrated in Zones 1 and 2

wolf conservation activity is focused. Within these two zones, den sites are protected, and public land agencies are encouraged to maintain forests with low road densities and provide adequate habitat for prey (deer and beavers). Reactive depredation control activities are focused within 1.6 km of depredation sites. Proactive control by government trappers will focus on pockets of agricultural land or areas of high interspersion of forest and farmland where livestock depredations are likely (Treves et al. 2004). Zone 3 represents marginal wolf habitat, and habitat management will focus mainly on maintaining adequate areas of forest cover to allow dispersing wolves to travel between Zones 1 and 2. Reactive depredation control activities are allowed up to 8 km from depredation sites, and liberal use of proactive control will be used on problem wolves. Zone 4 represents areas of poor wolf habitat, and liberal control will be applied to any problem wolves that enter the zone.

6.7   Future of Wisconsin's Wolf Population

Under the guidelines of the WDNR management plan, the wolf population in Wisconsin is expected to begin to stabilize (assuming lawsuits do not cause wolves to be relisted by the federal government), and should decline in areas of mixed forest and farmland that would be considered marginally suitable wolf habitat. The wolf population should continue to spread into northeastern Wisconsin, and eventually saturate most areas of suitable habitat in the area. Wolf populations will mostly be allowed to fluctuate naturally with prey populations within areas of public forest in northern and central Wisconsin. In agricultural areas, wolf depredations will be controlled through trapping and shooting by government trappers, and shooting permits for landowners; these controls are likely to have a dampening effect on wolf populations in agricultural regions. The Central Forest wolf population may eventually become more isolated by increased human developments and traffic. Overall, suitable habitat may also decline in northern Wisconsin due to extensive development of secondary homes in forests, especially in areas near lakes (Radeloff et al. 2005).

6.8   Summary

Wolves were abundant in Wisconsin when European settlement began in the 1830s, but were extirpated by 1960 due to human attitudes and bounties. Wolves returned to Wisconsin about 1975, and the WDNR began a population-monitoring program in 1979. The late-winter wolf population grew from 25 wolves in 1979–1980 to 540 wolves in 2006–2007. During this period the range occupied by territorial wolves grew from <1,500 km^2 to >14,000 km^2. Mean pack size has generally averaged slightly less than 4, survival rates of pups to the end of the first winter averaged 29%, and about 32% of packs were unsuccessful raising pups. The Wisconsin management plan includes a population goal of 350 wolves outside of Indian reservations, and uses a zone system as well as landowner and government control to manage the population toward this goal.

Despite Wisconsin's reputation as an agricultural and heavily populated state, wolves have been able to successfully return after extirpation in the 1950s. This successful recovery was possible because of adequate habitat in portions of the state, a high prey base, proximity to a large source population, public education and changing public attitudes toward wolves, and adequate legal protection by federal, state, and tribal agencies. The wolf population is relatively secure in the state for the foreseeable future, but continued human developments and human population growth are likely threats. Intense population monitoring and protection of habitat will need to continue to assure that wolves remain secure in Wisconsin.

Acknowledgments We thank Sarah Boles, Larry Prenn, Pam Troxell, Bob Welch, Jim Halfpenny, WDNR pilots, WDNR wildlife biologists and technicians, volunteer trackers, tribal biologists, USGS Wildlife Health Center, WDNR Wildlife Health Team, and Great Lakes Indian Fish and Wildlife Commission. Funding was provided through Federal Aid in Wildlife Restoration (PR Funds), USFWS Endangered Species grants, Wisconsin Endangered Resources funds, US Forest Service, Timber Wolf Alliance, Timber Wolf Information Network, Defenders of Wildlife, National Wildlife Federation, and private donations.

References

Fuller, T. K. 1989. Population dynamics of wolves in north central Minnesota. Wildlife Monographs 105:1–41

Fuller, T. K., Berg, W. E., Radde, G. L., Lenarz, M. S., and Joselyn, G. B. 1992. A history and current estimate of wolf distribution and numbers in Minnesota. Wildlife Society Bulletin 20:42–55.

Fuller, T. K., York, E. C., Powell, S. M., Decker, T. A., and DeGraff, R. M. 2002. An evaluation of territory mapping to estimate fisher density. Canadian Journal of Zoology 79:1691–1696.

Fuller, T. K., Mech, L. D., and Cochrane, J. F. 2003. Wolf population dynamics. In Wolves: Behavior, Ecology and Conservation, eds. L. D. Mech and L. Boitani, pp. 161–191. Chicago: University of Chicago Press.

Halfpenny, J. 1986. A Field Guide to Mammal Tracking in North America. Boulder: Johnson Publishing.

Harrington, F. H. and Mech, L. D. 1982. An analysis of howling response parameters useful for wolf pack censusing. Journal of Wildlife Management 46:686–693.

Harrington, F. H. and Asa, C. S. 2003. Wolf communication. In Wolves: Behavior, Ecology and Conservation, eds. L. D. Mech and L. Boitani, pp. 66–103. Chicago: University of Chicago Press.

Heilhecker, E. In Press. Wolf, *Canis lupus*, behavior in areas of frequent human activity. Canadian Field-Naturalist.

Jackson, H. H. T. 1961. Mammals of Wisconsin. Madison: University of Wisconsin Press.

Kreeger, T. J. 2003. The internal wolf: Physiology, pathology, and pharmacology. In Wolves: Behavior, Ecology and Conservation, eds. L. D. Mech and L. Boitani, pp. 192–217. Chicago: University of Chicago Press

Kuehn, D. W., Fuller, T. K., Mech, L. D., Paul, W. J., Fritts, S. H., and Berg, W. E. 1986. Trap related injuries in gray wolves in Minnesota. Journal of Wildlife Management 50:90–91.

Mech, L. D. 1970. The Wolf: The Ecology and Behavior of an Endangered Species. Garden City: Natural History Press.

Mech, L. D., and Nowak, R. M. 1981. Return of the gray wolf to Wisconsin. American Midland Naturalist 105:408–409.

Mladenoff, D. J., Sickley, T. A., Haight, R. G., and Wydeven, A. P. 1995. A regional landscape analysis and prediction of favorable gray wolf habitat in the northern Great Lakes region. Conservation Biology 9:279–294.

Mladenoff, D. J., Haight, R. G., Sickley, T. A., and Wydeven, A. P. 1997. Causes and implications of species restoration in altered ecosystems. BioScience 47:21–31.

Mladenoff, D. J., Sickley, T. A., and Wydeven, A. P. 1999 . Predicting gray wolf landscape recolonization: Logistic and regression models vs. new field data. Ecological Applications 9:37–44.

Mohr, C. O. 1947. Table of equivalent animal populations of North American small mammals. American Midland Naturalist 37:233–249.

Olson, J. F., and Tischaefer, R. 2004. Cable restraints in Wisconsin: A guide for responsible use. Madison: Wisconsin Department of Natural Resources, PUBL-WM-443.

Peters, R. P., and Mech, L. D. 1975. Scent-marking in wolves: A field Study. American Scientist 63:628–637.

Pollock, K. H., Winterstein, S. R., Bunck, C. M., and Curtis, P. D. 1989. Survival analysis in telemetry studies: The staggered entry design. Journal of Wildlife Management 53:7–15.

Radeloff, V. C., Hammer, R. B., and Stewart, S. I. 2005. Sprawl and forest fragmentation in the U. S. Midwest from 1940 to 2000. Conservation Biology 19:793–805.

Rothman, R. J., and Mech, L. D. 1979. Scent marking in lone wolves and newly formed pairs. Animal Behavior 17:750–760.

Thiel, R. P. 1978. The status of the timber wolf in Wisconsin in 1975. Transactions of the Wisconsin Academy of Science, Arts and Letters 66:186–194.

Thiel, R. P., and Welch, R. J. 1981. Evidence of recent breeding activity in Wisconsin wolves. American Midland Naturalist 106:401–402.

Thiel, R. P. 1993. The Timber Wolf in Wisconsin: The Death and Life of a Majestic Predator. Madison: University of Wisconsin Press.

Treves, A., Naughton-Treves, L., Harper, E. L., Mladenoff, D. J., Rose, R. A., Sickley, T. A., and Wydeven, A. P. 2004. Predicting human-carnivore conflict: A spatial model based on 25 years of wolf depredation on livestock. Conservation Biology 18:114–125.

United States Fish and Wildlife Service. 1992. Recovery Plan for the Eastern Timber Wolf. Twin Cities: U.S. Fish and Wildlife Service.

Wisconsin Department of Natural Resources. 1989. Wisconsin Timber Wolf Recovery Plan. Madison: Wisconsin Endangered Resources Report 50. Wisconsin Department of Natural Resources.

Wisconsin Department of Natural Resources. 1999. Wisconsin Wolf Management Plan. Madison: PUBL-ER-099 99, Wisconsin Department of Natural Resources.

Wisconsin Legislative Reference Bureau. 2003. State of Wisconsin 2003–2004 Blue Book. Joint Committee on Legislative Organization, Wisconsin Legislature, Madison, Wisconsin.

Wydeven, A. P. 1993. Wolves in Wisconsin: Recolonization underway. International Wolf 3(1):18–19.

Wydeven, A. P., Schultz, R. N., and Thiel, R. P. 1995. Monitoring of a recovering gray wolf population in Wisconsin, 1979–1991. In Ecology and Conservation of Wolves in a Changing World, eds. L. N. Carbyn, S. H. Fritts, and D. R. Seip, pp. 147–156. Edmonton: Canadian Circumpolar Institute.

Wydeven, A. P., Schultz, R. N., and Megown, R. A. 1996. Guidelines for carnivore track surveys during winter in Wisconsin. Madison: Wisconsin Endangered Resources Report 112. Wisconsin Department of Natural Resources.

Wydeven, A. P., Jurewicz, R. L., Van Deelen, T. R., Erb, J., Hammill, J. H., Beyer, D. E., Roell, B., Wiedenhoeft, J. E., and Weitz, D. A. in press. Gray wolf conservation in the western Great Lakes region of the United States. In World of Wolves: New Perspectives on Ecology, Behavior and Policy, eds. M. Musiani, L. Boitani, and P. C. Paquet. Calgary: University of Calgary Press.

Chapter 7
A Disjunct Gray Wolf Population in Central Wisconsin

Richard P. Thiel, Wayne Hall, Jr., Ellen Heilhecker, and Adrian P. Wydeven

7.1 Introduction

Wisconsin's Central Forest Region (CFR) is a 7,155-km^2 L-shaped area in west-central Wisconsin extending from Chippewa Falls and Eau Claire to Tomah, Adams-Friendship, and Wisconsin Rapids (Fig. 7.1; Curtis 1959; Finley 1976). The CFR lies within the unglaciated driftless area and consists of flat, sandy, late Pleistocene glacial lake sediments and occasional Cambrian sandstone or Precambrian igneous outliers. Extreme western portions of the CFR consist of ridges and deeply incised valleys of Cambrian sandstones (Martin 1965; Schultz 1985). This region was logged between 1850 and 1920. In the past century its marshes were drained, its uplands and lowlands farmed, and much of it was abandoned by the time of the Great Depression of the 1930s (Grange 1948).

Humans sparsely inhabit the CFR. Economic activities are mainly forestry, outdoor recreation, and cranberry (*Vaccinium*) agriculture. The region consists of forests of oak (*Quercus*), aspen (*Populus*), pine (*Pinus*) and a variety of wetlands ranging from tamarack (*Larix laracina*) and black spruce (*Picea mariana*) swamps to sedge and sphagnum bogs. Floristically, the CFR resembles the northern forested region of Wisconsin, but is isolated from it by a 22–72-km wide zone of intense agriculture (primarily dairy, grain, and forage crops). About 2,574 km^2 in the central CFR is publicly owned, consisting of a mixture of county forests (Adams, Clark, Eau Claire, Jackson, Juneau, Monroe, and Wood counties), state forests (Black River State Forest and several Wildlife Areas,) and federal properties (Necedah National Wildlife Refuge, Meadow Valley State Wildlife Area, Fort McCoy Military Reservation). These forests primarily are managed for forestry, recreation, and wildlife conservation.

Gray wolves (*Canis lupus*) ranged throughout the CFR prior to European settlement and were probably extirpated as a breeding population by 1920 (Thiel 1993). Some dubious bounty data and local accounts suggested that individual wolves may have survived within the CFR until the 1950s when the species disappeared from all of Wisconsin.

Initially wolves recolonized several small, isolated areas within northwestern and north-central Wisconsin during the mid- to late-1970s (Mech and Nowak 1981;

A.P. Wydeven et al. (eds.), *Recovery of Gray Wolves in the Great Lakes Region of the United States*,
DOI: 10.1007/978-0-387-85952-1_7, © Springer Science+Business Media, LLC 2009

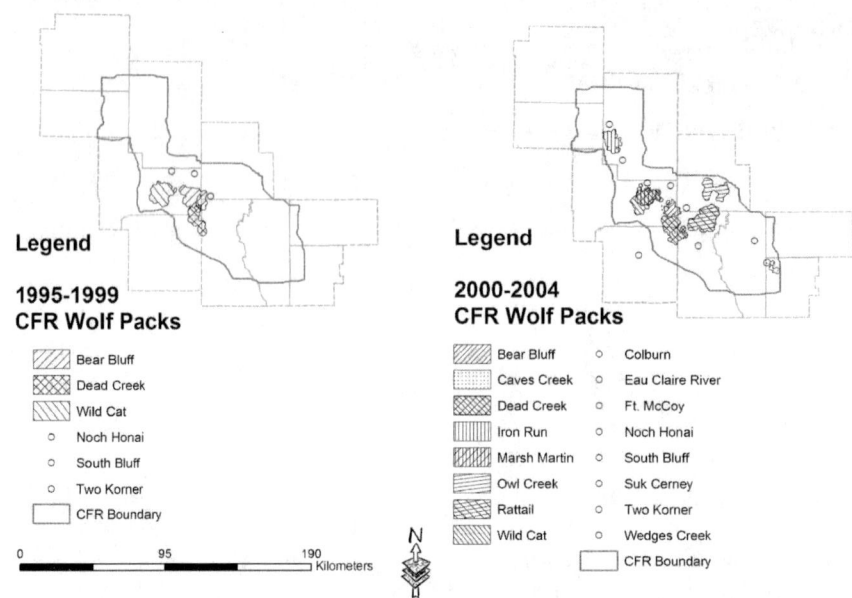

Fig. 7.1 Map of pack distribution in the Central Forest Region (*CFR*) in 1995–1999 and 2000–2004

Thiel and Welch 1981; Thiel 1993; Wydeven et al. 1995). By 1990, wolves began expanding their range in Wisconsin (Wydeven et al. 1995). Citizens' reports of large canids in CFR surfaced in 1992. In November 1994, a radiocollared yearling male wolf was killed near Oakdale on Interstate Highway 90/94 just south of the CFR. Surveys and monitoring for gray wolves began during winter 1994–1995 in the CFR. This chapter summarizes monitoring of gray wolves in the CFR between the winter of 1994–1995 and 2005–2006. We also address concerns for the long-term maintenance of the CFR wolf population because of its geographic separation from the larger Lake Superior basin wolf population located further north.

7.2 Methods for Monitoring Wolves in Wisconsin's CFR

As employees of Wisconsin's Department of Natural Resources (WDNR), we maintained records of citizen and staff reports of wolf sightings and tracks. Initially, we tried to confirm presence by searching areas of sightings after snowfall. Following discovery of wolves, department staff and cooperators from other agencies (United States [US] Fish and Wildlife Service, Ft. McCoy Military Reservation, Jackson and Wood County Forestry, and the Ho-Chunk Nation's Department of Natural Resources) and trained volunteers conducted winter-track surveys annually to count wolves throughout the CFR.

Wolves were captured in foot-hold traps from May through mid-September, or radioed when coyote trappers accidentally captured wolves during autumn and winter trapping. Methods for sedating, handling and radiocollaring, and aerially locating CFR wolves are described by Wydeven et al. (this volume). Additional information came from a study of mortality of wolf pups outfitted with ear-tag radio transmitters (Heilhecker et al. in press).

We determined number of packs, number of wolves per pack, and total population size in the CFR through aerial observations of wolves accompanying radioed individuals, and by counting tracks while conducting winter-track surveys annually (Wydeven et al. 1995; Thiel and Welch 1981). We defined a pack as a group of ≥2 wolves. Dispersal and mortality data are limited to CFR wolves radiocollared between 1995 and 2004.

We used the minimum convex polygon and fixed kernel methods to determine the sizes of wolf pack territories in the CFR (Millspaugh and Marzluff 2001). Road densities (RDs) were calculated within pack territories using criteria described by Mladenoff et al. (1995, 1999). Wydeven et al. (2001) found that RDs were related to human-caused mortalities. The lowest RDs in CFR are above the 0.45 km/km^2 threshold used by Mladenoff et al. (1995, 1999) to separate suitable from nonsuitable habitat for wolves. We tested whether pack sizes differed and whether rates of human-caused mortality were positively associated with RD for CFR packs inhabiting the lowest RD, or "core" habitat areas (RD < 0.80 km/km^2), and relative to wolves inhabiting higher RD, or "marginal" habitat areas (RD > 0.79 km/km^2). Wolf population density was estimated by dividing the sum of the numbers of wolves observed in winter in radioed packs by the summed sizes of their respective pack territories (Fuller et al. 2003).

We recovered dead wolves and transported them to the US Geological Survey Wildlife Health Center or the WDNR Wildlife Health lab for necropsy to determine cause of death. For incomplete diagnostic cases we relied on field evidence to estimate cause of death. Mortality information was based on two classifications of dead wolves: radioed individuals and all recovered animals. Radioed wolves provided the least biased determination of mortality, whereas all recoveries provided evidence of the breadth of the more minor causes of death among wolves. Annual survival rates were calculated from radioed wolves following Heisey and Fuller (1985).

We conducted howl surveys annually between June and September to estimate numbers of CFR packs that produced pups (Fuller and Sampson 1988; Harrington and Mech 1982; Wydeven et al. this volume). Lack of response from pups in surveyed packs cannot be interpreted as an absence of pups. Therefore, percent of packs with pups detected represents a minimum estimate.

Seasons were defined as winter: December–February; spring: March–May; summer: June–August; and fall: September–November.

7.3   Population Trends and Ecology of CFR Wolves

Founder wolves probably arrived in the CFR by 1992–1993. During winter 1994–1995, track surveys detected ten CFR wolves, organized as one reproductive pair and a pup, three colonizing pairs, and one single wolf (Thiel et al. 1997). An additional pair of wolves was likely present but was not found until the following winter.Thus, the CFR wolf population consisted of 12 wolves in 1994–1995.

Between 1995 and 2004 we accumulated 2,798 trap nights and captured 54 wolves, including 18 pups captured as part of a discrete study of mortality of wolf pups conducted in collaboration with the University of Wisconsin-Stevens Point (UWSP, Heilhecker et al. in press). Our combined capture rate was 52 trap nights per wolf. We radiocollared 12 females (15 total captures of 2 pups, 4 yearlings, and 6 adults) and 18 males (19 total captures of 6 pups, 3 yearlings, and 9 adults). Our data consisted of 15,253 radio days from radiocollared wolves and 1,359 radio days from ear-transmittered pups.

Wolves rapidly colonized the CFR, expanding to >50 wolves in 16 packs by 2003–2004 (Fig. 7.1; Table 7.1).The observed mean annual finite rate of increase during the 9-year period was 1.22. The annual survival rate of radiocollared wolves, excluding animals in the UWSP pup study, was 0.73 (CI: 0.62–0.89, $n = 15,252$ transmitter days).

Territory sizes of radioed wolf packs averaged 144 km^2 and varied from 71 to 233 km^2. Territory size was relatively stable over the years. We calculated a density of 24 wolves/1,000 km from eight radioed wolf packs monitored between 2000 and 2004.

Three population pulses were evident between 1994–1995 and 2005–2006 as the CFR wolf population expanded and dispersers established additional territories (Table 7.1). Mean pack size when more than one new territory was detected annually was 3.1 wolves/pack versus 4.2 wolves/pack when less than two new territories were detected annually. This pattern was related to varying numbers of founding pairs as colonization proceeded.

We estimated that litter size at birth was about seven pups based on placental scars from necropsies of four adult CFR females. The average observed litter size during summer was 4.8 pups ($n = 9$; range 3–6 during 7 pack-years). Howling surveys to detect presence of pups in CFR packs during summer months indicated that pups were present in an average of 72% of packs surveyed each year (range: 57–89%) from 1996 to 2004.

Six of 16 ear-transmittered pups (37%) died during surveillance between July 15 and January 15 in 2002 and 2003. Five died between 91 and 143 days of age from

Table 7.1 Population trends of wolves in the Central Forest Region (CFR) of Wisconsin, 1994–1995 through 2005–2006

Year	1994–1995	1995–1996	1996–1997	1997–1998	1998–1999	1999–2000	2000–2001	2001–2002	2002–2003	2003–2004	2004–2005	2005–2006
No. of packs	4	4	5	8	9	9	9	13	14	16	14	14
No. of wolves	12	18	27	27	29	44	37	35	41–45	52–53	48–50	54–57
Mean pack size	3.0	4.5	5.4	3.4	3.2	4.9	4.1	2.7	3.0	3.3	3.5	4.0

a combination of mange and severe emaciation. One of these was also diagnosed with canine distemper. Four of the mortalities were members of a single litter born in 2002, and a fifth was from a litter born in 2003. A sixth pup was shot illegally by a deer hunter at ~224 days of age. The 6-month survival rate of the ear-tag-transmittered pups, (ages 3–9 months), was 0.20 (95% CI: 0.05–0.72).

Between 1999 and 2004, 33 dead CFR wolves were recovered. Twenty of these were radiocollared, of which 13 had functional radios at the time of death. Age was determined for 29 individuals and included 12 pups (41%), 2 yearlings (7%), and 15 adults (52%). Seasonal distribution was 6 deaths during spring (18%), 7 during summer (21%), 13 during autumn (40%), and 7 during winter (21%).

Fifty-three percent of the radiocollared subsample and 55% of all recovered wolves were killed by humans (e.g., gunshot, vehicle collisions) followed by 23% of radioed and 18% of all wolves, respectively, dying from disease/parasitism (Fig. 7.2). One radioed wolf was euthanized after Department of Natural Resources (DNR) officials received reports of a wolf dragging its hind legs along a state highway. Necropsy was inconclusive, but this wolf had not been hit by a vehicle as we suspected.

Between 1996 and 2006, 52 complaints of gray wolf-livestock depredations, pet attacks, and nuisance wolves were reported within the CFR. Fourteen (27%) of these complaints were caused by wolves. Coyotes (25%), dogs (11%), and other causes (29%, includes still-births, car-killed livestock or pets, etc.) predominated. Wolves depredated two calves at a single farm, killed or injured seven dogs, and

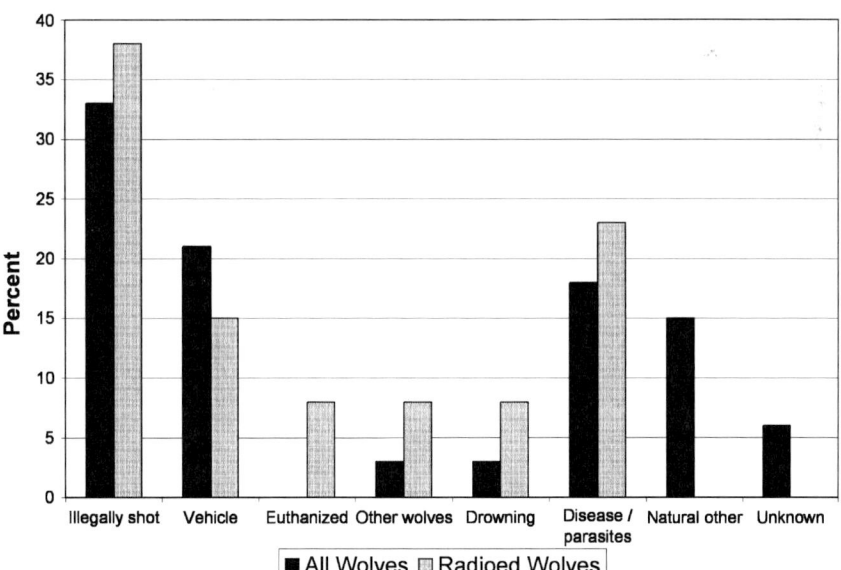

Fig. 7.2 Causes of death for radio-equipped wolves ($n = 20$) and all wolves ($n = 33$) recovered in the Central Forest Region (*CFR*) of Wisconsin, 1999–2004. Radio-equipped wolves are a subset of all wolves

were responsible for five cases exhibiting human-habituated behaviors or threatening domestic animals.

Twelve radiocollared (6 males, 6 females) and 3 ear-transmittered wolves (1 male, 2 females) dispersed. One female made 2 predispersal movements at 16 and 20 months of age prior to dispersing at 28 months. Likewise 2 males made single predispersal forays at 31 and >94 months before dispersing at 42 and >128 months, respectively. Males tended to disperse at an older age (mean = 51 months; range:11–128 months) than females (mean = 20 months; range: 9–31 months, $P = 0.07$), although this is mostly due to a male that dispersed at >128 months.

Dispersal times, measured as the number of days between leaving a territory to settling into a new territory, did not differ between males (mean = 67 days; range: 1–127 days) and females (mean = 44 days; range: 1–109 days, $P = 0.29$). Dispersing males traveled an average of 186 km (range: 124–312 km), whereas females covered 104 km (range: 99–117 km, $P = 0.56$). Straight-line distances between the centers of vacated territories and new territories colonized or integrated into, or sites of death prior to settlement did not differ for males (mean = 83 km; range 22–288 km) and females (mean = 67 km; range: 50–82 km, $P = 0.67$).

Five dispersers (4 males, 1 female) successfully colonized new territories. One dispersing male and 4 dispersing females integrated into existing pack territories. Seven of the 12 dispersers (3 males, 4 females, 58%) were known or suspected to have reproduced. Eight (3 males, 5 females) settled within the CFR.

Five radioed wolves (four males, one female) dispersed out of the CFR. Two males and a female dispersed north-northeast between 42 and 288 km. The greatest northerly distance was traveled by male wolf 307 who disappeared on March 28, 2000 and whose skull and radiocollar were found near L'Anse, Michigan, during spring 2007. Two males dispersed south-southeast distances of 119 and 689 km. The greatest southerly distance traveled was of a male pup 409, who left its natal territory sometime after January 15, 2003. Its carcass was recovered 19 km west of the Ohio/Indiana state line on June 20, 2003, having moved 689 km within 6 months (Heilhecker et al. in press).

Wolves from other populations dispersed into the CFR as well. A yearling male wolf radiocollared in Lincoln County, Wisconsin, was killed by a car 10 km south of the CFR in November 1994. A radiocollared female wolf pup dispersed from its Gogebic County, Michigan, pack after December 20, 1999 and was found shot on January 15, 2000, ~15 km northeast of CFR, having dispersed 173 km in a 37-day period.

7.4 Discussion

Analysis of habitat preferences of radioed wolves in northwestern Wisconsin classified wolf habitat within the northern half of Wisconsin, upper Michigan, and northern Minnesota (Mladenoff et al. 1995, 1999). The best landscape filter for identifying wolf habitat used RDs with a threshold of <0.45 km/km² corresponding to a >0.50 probability of wolf occupancy (Mladenoff et al. 1995, 1999). One outlier of potentially favorable wolf habitat identified in Mladenoff et al.'s (1995, 1999) analysis was Wisconsin's CFR, considered wolf habitat despite its small size and marginally suitable RDs (RDs >0.45 km/km^2). The CFR is separated from the more extensive contiguous forest region of northern Wisconsin, Michigan's Upper Peninsula, and Minnesota by a 22–72-km swath of intensive agriculture and a buffer of surrounding forest–farm mix that is considered marginal wolf habitat (Wisconsin DNR 1999).

The CFR wolf population increased from 12 wolves in 4 packs to more than 50 wolves in 16 packs within 9 years (Table 7.1). The high annual finite rate of increase of 1.22 exhibited by the CFR wolf population is typical of growth rates observed in naturally colonizing populations studied in northwestern Minnesota and in northern Wisconsin (Fritts and Mech 1981; Wydeven et al. 1995, this volume; Van Deelen this volume), and by populations recovering from intensive experimental control actions (Ballard et al. 1987; Hayes and Harestad 2000; Fuller et al. 2003).

Pack territory size is usually larger in colonizing populations relative to established or saturated populations (Fritts and Mech 1981; Hayes and Harestad 2000; Mech and Boitani 2003). Territory sizes of CFR packs ranged from 71 to 233 km^2, and averaged 144 km^2, smaller than most reported wolf territories where primary prey are white-tailed deer (e.g., 116–344 km^2; Fuller et al. 2003). Territories of CFR wolves were also smaller than mean territory sizes (179 km^2; range 47–179 km^2) observed in a recolonizing population of wolves studied in northern Wisconsin between 1980 and 1990 (Wydeven et al. 1995).

Wydeven et al. (1995) observed an inverse relationship between sizes of wolf pack territories and white-tailed deer density. Densities of deer in CFR were higher (10–19 deer/km^2) than those observed by Wydeven et al. (1995). The smaller territory sizes for CFR wolf packs are consistent with observations of an inverse relationship between territory size and deer densities as a general feature of wolf–deer systems (Wydeven et al. 1995; Fuller et al. 2003).

Territories of founder CFR wolf packs decreased in size between 1995–1999 and 2000–2004, but not significantly so. Yearling wolves occasionally established territories that usurped edges of their natal pack territories, accounting for some of the observed reductions in sizes of territories of founder packs over time. Similar observations were made by Fritts and Mech (1981) and Hayes and Harestad (2000), but in those studies the decreases in territory sizes over time was significant.

High reproductive rates probably contributed to the population increase observed in the CFR. Fecundity among breeding females in the CFR, as measured by uterine placental scars, was high relative to those reported in Fuller et al.'s (2003) review, but this may be an artifact of the small sample size in the CFR data set. Similarly, we detected pups in most packs surveyed during summer howl routes. A high percentage of CFR packs reproduced annually, and surviving offspring likely colonized new CFR territories. Similarly, a significant number of dispersing radiocollared wolves either established new territories in the CFR or integrated into existing CFR packs. Thus, internal dynamics of the CFR population likely facilitated its rapid growth.

Mean yearly pack size was highly variable over the 10-year study but seemed related to periodic expansion of the CFR population. Years with larger mean pack sizes were associated with years when fewer new territories were detected. Mean pack size was also affected by pack location relative to RDs.

Changes in patterns of human activity, densities of rural residences, variations in human tolerance of wolves, and wolves'-demonstrated adaptability to humans alter the effect that RDs have on wolf habitat (Thiel et al. 1998; Kohn et al. this volume). Nonetheless, RDs seem to define wolf habitat in the Great Lakes region. Human-caused mortality accounted for 17–31% of wolf deaths in Minnesota (Fuller et al. 2003) where RDs are lowest. In northern Wisconsin, with intermediate RDs, 39–72% of wolf deaths were caused by humans from 1979 to1992 (Mladenoff et al. 1995, 1999, 2006; Wydeven et al. 1995). Wydeven et al. (2001) reported that 60% of human-caused wolf mortality occurred at RDs >0.63 km/km², and "most shootings and vehicle collisions occurred at RDs of 0.84–1.14 km/km²." Our data on the CFR wolf population support the notion that RDs affect survival of wolves and shape wolf distribution (Mladenoff et al. 1995, 1999, 2006; Kohn et al. this volume).

As expected, human-caused wolf mortality predominated in the CFR. Packs within core habitat areas (mean RD = 0.76 km road/km²) held 47% of the CFR population (172/365 wolves) over the 10-year census period, but sustained only 17% (3/18 wolves) of known human-caused mortality (shot or vehicle killed). Eighty-three percent of the human-caused mortality occurred in marginal habitat areas (mean RD = 0.94 km road/km²). This corroborates Wydeven et al. (2001), who observed that 75% of wolf mortality in northern Wisconsin was human-caused in areas with RDs between 0.84 and 1.14 km/km². During our study, larger mean pack sizes tended to be in core habitat areas where RDs were lower (4.2 wolves/pack; 47 pack-years; 195 wolves) relative to marginal habitats where RDs were higher (3.1 wolves/pack; 68 pack-years; 211.5 wolves), though this difference was not significant ($P = 0.20$). Differences in human-caused mortality in the core (3/18 wolves) relative to marginal (15/18 wolves) habitats were likely the cause for differences in observed pack sizes. Higher human-caused mortality may suppress mean pack size in areas with higher RDs through higher turnover of both adult breeders and their offspring.

Mladenoff et al. (1995, 1999) determined that RDs were the best predictor of habitats capable of sustaining viable wolf populations, with a >0.50% probability of wolf occupancy associated with areas supporting RDs <0.45 km/km². Mech (2006), however, argued that RD thresholds were a poor predictor of wolf habitat suitability. RDs within the ranges of radioed CFR wolf packs ranged from 0.55 to 1.16 km/km² and, as a consequence, very little of the CFR has RD values below the Mladenoff et al. (1995, 1999) threshold. However, the CFR wolf population first occupied areas within CFR with lower RD (<0.80 km/km²), as was also predicted by Mladenoff et al. (1995, 1999). Our data on patterns of wolf colonization and different mortality rates in regions with varying RDs support Mladenoff et al.'s (1995, 1999, 2006) general predictions that RDs do affect patterns of wolf colonization and wolf population demographics.While the Mladenoff et al. (1995, 1999)

RD threshold may be conservative in areas with established wolf populations and needs reevaluation, we feel it remains a very powerful predictor of geographic areas likely to be colonized by wolves.

Since 2000, CFR wolves have increasingly inhabited areas near humans. The edge of one pack's territory (Seneca pack) abuts the city of Wisconsin Rapids with a population of about 20,000 people. Other packs regularly establish rendezvous sites in cranberry beds within daily view of agricultural workers, and one pack established a territory on the Fort McCoy Military Reservation. CFR wolves have developed a tolerance toward humans that may necessitate a reanalysis of Mladenoff et al.'s (1995, 1999) RD thresholds as human-tolerant wolves continue to disperse and perhaps successfully colonize landscapes with even greater human activity (Thiel et al. 1998).

Survival of pups in the CFR was estimated from a single study (Heilhecker personal communication), and is low compared to other studies (Fuller et al. 2003). We feel our observed annual survival rate of 0.20 for pups is probably biased because of small sample size ($n = 16$), small numbers of packs, short duration of the study (2 years), and a high premature failure rate among ear transmitters in the second year of the project (Heilhecker personal communication). Our estimated annual suvival rate of 0.72 for adults was well above the threshold rate of 0.65 associated with stable populations (Fuller et al. 2003). These survival values were comparable to rates observed in northern Wisconsin during the late-1980s and early-1990s (Wydeven et al. 1995).

CFR pups less than 6-month old died from disease and parasitism, whereas more than 50% of mortalities in adult-sized wolves were caused by humans. This is not surprising since pups are less mobile and, therefore, less likely to encounter mortality risks associated with humans. Older wolves, by contrast, are more likely to encounter humans, given the relatively small size of the CFR, its proximity to relatively dense human population centers (including three cities with >40,000 residents within a 1-h drive of CFR), and the high RDs that provide access to humans.

The low depredation rates observed in CFR may be explained in part by an initial unawareness of wolves by residents, unfamiliarity in reporting procedures, and a paucity of livestock throughout most of the areas inhabited by wolves. Mech et al. (1988) reported an inverse relationship between rates of depredation and severity of the previous winter in northern Minnesota. They suggested that depredation rates were in part explained by the availability of vulnerable fawns, with mild winters leading to greater numbers of fawns and decreased depredations. Perhaps, high deer densities in CFR effectively provide wolves with a readily available food source and diminishes depredation rates.

Although the CFR is somewhat isolated geographically, two radioed wolves from farther north – one from Lincoln County, Wisconsin, and the other from Gogebic County, Michigan – were recovered dead within 20 km of CFR during this study. These dispersers demonstrated that wolves from the Lake Superior region can and, likely, do reach CFR regularly. Our data on 11 dispersers from CFR similarly documented 3 dispersers that left the CFR in northerly routes, one that reached

Michigan's Upper Peninsula, and one that moved southeasterly, reaching east-central Indiana. During our study, the CFR appeared connected to the more substantive Lake Superior basin wolf population by dispersal, and was a source for wolves moving or settling further to the south. The agricultural belt surrounding CFR, though consisting of relatively open, nonforested terrain, does not appear to be a barrier to wolf movement in either direction.

7.5   Conclusion

The CFR wolf population exhibited robust growth during our study. Once established, wolves rapidly colonized available core and marginal habitats. These wolves exhibited high reproductive and adequate survival rates. Dispersers have colonized local patchy habitats and replaced breeders within CFR, and wolves from CFR provided a source of dispersers that may reach distant suitable habitats.

We recognize that changes in human occupancy and use of CFR is inevitable and unpredictable. To ensure continued survival of wolves in the CFR we recommend annual winter-track surveys to estimate year-to-year population trends. We also recommend that more intensive radiotelemetry monitoring of CFR wolves be conducted over 2-year periods, every 5 years, to assess demographic trends, and to assess changes in territorial-spacing mechanisms and mortality rates otherwise not available from chance carcass recoveries.

Acknowledgments   We acknowledge the work of numerous summer howl and winter track survey volunteers, various personnel of the US Fish and Wildlife Service at Necedah National Wildlife Refuge, US Department of the Army personnel, officials at Wood County and Jackson County Forestry, numerous Bureau of Wildlife Management staff within the Wisconsin Department of Natural Resources, and Matthew Schuler who performed the statistical calculations.

References

Ballard, W. B., Whitman, J. S., and Gardner, C. L. 1987. Ecology of an exploited wolf population in south-central Alaska. Wildlife Monographs. 98: 1–54.

Curtis, J. T. 1959. The vegetation of Wisconsin. Madison, WI: The University of Wisconsin Press.

Finley, R. W. 1976. Geography of Wisconsin: a content outline. Madison, WI: Regents of the University of Wisconsin System.

Fritts, S. H. and Mech, L. D. 1981. Dynamics, movements, and feeding ecology of a newly protected wolf population in northwestern Minnesota. Wildlife Monograph. 80: 1–79.

Fuller, T. K. and Sampson, B. A. 1988. Evaluation of a simulated howling survey for wolves. Journal of Wildlife Management. 52: 60–63.

Fuller, T. K., Mech, L. D., and Cochrane, J. F. 2003. Wolf population dynamics. In Wolves: behavior, ecology, and conservation, eds. L. D. Mech and L. Boitani, pp. 161–191. Chicago, IL: University of Chicago Press.

Grange, W. B. 1948. Wisconsin grouse problems. Madison, WI: Wisconsin Conservation Department.

Harrington, F. H. and Mech, L. D. 1982. An analysis of howling response parameters useful for wolf pack censusing. Journal of Wildlife Management. 46: 686–693.

Hayes, R. D. and Harestad, A. S. 2000. Demography of a recovering wolf population in the Yukon. Canadian Journal of Zoology. 78: 36–48.

Heilhecker, E., Thiel, R. P., and Hall, W., Jr. In Press. Wolf, *Canis lupus*, behavior in areas of frequent human activity. Canadian Field-Naturalist.

Heisey, D. M. and Fuller, T. K. 1985. Evaluation of survival and cause-specific mortality rates using telemetry data. Journal of Wildlife Management. 49: 668–674.

Martin, L. 1965. The physical geography of Wisconsin. Madison, WI: The University of Wisconsin Press.

Mech, L. D. 2006. Prediction failure of a wolf landscape model. Wildlife Society Bulletin. 34: 874–877.

Mech, L. D. and Boitani, L. 2003. Wolf social ecology. In Wolves: behavior, ecology, and conservation, eds. L. D. Mech and L. Boitani, pp. 1–34. Chicago, IL: University of Chicago Press.

Mech, L. D. and Nowak, R. M. 1981. Return of the gray wolf to Wisconsin. American Midland Naturalist. 105: 408–409.

Mech, L. D., Fritts, S. H., and Paul, W. J. 1988. Relationship between winter severity and wolf depredations on domestic animals in Minnesota. Wildlife Society Bulletin. 16: 269–272.

Millspaugh, J. and Marzluff, J. M. 2001. Radio tracking and animal populations. San Diego, CA: Academic Press.

Mladenoff, D. J., Sickley, T. A., Haight, R. G., and Wydeven, A. P. 1995. A regional landscape analysis and prediction of favorable gray wolf habitat in the Northern Great Lakes Region. Conservation Biology. 9: 279–294.

Mladenoff, D. J., Sickley, T. A., and Wydeven, A. P. 1999. Predicting gray wolf landscape recolonization: logisitic regression models vs new field data. Ecological Applications. 9: 37–44.

Mladenoff, D. J., Clayton, M. K., Sickley, T. A., and Wydeven, A. P. 2006. L. D. Mech critique of our work lacks scientific validity. Wildlife Society Bulletin. 34: 878–881.

Schultz, G. 1985. Wisconsin's foundations: a review of the state's geology and its influence on geography and human activity. Cooperative Extension Service, University of Wisconsin. Dubuque, IA: Kendall/Hunt Publishing Company.

Thiel, R. P. 1993. The timber wolf in Wisconsin: life and death of a majestic predator. Madison, WI: The University of Wisconsin Press.

Thiel, R. P. and Welch, R. J. 1981. Evidence of recent breeding activity in Wisconsin wolves. American Midland Naturalist. 106: 401–402.

Thiel, R. P., Hall, W. H., and Schultz, R. N. 1997. Early den digging by wolves, *Canis lupus*, in Wisconsin. Canadian Field-Naturalist. 111: 481–482.

Thiel, R. P., Merrill, S., and Mech, L. D. 1998. Tolerance by denning wolves, *Canis lupus*, to human disturbance. Canadian Field-Naturalist. 112: 340–342.

Wisconsin Department of Natural Resources. 1999. Wisconsin wolf management plan. Madison, WI: Author.

Wydeven, A. P., Schultz, R. N., and Thiel, R. P. 1995. Monitoring of a recovering gray wolf population in Wisconsin, 1979–1991. In Ecology and conservation of wolves in a changing world, eds. L. N Carbyn, S. H. Fritts, and D. R. Seip, pp. 147–156. Edmonton, AB: Canadian Circumpolar Institute.

Wydeven, A. P., Mladenoff, D. J, Sickley, T. A, Kohn, B. E, Thiel, R. P., and Hansen, J. L. 2001. Road density as a factor in habitat selection by wolves and other carnivores in the Great Lakes region. Endangered Species Update. 18: 110–114.

Chapter 8
Change in Occupied Wolf Habitat
in the Northern Great Lakes Region

**David J. Mladenoff, Murray K. Clayton, Sarah D. Pratt,
Theodore A. Sickley, and Adrian P. Wydeven**

8.1 Introduction

The concept of wolf habitat and relative suitability has changed significantly over the past several decades. In large part, this occurred because of insights gained during expansion of the wolf population in the northern Great Lakes states (Mech 1970; Erb and DonCarlos, this volume; Beyer et al., this volume; Wydeven et al., this volume). Protection from intentional killing of wolves since 1974, under the Endangered Species Act of 1973, began the process of wolf population growth and expansion in northeastern Minnesota, with eventual recolonization of northern Wisconsin and upper Michigan (Beyer et al., this volume; Wydeven et al., this volume).

In 1955, Wisconsin game manager John Keener wrote about wolves, "This animal is a symbol of the true wilderness. He cannot tolerate the advancing civilization of his wild home" (Keener 1955: 22). As late as the 1980s it was still generally believed that wolves required wilderness to survive, though research was beginning to show otherwise (Mech et al. 1988; Mech 1989). This concept lasted for so long in part because wolves had persisted only in the Boundary Waters Canoe Area Wilderness and adjacent areas of the Superior National Forest in northeastern Minnesota (Erb and DonCarlos, this volume), as well as Isle Royale National Park in Lake Superior (Vucetich and Peterson, this volume). Gradually, it became clearer that the role of wilderness was largely one of protection for wolves from killing through reduced human accessibility, rather than any innate requirements of wolves and their behavior (Mech 1995).

With protection, wolves colonized areas with greater human presence. At the same time, remoteness clearly has a positive effect on wolf survival because of reduced conflict with humans, reduced accidental killing of wolves (such as by vehicles), and perhaps less disease, as well as less intentional illegal killing. Remoteness provides one relative factor in defining degrees of habitat suitability for wolves. The other important factor is prey abundance. Ironically, in today's human-dominated landscape, these factors are often in conflict. Human-dominated landscapes, both forests subject to harvesting and re-growth and agricultural lands, support high levels of prey (white-tailed deer, *Odocoileus virginianus*) abundance.

A.P. Wydeven et al. (eds.), *Recovery of Gray Wolves in the Great Lakes
Region of the United States*,
DOI: 10.1007/978-0-387-85952-1_8, © Springer Science + Business Media, LLC 2009

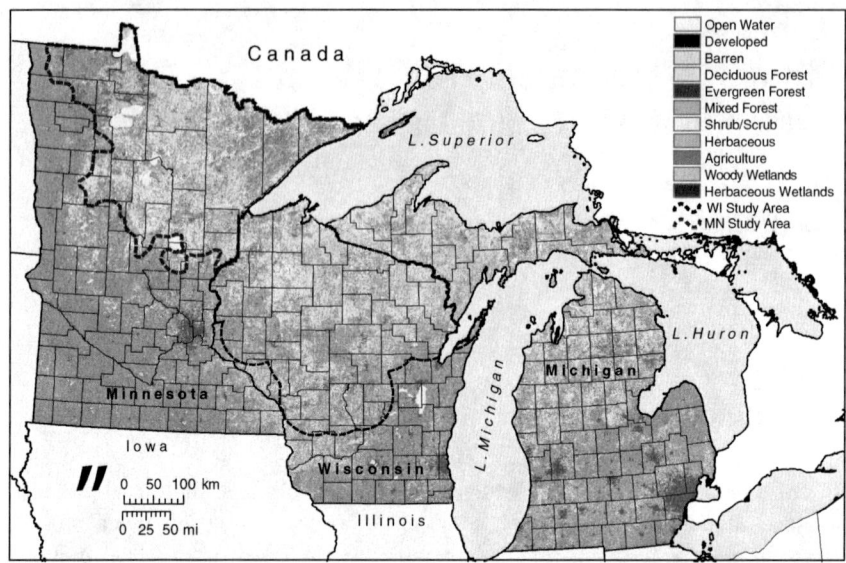

Fig. 8.1 Current land cover in Minnesota, Wisconsin, and Michigan. Data source: USGS National Land Cover Database 2001 (2006). Map created: Forest Landscape Ecology Laboratory, Department of Forest and Wildlife Ecology, University of Wisconsin-Madison. © David J. Mladenoff (*see Color Plate 1*)

But by supporting higher abundance of prey, these human-dominated landscapes can also support a higher wolf population. Greater wolf–human conflict ultimately often results (Mladenoff et al. 1997).

Areas with more intensive forest management and agriculture provide highly productive deer habitat. In the Great Lakes states, this gradient of productivity is controlled primarily by climate and soils, and declines generally from north to south. This gradient today largely follows the increasing area and productivity of agriculture from north to south (Fig. 8.1). Agriculture maintains productive browse and grazing within reach of deer, and the intensive nutrient inputs to crops and pasture accentuate the attractiveness of crops to deer, as well as providing livestock (prey) availability to wolves (Treves et al. 2002).

8.2 Historical Vegetation and Habitat

Prior to widespread Euro-American settlement in the 1800s, all ecosystems in the region were wolf habitat. Indeed, this is the case nearly worldwide. Wolves occur, or occurred, wherever there was prey (Mech 1995).

For a top predator like the wolf, the most important component of habitat is adequate prey, and for wolves the major prey species are ungulates. Where prairie and open savanna occurred, bison (*Bison bison*) and elk (*Cervus elaphus*) would have been most abundant. Furthest north, woodland caribou (*Rangifer tarandus*) and moose (*Alces alces*) were most abundant. In the mixed conifer-deciduous forests in between, white-tailed deer were most abundant (DelGiudice et al., this volume). These general relationships were similar even as vegetation zones oscillated with climate over recent millennia. Prey species and their relative abundance varied as influenced by climate and productivity, as well as hunting by Native Americans.

Our best picture of the distribution of regional ecosystems before modern levels of change is from the mid- to late 1800s, the period of the US General Land Office Survey. While the main purpose of this survey was to facilitate sale of the public domain and settlement, surveyors also recorded valuable information on the natural systems present. Also, because of the land survey, this additional information is well referenced spatially, and can be transcribed and mapped (Schulte and Mladenoff 2005). To varying degrees, this occurred in the three states of the Great Lakes region. More detailed state efforts (Schulte et al. 2002) generalize to a regional picture of ecosystems present before the main influx of Euro-American settlers, logging, agriculture, and wolf extermination (Fig. 8.2).

The Great Lakes region was dominated by prairie and oak (*Quercus*) savanna to the west and south, mixed evergreen-deciduous forest in northern Wisconsin

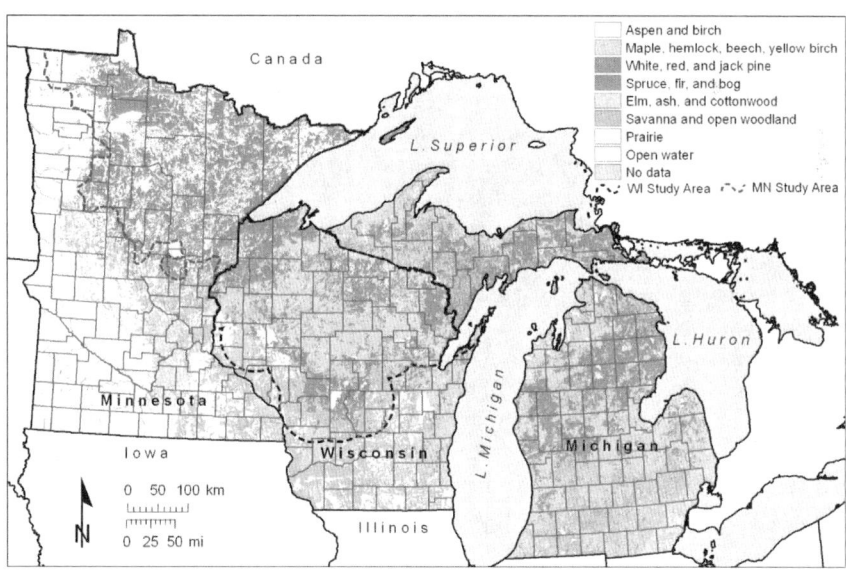

Fig. 8.2 Pre-European land cover (1800s) in the Great Lakes region. Data compiled for the three states from the US Government Land Office survey of the 1800s. Data source: US Forest Service Great Lakes Assessment. Land cover classification and map: Forest Landscape Ecology Laboratory, Department of Forest and Wildlife Ecology, University of Wisconsin-Madison. © David J. Mladenoff (*see Color Plate 2*)

and upper Michigan on heavier moraine soils, and pine (*Pinus*) on sandy outwash plains (Curtis 1959). Northeastern Minnesota was distinctive with a mixture of pine- and aspen- (*Populus*) dominated uplands, and abundant spruce-fir (*Picea Abies*) in lowlands, and in some uplands (Heinselman 1996). The possible patterns of ungulate and wolf abundance that occurred over time were unclear and potentially complex. They also would have been dynamic. Ungulate, and therefore wolf, biomass generally should have followed patterns of vegetation productivity, which because of climate (higher temperature and precipitation) and better soils, would have been most abundant in the southeastern Great Lakes states region. However, to the east, increasing vegetation productivity went largely into woody plant tissue (trees). In the western portion of the region in the prairie and savannas, also very productive on good soils, much above-ground plant productivity was within the reach of ungulates, but only seasonally. Also, the higher productivity in the southeast would have been tempered by a similarly higher Native American population and their impact on ungulate populations (Cleland 1983). Native American agriculture also was more important in the southeastern Great Lakes region and therefore the net level of competition for prey with humans is not obvious. Competition with Native Americans for ungulate prey may have been greater to the north, where hunter-gatherer cultures were more prevalent (Cleland 1983).

In the northern Great Lakes region, the most productive soils were dominated by mixed forest, and in particular in northern Wisconsin and upper Michigan closed canopy forests of sugar maple (*Acer saccharum*), hemlock (*Tsuga canadensis*), and yellow birch (*Betula alleghaniensis*). These forests have low light under their canopies and relatively low understory plant productivity without canopy-opening disturbance (Curtis 1959). While windthrow was not infrequent (Schulte and Mladenoff 2005), this mesic forest ecosystem, the most extensive in Wisconsin and upper Michigan, likely contained the lowest prey biomass per unit area, and thus likely the lowest density of wolves. The pine-dominated sandy outwash plains in the north would have had lower absolute productivity, but because of their drier soils and higher fire frequency, these forests had more open canopies, grading to savanna, and thus greater productivity in the understory.

Northern Minnesota, despite lower productivity based on climate and soil, had higher rates of fire that kept much of the forest in pine and aspen, resulting in higher light levels and more plant productivity within reach of ungulates. Furthermore, the likely higher population of moose relative to deer in northern Minnesota also meant that larger woody browse was available because of the ability of moose to down saplings and feed on bark (Peek 1974).

Given all of this natural variability, it is difficult to conclude what the actual relative prey biomass and therefore wolf population might have been. Jackson (1961) estimated wolf numbers in Wisconsin at the beginning of Euro-American settlement at 20,000–25,000, but this now seems doubtful. A more reasonable estimate based on recent data of wolf densities suggests a possible range of 3,000–5,000 (Wydeven et al. 1995).

8.3   Logging and Post-Logging Era

The era of massive logging, land cover change, and settlement in the northern Great Lakes states occurred from the mid-1800s in the south to the early 1900s in the far north (Mladenoff et al. 1993; Schulte et al. 2007). At the same time, wolves were being extirpated deliberately through bounty programs (Thiel 1993), and indirectly because of prey elimination due to habitat destruction and unregulated hunting (DelGiudice et al., this volume). During the 1900s, habitat for prey improved in reforesting areas following effective fire suppression beginning in the 1930s (Rhemtulla and Mladenoff 2007), and more effective hunting regulation resulted in a gradual resurgence of white-tailed deer especially, and an eventual reestablishment of deer into the southern parts of the region by the 1960s (DelGiudice et al., this volume). In the north, much of this post-fire reforestation was aspen, preferred browse for deer. Since the mid-1900s aspen has declined in Wisconsin and upper Michigan, but not in Minnesota (Schulte et al. 2007).

This change in habitat and prey was little benefit to wolves, since by the 1950s they were extirpated in the region except for wilderness in far northeastern Minnesota (Erb and DonCarlos, this volume). By the 1970s, with wolf protection under the federal Endangered Species Act of 1973, wolves began recolonizing more of northeastern Minnesota, challenging beliefs about the kind of habitat they actually needed. Wolf protection, changing human attitudes, and very high prey (deer) populations complicated examination of this picture, since this was a different landscape in all ways that wolves were re-invading.

8.4   Present Landscape

The land cover composition and pattern of today generally reflects what has existed in the last half century (Fig. 8.1) and the landscape that wolves have recolonized over the past three decades. While land cover alone cannot be read as a habitat map for wolves, it provides key elements. Apart from eradication efforts, human land use activities have drastically altered the 1800s landscape and in fact generally provided better habitat for wolves in the north. In the south, prairie and oak savanna and open woodland have been completely eliminated and replaced by agriculture, with remnant patches of forest. Remnants have become closed canopy forest woodlots due to fire suppression. In the north, the dominant large-statured conifers, pine and hemlock, have also been dramatically reduced and replaced by aspen where slash fires followed the original logging, or younger maple-dominated hardwoods (Rhemtulla et al. 2007). In response, prey, especially deer, have increased dramatically across Wisconsin and the region in the 1900s (DelGiudice et al., this volume).

The regional road network also provides another key layer of data. In the 1980s, knowledge of wolf habitat needs was changing. Legal protection showed that wolves did not need "wilderness" as conceived by humans, but refuge from human contact and its accidental and intentional killing of wolves (Thiel 1985; Mech et al.

1988). Protection and attitude changes reduced intentional killing, and high deer populations may have boosted wolf fecundity. It also became clear that wolves would use roads and snowmobile trails, rather than always avoid them. Clearly, what made up the matrix of favorable wolf habitat was more complex than originally thought at the beginning of wolf recovery.

8.5   Original Great Lakes States Wolf Habitat Model

Thiel (1985) and Mech et al. (1988) identified road density as an important predictor of where wolves would live. By the early 1990s, the first quantitative model of preferred, regional wolf habitat showed that the most important predictor was road density (Mladenoff et al. 1995). Our analysis was based on a logistic regression model of recolonizing wolves in northern Wisconsin:

$$\text{logit}\,(p) = 14.61R - 6.5988$$

where R = road density (km/km^2).

Areas of lower road density (at that time, a threshold of 0.7 km/km^2) were more likely to be occupied by wolves. Since the population was apparently still continuing to recolonize new areas with these characteristics, a more correct interpretation of this pattern would be that wolves were more likely to survive in such areas, and thus tended to establish in these areas first. Occupancy did not have a linear relationship with road density, suggesting that other factors were involved, and that this relationship could continue to change over time (Mladenoff et al. 1997). But mapping these first regional habitat quality classes region-wide and extrapolating estimates of potential wolf numbers (based on the method of Fuller et al. 1992) revealed some starting predictions for potential wolf numbers in the northern Great Lakes states (Mladenoff et al. 1997). At the time, the Wisconsin goal (set in 1989) for downlisting wolves from endangered to threatened status was for a population of 80 wolves, and few thought the population would rise much beyond this level because of the lack of wilderness in the state (Wisconsin DNR 1989). Our estimates suggested that in fact Wisconsin could accommodate 262–662, and upper Michigan 581–1,357 wolves (Mladenoff et al. 1997). In fact, the lower limit of this range has been well surpassed in Wisconsin, with 540 wolves in 2007 (Wydeven et al., this volume). Recent more detailed work in upper Michigan (Potvin et al. 2005) suggested that lower deer densities in high snow areas of upper Michigan may limit wolf numbers in upper Michigan to 10% less than our initial projection. In general, work during the mid-1990s emphasized that managers would likely need to re-assess their estimates and plans for the size of the potential wolf population for the Great Lakes region.

In the late 1990s, we re-assessed our statistical model developed in the early 1990s, using new data from additional recolonizing wolves. This analysis showed that the most favorable classes of habitat with lowest road densities were still being occupied preferentially (Mladenoff et al. 1999). After 2002, wolf numbers approached the lower bounds we had predicted, and were clearly increasing. This was an opportunity to re-examine how our original roads-only-based model described how wolves continued to recolonize. Changes were expected, as many of the most favorable areas were occupied. We did this by recalculating the original roads-based model for the wolf packs occupying Wisconsin in each year from 1979 to 2003. The pattern revealed was a complex one of changes over time and space (D. Mladenoff, unpublished data). Wolves continued preferentially to occupy the most favorable (lowest road density) class, even as this class became increasingly frag-mented and reduced. Wolf packs are increasingly occupying less preferred classes, usually by also including portions of the better classes in their territories. Further, they have begun to occupy the lowest quality class, but at very low rates, given that this class constitutes 70% of the landscape (Fig. 8.3; Mladenoff et al., unpub-lished data). But overall, the original relationship still holds: the wolf population has continued to disproportionately favor the better habitat classes beyond random expectations and availability (Mladenoff et al., unpublished data). As we hypothe-sized previously, we suspect that the now higher wolf population in better areas can subsidize higher mortality in poorer areas. In such source–sink dynamics (Pulliam 1988), the poorer habitat areas are neither truly biologically suitable in all ways, nor can they be sufficient for a truly sustainable population.

After assessing our original modeling, it is logical to conduct a new model-building exercise based on the current distribution of wolf packs on the current landscape. To do this we used a procedure similar to Mladenoff et al. (1995).

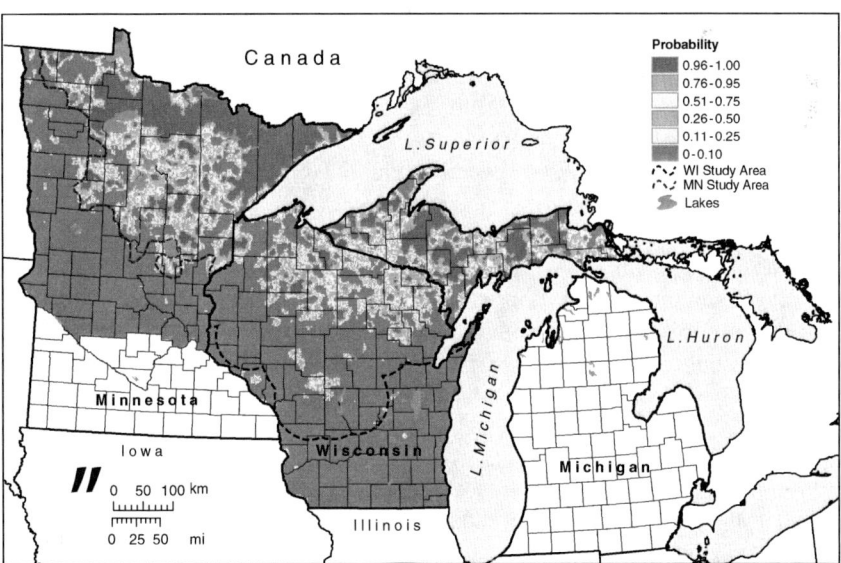

Fig. 8.3 Map of northern Great Lakes states wolf habitat classes from Mladenoff et al. (1995). Map created: Forest Landscape Ecology Laboratory, Department of Forest and Wildlife Ecology, University of Wisconsin-Madison. © David J. Mladenoff (*see Color Plate 3*)

8.6   Development of a New Model of Wolf Occupancy

8.6.1   Study Region

Our study region is located in northern and central Wisconsin and covers two-thirds of the land area of the state, ~97,000 km² (Fig. 8.4). When our original model was developed, the study area was defined by the boundaries of the wolf recovery area as established the U.S. Fish and Wildlife Service (1992). Wolves have expanded beyond the original study area so here we used the existing wolf pack locations, buffered by 50 km, to define a new study area. It is possible that the absence of wolf packs beyond this area may not represent less desirable habitat, but may be areas that wolves had not yet reached, and have not had the opportunity to be selected as habitat. In fact, this likely is not true, and is a very conservative assumption. Data suggest that wolves have thoroughly moved throughout the region.

8.6.2   Data Preparation

We acquired the locations of Wisconsin wolf packs for 2007 from the Wisconsin Department of Natural Resources (DNR). Data included polygons of pack territories for 143 packs. Pack territories were mapped using radio-locations from collared wolves or from winter track surveys and reported observations (Wydeven et al. 1995). We then created 143 non-pack areas by randomly locating polygons with an area equal to the mean pack territory size (135 km²) throughout the study area. These non-pack areas were constrained so they would not overlap with any existing packs or with each other. We randomly selected 95 pack and 95 non-pack areas for use in model development (Fig. 8.4).

We collected and processed several spatial datasets of landscape variables for use as predictor variables in the model including land cover, stream density, and three different metrics based on roads. We used the 2001 National Land Cover Dataset (NLCD), which maps land cover at a 30-m pixel resolution using LANDSAT satellite data (Homer et al 2004). To represent the urban land cover class we used the WISCLAND dataset developed by the Wisconsin DNR (Reese et al 2002). For each pack and non-pack polygon, we calculated the percentage of that area that was occupied by each land cover class. We also aggregated some land cover classes to form additional predictor variables including an agriculture class (crops and pasture) and a forest class (deciduous, mixed, and coniferous forests).

We obtained the 1:1,000,000 scale National Hydrography Dataset (NHD) and created a line shapefile of streams by extracting all streams and artificial paths which did not intersect water bodies (US Geological Survey 1999). We then calculated the density of streams in km/km² for each pack and non-pack area. Finally, we used the US Census Bureau 2000 Tiger line data to create a roads layer that included all road categories usable by ordinary cars and trucks but not roads or trails passable only by four-wheel drive vehicles (categories A1, A2, A3, A4, and A6 except A6.5;

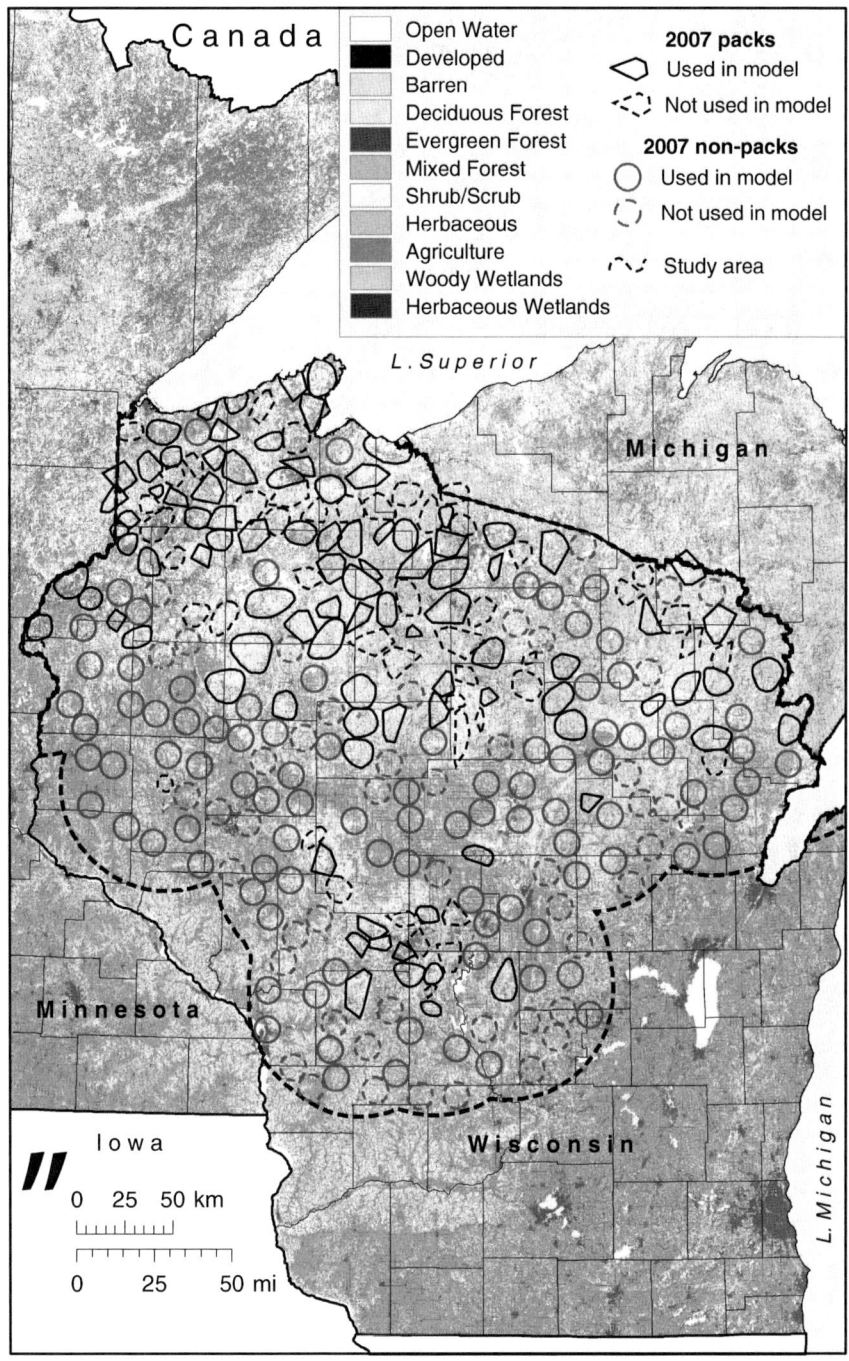

Fig. 8.4 Map of 2006–2007 wolf pack areas and randomly selected non-pack areas used in the analysis for Wisconsin. Map created: Forest Landscape Ecology Laboratory, Department of Forest and Wildlife Ecology, University of Wisconsin-Madison. © David J. Mladenoff (*see Color Plate 4*)

U. S. Census Bureau 2001). From this roads layer, we also created a highway layer that included interstate, USA, and major state and county highways (categories A1 and A2). Similar to streams, we calculated road density and highway density in km/km^2 for each pack and non-pack area. We also calculated the distance, in kilometers, to the nearest highway for each area.

8.6.3   Analysis

We used SAS 9.1 (SAS Institute, Cary, North Carolina) to analyze the data. We calculated summary statistics for the pack and non-pack areas as well as for the Wisconsin study region for each of the landscape variables. We tested for differences in the pack and non-pack areas using the nonparametric Kruskal-Wallis test. A Pearson Correlation Coefficient was calculated between all independent variables to examine multicollinearity. We then used a stepwise logistic regression procedure with all possible variables to find the best set of predictor variables. The resulting model was evaluated based on Akaike's information criterion (AIC, Akaike 1974), the Wald test for maximum likelihood estimates for individual model parameters, and model validation using hold-out pack and non-pack areas. After our initial model fitting, we removed some land cover classes from the stepwise procedure based on analysis of preliminary results and then refit the model. We assessed spatial autocorrelation of the pack locations by calculating semi-variograms of regression residuals, and also by mapping them for visual inspection. Autocorrelation was present but low, and comparison showed that it did not affect the regression results. We therefore used the non-spatial models in the final analysis.

8.7   Results

8.7.1   Landscape Composition

Current landscape composition in pack areas differed between occupied and unoccupied areas (Table 8.1). The greatest differences were in use of agriculture land, which was most negatively correlated with pack locations and constituted 5.3% of pack areas and 41.0% of non-pack areas. The total study area landscape was 27.2% agriculture (Table 8.1, Fig. 8.1). Deciduous forest, by far the largest forest class, was most positively related to pack areas, comprising 49.2% of pack areas and 33.8% of non-pack areas, with the overall study area containing 39.7% deciduous forest. Road density also continued to be negatively associated with packs, with the threshold remaining below 1 km/km^2 for wolf packs. Forested

Table 8.1 Summary of and use and land cover data derived from satellite classification. Means (and standard deviations) for occupied pack territories and randomly located non-pack areas (see modeling description), and for the defined Wisconsin study region overall. Kruskal-Wallis test of differences between pack and non-pack areas are significant at $P < 0.01$ for all classes, except water and shrub/scrub

Variable	Pack territories	Non-pack areas	Study area
Land cover (%)			
Water	2.51 (3.22)	3.18 (4.96)	3.22
Developed	0.03 (0.18)	0.76 (2.84)	0.53
Barren	0.09 (0.15)	0.24 (0.36)	0.26
Deciduous forest	49.24 (17.18)	33.77 (14.46)	39.70
Evergreen forest	8.58 (9.16)	3.98 (5.07)	5.42
Mixed forest	6.84 (7.73)	3.36 (5.67)	5.16
All forest	64.66 (15.25)	41.11 (17.50)	50.28
Shrub/scrub	1.20 (2.20)	1.05 (1.17)	1.20
Herbaceous	3.08 (5.08)	2.66 (2.09)	2.83
Agriculture	5.30 (7.68)	40.98 (23.60)	27.16
Woody wetlands	16.01 (13.05)	7.60 (8.28)	11.00
Emergent wetlands	6.69 (7.64)	2.43 (3.25)	3.51
Road density (km/km²)	0.93 (0.35)	1.42 (0.48)	1.31

wetlands and forest cover overall also were major land cover classes that showed a strong positive association with wolf pack territories (Table 8.1).

Correlations among the variables reflected these positive and negative associations, with the forest variable being most strongly negatively correlated with agriculture and roads, and positive correlations within the variables of these two groups (Table 8.2).

8.7.2   New Model

Our new, current model of wolf occupancy revealed how behavior has changed since the early 1990s. In this model (Figs. 8.5 and 8.6), the broader extent of wolf occupancy is reflected in a change in the significant variables in the model. While road density was the only significant variable in the 1995 model, in the current analysis model results were more varied. For direct comparison with the 1995 model, we calculated a roads-only model from the current data. This model remained a significant predictor of wolf presence, though it was not among the best models found.

More comprehensive analyses (including other predictors) resulted in several models of nearly identical AIC values and validation, but varying in their interpretability. One included road density, mixed forest, and agriculture, with all having negative coefficients for wolf prediction. This result was unexpected for mixed forest, which by itself was highly correlated with wolf presence (Table 8.1). The negative coefficients for roads and agriculture are expected, because these are variables that indicate high human presence on the landscape (Mladenoff et al. 1995). This complex

Table 8.2 Pearson correlation matrix for land use and land cover variables within the occupied pack areas for wolf habitat in Wisconsin

	Open water	Developed	Barren	Deciduous forest	Conifer forest	Mixed forest	All forest	Agriculture	Shrub/scrub	Herbaceous	Woody wetland	Herbaceous wetland	Road density	Highway density	Distance to highway	Stream density
Open water	1.00	0.21	0.09	-0.21	0.18	0.38	0.02	-0.28	-0.06	-0.04	0.13	0.09	0.28	-0.07	-0.04	-0.06
Developed	0.21	1.00	0.21	-0.16	-0.04	-0.07	-0.18	0.07	-0.04	0.10	-0.09	-0.05	0.65	0.23	-0.13	0.00
Barren	0.09	0.21	1.00	-0.32	-0.11	0.01	-0.32	0.32	-0.07	0.03	-0.15	-0.13	0.32	0.33	-0.27	-0.11
Deciduous forest	-0.21	-0.16	-0.32	1.00	0.12	-0.21	0.85	-0.54	-0.08	-0.10	-0.12	-0.07	-0.42	-0.25	0.24	0.22
Conifer forest	0.18	-0.04	-0.11	0.12	1.00	0.07	0.51	-0.42	-0.07	0.28	-0.20	0.12	0.14	-0.12	0.15	0.02
Mixed forest	0.38	-0.07	0.01	-0.21	0.07	1.00	0.19	-0.46	0.00	-0.22	0.62	-0.03	-0.10	-0.13	0.02	0.20
All forest	0.02	-0.18	-0.32	0.85	0.51	0.19	1.00	-0.79	-0.09	-0.05	0.03	-0.03	-0.35	-0.31	0.28	0.27
Agriculture	-0.28	0.07	0.32	-0.54	-0.42	-0.46	-0.79	1.00	-0.07	-0.04	-0.49	-0.31	0.40	0.38	-0.35	-0.29
Shrub/scrub	-0.06	-0.04	-0.07	-0.08	-0.07	0.00	-0.09	-0.07	1.00	0.39	0.02	0.08	-0.06	-0.04	0.00	-0.02
Herbaceous	-0.04	0.10	0.03	-0.10	0.28	-0.22	-0.05	-0.04	0.39	1.00	-0.29	0.15	0.22	0.04	-0.06	-0.21
Woody wetland	0.13	-0.09	-0.15	-0.12	-0.20	0.62	0.03	-0.49	0.02	-0.29	1.00	0.10	-0.40	-0.24	0.19	0.23
Herbaceous wetland	0.09	-0.05	-0.13	-0.07	0.12	-0.03	-0.03	-0.31	0.08	0.15	0.10	1.00	-0.26	-0.12	0.29	0.01
Road density	0.28	0.65	0.32	-0.42	0.14	-0.10	-0.35	0.40	-0.06	0.22	-0.40	-0.26	1.00	0.36	-0.31	-0.18
Highway density	-0.07	0.23	0.33	-0.25	-0.12	-0.13	-0.31	0.38	-0.04	0.04	-0.24	-0.12	0.36	1.00	-0.42	-0.10
Distance to highway	-0.04	-0.13	-0.27	0.24	0.15	0.02	0.28	-0.35	0.00	-0.06	0.19	0.29	-0.31	-0.42	1.00	0.06
Stream density	-0.06	0.00	-0.11	0.22	0.02	0.20	0.27	-0.29	-0.02	-0.21	0.23	0.01	-0.18	-0.10	0.06	1.00

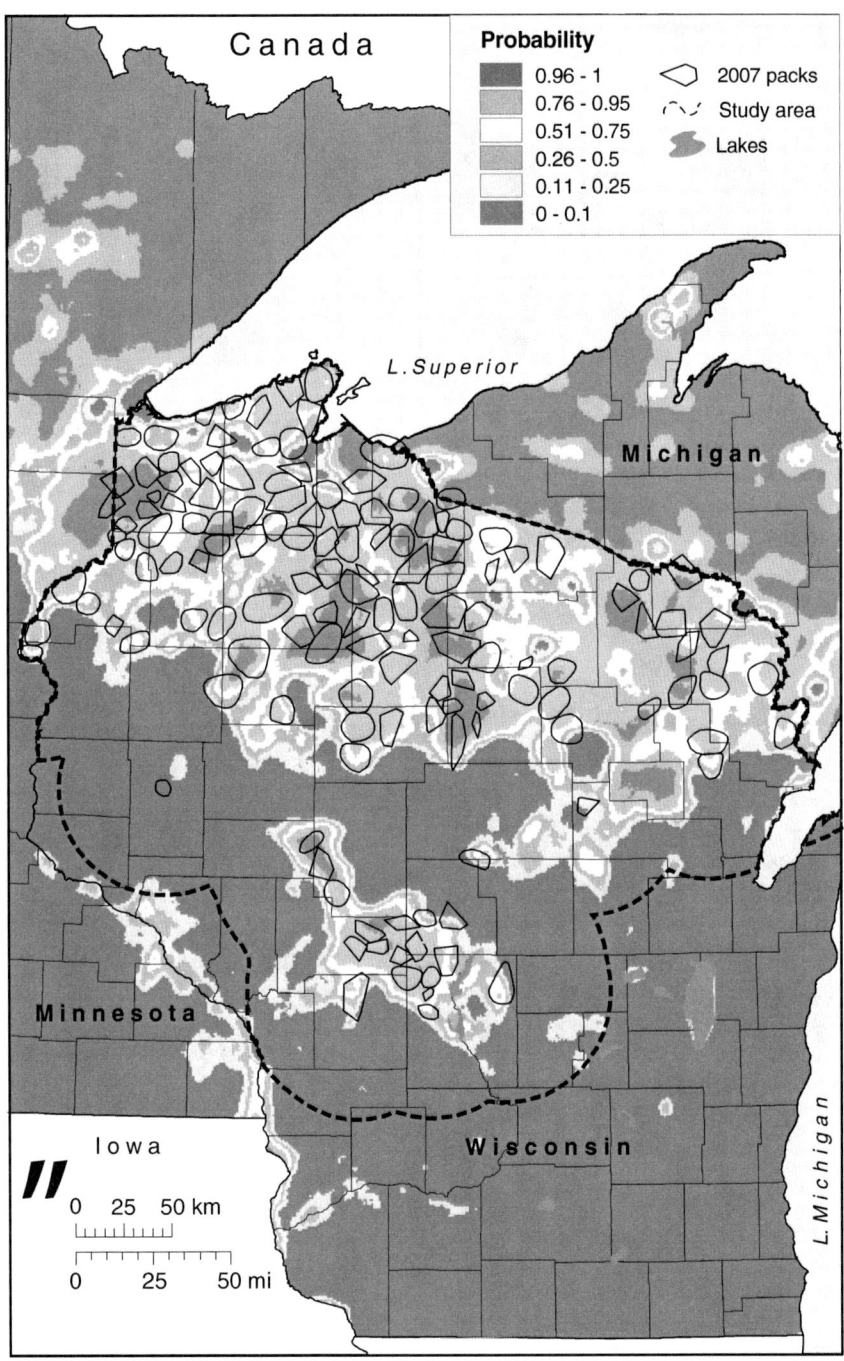

Fig. 8.5 Map of wolf habitat probability classes in Wisconsin based on new analysis and wolf pack occupancy in winter 2006–2007. Map created: Forest Landscape Ecology Laboratory, Department of Forest and Wildlife Ecology, University of Wisconsin-Madison. © David J. Mladenoff (*see Color Plate 5*)

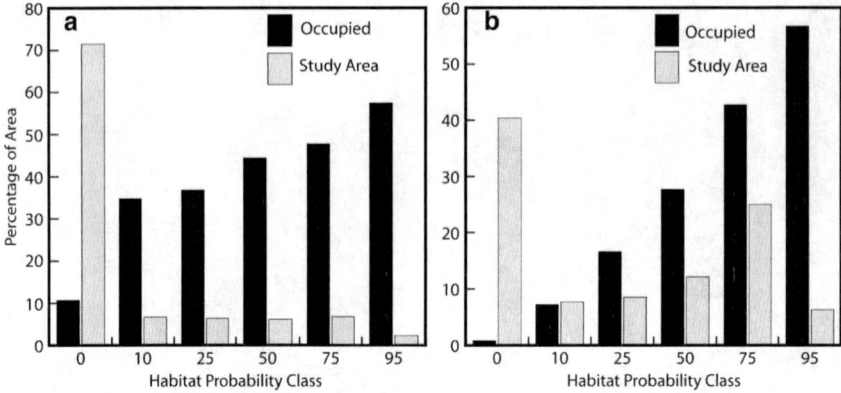

Fig. 8.6 Percentage of habitat area classes in the Wisconsin study area and occupied percentage for (**a**) the new model (map Fig. 8.7) and (**b**) the original 1995 model (map Fig. 8.3).

model result may occur because mixed forest is a small forest class in the landscape, and co-occurs with a relatively high road density area in the north central lakes area (Table 8.1, Fig. 8.1). Hence, complex interactions among variables resulted in a low but negative coefficient for mixed forest. To address this, we merged the three forest classes into one forest class and repeated the model-building exercise. The most interpretable model with equally high AIC and validation to the roads-ag-mixed forest model included roads and agriculture only:

$$\text{Logit}\,(p) = 5.0018 - 11.7095A - 2.5655R,$$

where p is the probability of occurrence of a wolf pack, A is agricultural land use (%), and R is road density (km/km^2, Table 8.3).

Validation was nearly identical for the models, with a cut level at the $\geq$70% probability level resulting in 81.6% correct classification, with 40 pack and 40 non-pack areas (80) correctly classified, and eight each (16) misclassified. In this current model, much more of the landscape is in high occupancy classes (>75% probability, Fig. 8.6) compared to the 1995 roads-only model (Fig. 8.4). This reflects the fact that a high proportion of the forested landscape (Fig. 8.1) is now occupied, but wolves typically avoided agriculturally dominated areas. The highest probability habitat (>95% probability) remained strongly influenced by road density, as shown by the conjunction of this class in the model maps from the two periods (Figs. 8.3 and 8.5).

8.8 Discussion

Our analysis of wolf habitat and information on current distribution of wolves demonstrates that wolf packs are able to occupy areas well beyond wilderness across the Great Lakes region. Given adequate protection, wolves are able to occupy most

Table 8.3   Logistic regression model statistics for final model

Parameter	DF[a]	Estimate	SE[b]	Wald X^2	P
Intercept	1	5.00	0.95	27.49	<0.01
Road density	1	–2.57	0.74	12.07	<0.01
Agriculture	1	–11.71	1.99	34.75	<0.01

[a]Degrees of freedom
[b]Standard error

large tracts of forest and other wildland habitat. Similar selection for forest and wildlands, while avoiding anthropogenic features such as agricultural land, roads, and areas of high human density, was demonstrated through similar habitat modeling in the northern Rocky Mountains of the USA (Oakleaf et al. 2006), across Poland (Jędrzejewski et al. 2004, 2005), and in Italy (Corsi et al. 1999).

Mech (2006) questioned the value of this type of modeling for predicting wolf distributions in Wisconsin, though his earlier analysis also showed the value of using road densities to describe wolf distribution in Minnesota. Mech's critique (2006) of our modeling efforts was not based on sound analysis (Mladenoff et al. 2006). Our analysis demonstrated the usefulness of our earlier model for predicting initial colonization and direction of habitat occupancy, and it continues to be robust. Our new model probably more fully describes the extent of future pack occurrences under current human-caused mortality regimes. If human-caused mortality rates change in the future, areas of potential wolf habitat would likely change as well.

We suggest that our new model and the 1995 model reflect differences not only due to the changed number and location of wolves, but that the models also mean very different things and should be viewed and used in different ways. This is particularly true if we extrapolate the new Wisconsin model to the three-state region (Fig. 8.7).

Our original (Mladenoff et al. 1995) model reflected a small, new, colonizing wolf population in northern Wisconsin and showed how wolves were occupying the landscape in a very selective way. This pattern changed gradually over intervening years, as wolf numbers increased from 25 to >540 (Wydeven et al., this volume). Over this time, our original model was able to continue to identify the most preferred habitat (Mladenoff et al. 1999; Mladenoff et al., unpublished data). Though its predictability gradually declined in the most recent half decade, it remained a significant predictor of wolf distribution, and road density remains a key secondary variable in the new model, especially for indicating the highest quality habitat areas.

Our new model reflects different information. This model reflects current habitat occupancy, but likely under very different dynamics, not merely higher and more broadly dispersed wolf numbers. We suggest that our current model reflects the result of strong source–sink dynamics (Pulliam 1988) as suggested by our earlier work (Mladenoff et al. 1995, 1999, 2006). We hypothesize that our new model shows wolf occupancy under the influence of strong fecundity and available dispersers from core, more reliable habitat that is smaller and more fragmented in northern Wisconsin, and is more abundant in Minnesota and upper Michigan (Table 8.4, Figs. 8.3 and 8.7). These core areas are shown in red in the original model (Fig. 8.3),

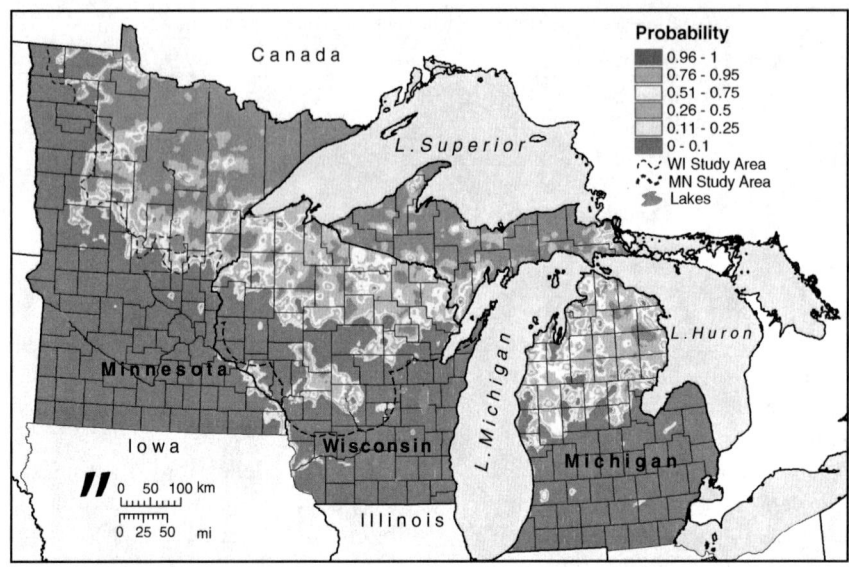

Fig. 8.7 Mapped extrapolation of the new Wisconsin model to the three states in the Great Lakes region. Map created: Forest Landscape Ecology Laboratory, Department of Forest and Wildlife Ecology, University of Wisconsin-Madison. © David J. Mladenoff (*see Color Plate 6*)

Table 8.4 Percentage of probability classes for the new wolf habitat model extrapolated to the three-state area

| Probability class (*p*) | Michigan | | | | Minnesota | | Wisconsin | |
| | Lower Peninsula | | Upper Peninsula | | Wolf range | | Study area | |
	Area (km²)	(%)	Area (km²)	(%)	Area (km²)	(%)	Area (km²)	(%)
≥0.95	2,143	2.0	29,818	68.9	52,927	42.2	6,121	6.3
0.75–0.94	15,513	14.5	9,775	22.6	20,401	16.3	24,161	25.0
0.50–0.74	10,609	9.9	1,826	4.2	7,280	5.8	11,735	12.1
0.25–0.49	7,246	6.8	1,082	2.5	5,812	4.6	8,220	8.5
0.10–0.24	5,645	5.3	442	1.0	5,033	4.0	7,432	7.7
<0.10	65,793	61.5	309	0.7	33,972	27.1	39,147	40.4
Total	106,949	100	43,252	100	125,425	100	96,816	100

and served as secure source and population expansion areas during the colonization years of the 1980s through early 2000s. Subsequent continued colonization into larger portions of the map (class ≥75% probability, Figs. 8.6 and 8.7) may have been dependent on those secure source areas. This may be why avoidance of agriculture (essentially, high-human-contact areas) is now the key variable in the model. Road density reflected similar information, but areas of moderate road density also occur with some areas of high forest cover that remain important in the new model.

This interpretation further suggests that our new model should not be seen as a map of favorable habitat that the 1995 model reflects. In particular, the new model should not be interpreted in that way in disjunct areas such as lower Michigan (Gehring and Potter 2005), where there is only limited access to the nearby source populations (Fig. 8.7). Our new map may better reflect potential habitat in upper Michigan and Minnesota, especially in the large expanses of high-quality source habitat (Fig. 8.7). It is unlikely that wolf occupancy will change again in the near future as significantly as it has in the last decade. There is evidence in Minnesota that expansion has slowed, possibly reflecting an equilibrium between the source population colonizers and increasing mortality in increasingly less favorable areas (Erb and DonCarlos, this volume), and population growth has declined in Wisconsin and Michigan, possibly due to a similar mechanism (Van Deelen, this volume). Also, wolf control has increased gradually as the growing population has spread into areas with more human contact. Federal delisting and more flexible controls applied by the states may further cause wolf range to stabilize at the agricultural–forest fringe. These factors may make it difficult to separate causes of any observed reduction in expansion of occupied wolf range.

Our earlier model of favorable habitat (Mladenoff et al 1995), and projected wolf populations (Mladenoff et al. 1997), were used extensively for wolf management planning in Wisconsin (Wisconsin DNR 1999). Although wolf pack distribution has extended beyond the original areas predicted to become occupied by packs, the models were used for creating management zones, establishing population goals, planning surveys, and planning for future population expansion. Our new model should also provide useful information for future habitat and population management for wolves in the Great Lakes region and will generate important research opportunities for understanding eventual limits to wolf population growth and range expansion.

8.9   Summary

Occupied wolf habitat in the Great Lakes region declined from region-wide at the time of Euro-American settlement to remote wilderness areas by the mid-1900s. By the time wolves were listed as a federal endangered species in 1974, they were perceived mainly as a wilderness species. Through protection of the Endangered Species Act, wolves were able to expand to forests and other wildland areas across the states of Minnesota, Michigan, and Wisconsin. Our modeling of wolf habitat in the 1990s showed that road densities were the best predictors of areas of initial colonization and direction of population spread. As core habitat areas have been mostly occupied and wolves have spread across much of the region's large expanses of forest, deriving a new model has shown both the changes and persistence of wolf preferences on the landscape. In our new model, the best predictors of wolf habitat are lack of agricultural land and low road density. Our mapping of wolf habitat shows extensive areas of suitable wolf habitat across the region, but core habitat is more fragmented in Wisconsin than in northern Minnesota and upper

Michigan. We anticipate that potential for further range expansion will be limited in the region, except possibly if wolves are able to expand to lower Michigan (Gehring and Potter 2005).

Conservationists should consider carefully our two modeling results in assessing future wolf population dynamics and management. Most importantly, our new model should be seen as more descriptive of wolf occupancy, and more dependent on complex population dynamics, than our simpler 1995 model. We suggest that the 1995 model is the best conservative indicator of preferred, most-critical habitat, especially in Wisconsin where both models show that core habitat remains more limited. This suggests that maintaining population security in Wisconsin will continue to require habitat management that limits road development and maintains areas of low road density on public forest lands, so that these areas can continue to function as core habitat. Long-term viability of Wisconsin wolves will likely require maintaining connectivity to the abundant high-quality source areas in Minnesota and upper Michigan.

Acknowledgments Funding was provided by the Wisconsin Department of Natural Resources, Endangered Resources Funds. Funding for wolf surveys were from Federal Aid in Wildlife Conservation Projects (Pittman-Robertson Funds), US Fish and Wildlife Service, Wisconsin Endangered Resources Funds, Chequamegon-Nicolet National Forest, and private donations. We are grateful to J. Wiedenhoeft for providing recent data. Major assistance was also provided on wolf surveys by R. Schultz, R. Thiel, B. Kohn, S. Boles, J. Wiedenhoeft, Wisconsin DNR pilots, other DNR staff, and volunteers.

References

Akaike, H. 1974. A new look at the statistical model identification. IEEE Transactions on Automatic Control 19:716–723.

Cleland, C. E. 1983. Indians in a changing environment. In S. L. Flader, ed., The Great Lakes Forest: An environmental and social history. pp. 83–95. Minneapolis: University of Minnesota Press.

Corsi, F., Dupre, E., and Boitani, L. 1999. A large-scale model of wolf distribution in Italy for conservation planning. Conservation Biology 13:150–159.

Curtis, J. T. 1959. Vegetation of Wisconsin: An Ordination of Plant Communities. Madison: University of Wisconsin Press.

Fuller, T. K., Berg, W. E., Radde, G. L., Lenarz, M. S., and Joselyn, G. B. 1992. A history and current estimate of wolf distribution and numbers in Minnesota. Wildlife Society Bulletin 20:42–55.

Gehring, T. M., and Potter, B. A. 2005. Wolf habitat analysis in Michigan: An example of the need for proactive land management for carnivore species. Wildlife Society Bulletin 33:1237–1244.

Heinselman, M. L. 1996. The Boundary Waters Wilderness Ecosystem. Minneapolis: University of Minnesota Press.

Homer, C., Huang, C., Yang, L., Wylie, B., and Coan, M. 2004. Development of a 2001 national land-cover database for the United States. Photogrammetric Engineering & Remote Sensing 70:829–840.

Jackson, H. H. T. 1961. Mammals of Wisconsin. Madison: University of Wisconsin Press.

Jędrzejewski, W., Niedziałkowska, M., Nowak, S., and Jędrzejewska, B. 2004. Habitat variables associated with wolf (*Canis lupus*) distribution and abundance in northern Poland. Diversity and Distributions 10:225–233.

Jędrzejewski, W., Niedziałkowska, M., Mysłajek, R. W., Nowak, S., and Jędrzejewska, B. 2005. Habitat selection by wolves, *Canis lupus* in the uplands and mountains of southern Poland. Acta Theriologica 50:417–425.

Keener, J. M. 1955. The case for the timber wolf. Wisconsin Conservation Bulletin 20:22–24.

Mech, L. D. 1970. The Wolf: The Ecology and Behavior of an Endangered Species. Garden City: Natural History Press.

Mech, L. D. 1989. Wolf population survival in an area of high road density. American Midland Naturalist 121:387–389.

Mech, L. D. 1995. The challenge and opportunity of recovering wolf populations. Conservation Biology 9:270–278.

Mech, L. D. 2006. Prediction failure of a wolf landscape model. Wildlife Society Bulletin 34:874–877.

Mech, L. D., Fritts, S. H., Radde, G. L., and Paul, W. J. 1988. Wolf distribution and road density in Minnesota. Wildlife Society Bulletin 16:85–87.

Mladenoff, D. J., Clayton, M. K., Sickley, T. A., and Wydeven, A. P. 2006. L. D. Mech critique of our work lacks scientific validity. Wildlife Society Bulletin 34:878–881.

Mladenoff, D. J., Haight, R. G., Sickley, T. A., and Wydeven, A. P. 1997. Causes and implications of species restoration in altered ecosystems: A spatial landscape projection of wolf population recovery. BioScience 47:21–31.

Mladenoff, D. J., Sickley, T. A., Haight, R. G., and Wydeven, A. P. 1995. A regional landscape analysis and prediction of favorable gray wolf habitat in the northern Great Lakes region. Conservation Biology 9:279–294.

Mladenoff, D. J., Sickley, T. A., and Wydeven, A. P. 1999. Predicting gray wolf landscape recolonization: Logistic regression models vs. new field data. Ecological Applications 9:37–44.

Mladenoff, D. J., White, M. A., Pastor, J., and Crow, T. R. 1993. Comparing spatial pattern in unaltered old-growth and disturbed forest landscapes. Ecological Applications 3:294–306.

Oakleaf, J. K., Murray, D. L., Oakleaf, J. R., Bangs, E. E., Mack, C. M., Smith, D. W., Fontaine, J. A., Jimenez, M. D., and Meier, T. J.. 2006. Habitat selection by recolonizing wolves in the northern Rocky Mountains of the United States. Journal of Wildlife Management 70:554–563.

Peek, J. M. 1974. A review of moose food habits in North America. Naturaliste Canadien 101:195–215.

Potvin, M. J., Drummer, T. D., Vucetich, J. A., Beyer, D. E., Jr., Peterson, R. O., and Hammill, J. H.. 2005. Monitoring and habitat analysis for wolves in Upper Michigan. Journal of Wildlife Management 69:1660–1669.

Pulliam, H. R. 1988. Sources, sinks, and population regulation. American Naturalist. 132:652–661.

Reese, H. M., Lillesand, T. M., Nagel, D. E., Stewart, J. S., Goldmann, R. A., Simmons, T. E., Chipman, J. W., and Tessar, P. A.. 2002. Statewide land cover derived from multiseasonal Landsat TM data: A retrospective of the WISCLAND project. Remote Sensing of the Environment 82:224–237.

Rhemtulla, J. M., and Mladenoff, D. J. 2007. Why history matters in landscape ecology. Landscape Ecology 22:1–3.

Rhemtulla, J. M., Mladenoff, D. J., and Clayton, M. K. 2007. Regional land-cover conversion in the U.S. upper Midwest: Magnitude of change and limited recovery (1850–1935–1993). Landscape Ecology 22:57–75.

Schulte, L. A., and Mladenoff, D. J. 2005. Severe wind and fire regimes in northern forests: Historical variability at the regional scale. Ecology 86:431–445.

Schulte L. A., Mladenoff, D. J., and Nordheim, E. V. 2002. Quantitative classification of a historic northern Wisconsin (USA) landscape: Mapping forests at regional scales. Canadian Journal of Forest Research 32:1616–1638.

Schulte, L. A., Mladenoff, D. J., Crow, T. R., Merrick, L. C., and Cleland, D. T. 2007. Homogenization of northern U.S. Great Lakes forests due to land use. Landscape Ecology 22:1089–1103.

Thiel, R. P. 1985. The relationship between road densities and wolf habitat in Wisconsin. American Midland Naturalist 113:404–407.

Thiel, R. P. 1993. The timber wolf in Wisconsin. Madison: University of Wisconsin Press.

Treves, A., Jurewicz, R. L, Naughton-Treves, L., Rose, R. A., Willging, R. C., and Wydeven, A. P. 2002. Wolf depredation on domestic animals in Wisconsin, 1976–2000. Wildlife Society Bulletin 30:231–241.

United States Census Bureau. 2001. Census 2000 TIGER/Line Files Technical Documentation. Washington, D.C.: U.S. Census Bureau.

United States Fish and Wildlife Service. 1992. Recovery Plan for the Eastern Timber Wolf. Twin Cities: U.S. Fish and Wildlife Service.

United States Geological Survey. 1999. Standards for National Hydrography Dataset: Reston, VA: U.S. Geological Survey, http://rockyweb.cr.usgs.gov/nmpstds/nhdstds.html

Wisconsin Department of Natural Resources 1989. Wisconsin Timber Wolf Recovery Plan. Wisconsin Endangered Resources Report 50, Madison: Wisconsin Department of Natural Resources.

Wisconsin Department of Natural Resources 1999. Wisconsin Wolf Management Plan. PUBL-ER-099 99, Madison: Wisconsin Department of Natural Resource.

Wydeven, A. P., Schultz, R. N., and Thiel, R. P. 1995. Monitoring of a recovering gray wolf population in Wisconsin, 1979–1991. In L. N. Carbyn, S. H. Fritts, and D. R. Seip, eds., Ecology and Conservation of Wolves in a Changing World. pp. 147–156. Alberta: Canadian Circumpolar Institute.

Chapter 9
Growth Characteristics of a Recovering Wolf Population in the Great Lakes Region

Timothy R. Van Deelen

9.1   Introduction

The northern forests of Wisconsin and Michigan's Upper Peninsula provide an extensive area of habitat favorable for gray wolves (*Canis lupus*) and are relatively isolated from established wolf populations in Ontario and Minnesota. Boundaries include Lakes Superior, Michigan, and Huron. The recent population of wolves in Wisconsin and Michigan evidently was founded by dispersing wolves that arrived in the mid-1970s. Growth of this population occurred while populations of white-tailed deer (*Odocoileus virginianus*) were at or near historic highs in abundance. Consequently, the re-colonizing wolves were entering a very high quality habitat where other major predators on adult deer such as grizzly bears (*Ursus arctos*) or cougars (*Puma concolor*) were absent. In addition to being an interesting ecological case study of density dependence and population growth in a population of large terrestrial predators, growth of this population has important conservation significance. Carrying capacity, and population growth rate with respect to carrying capacity, will determine impacts of wolves on other ecological, economic, and social factors of value to humans (e.g., deer populations, livestock depredation, wildlife viewing). In this chapter, I describe an attempt to derive empirical estimates of carrying capacity and growth rate for this population of wolves. I then discuss the potential for additional growth and what additional growth means for conservation and management.

Populations of gray wolves in Wisconsin and the Upper Peninsula of Michigan (hereafter, Upper Michigan) became listed as Federal Endangered in 1974, and State Endangered in 1975 (Wisconsin) and 1976 (Michigan). Formal monitoring of the recovering population began in 1979 (Wydeven et al. 1995). Biologists began radio-collaring adult wolves in Wisconsin in 1979 (Wydeven et al. 1995) and in Michigan in 1992 (Potvin et al. 2005) to facilitate the identity and location of wolf packs for aerial counts. These counts were supplemented by howling surveys and winter track surveys, but radio tracking has remained a centerpiece of population monitoring (Wydeven et al. 1995). Taken together, these efforts have provided rigorous annual counts of wolves.

A.P. Wydeven et al. (eds.), *Recovery of Gray Wolves in the Great Lakes Region of the United States*,
DOI: 10.1007/978-0-387-85952-1_9, © Springer Science + Business Media, LLC 2009

Wolves in Wisconsin and Upper Michigan are managed independently by their respective state natural resource agencies but both states share a common goal of population recovery with oversight and assistance provided by the US Fish and Wildlife Service under the direction of the Endangered Species Act. In addition, the border between Wisconsin and Upper Michigan is artificial with respect to wolf movement, comprising a remote and forested region of relatively good wolf habitat (Mladenoff et al. 1995, 1999). Thus, operationally and functionally the wolves occupying this region are properly considered members of a single biological population occupying the Southern Lake Superior (SLS) region.

The SLS wolves persisted at a relatively stable but low level (14–34 wolves) between 1979 and 1990 before entering a period of sustained growth (Fig. 9.1). This pattern of prolonged low population levels followed by an eventual robust growth phase is a common feature of recovering wolf populations (e.g., US Fish and Wildlife Service et al. 2007; Wabbakkan et al. 2000) and may be due to an Allee effect (Wabbakkan et al. 2000). Minimum population goals for recovery were 100 wolves for SLS region (total for both states) under the U.S. Endangered Species Act, and 450 wolves (200 in Michigan and 250 in Wisconsin) under the 2 states' endangered species laws (Wydeven et al. this volume; Beyer et al. this volume). In addition, Wisconsin established a statewide management goal of 350 wolves outside of Indian reservations. These goals were achieved during the 1990s and early 2000s (Wydeven et al. this volume). Recent (2006) population counts suggest roughly 1,000 wolves in the SLS. On 12 March 2007 the US Fish and Wildlife Service removed the western Great Lakes population of gray wolves from threatened status under the US Endangered Species Act. Federal delisting gives

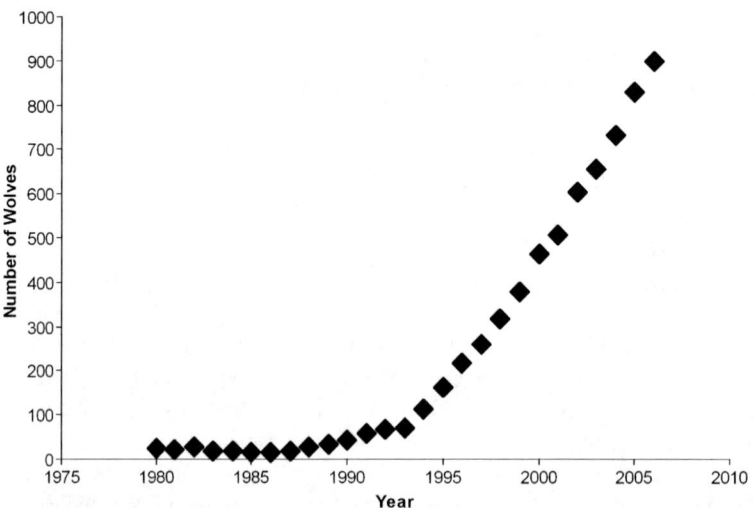

Fig. 9.1 Growth of the Southern Lake Superior (SLS) wolf population from 1980 to 2007 based on aerial counts of wolf packs containing radiocollared wolves, and winter track surveys. Data were obtained from the Michigan and Wisconsin Departments of Natural Resources

managers greater flexibility in managing wolf populations – especially with respect to problem wolves responsible for depredations on livestock and hunting dogs.

The record of continuous wolf counts over the 27-year period of recovery offers a uniquely valuable opportunity to study the population dynamics of a re-colonizing wolf population and may offer insights into the population growth that might be expected in other recovery efforts and with other large carnivores.

Most information about population dynamics of wolves comes from studies of radiocollared individuals (Fuller et al. 2003). Sibly and Hone (2003) argued that population growth is the organizing principle for the study of population dynamics. In this context, radiotelemetry studies contribute to a demographic or mechanistic paradigm for understanding population growth because telemetry-derived population parameters (e.g., individual-based age- and sex-specific mortality and productivity rates and their environmental drivers) are viewed as components of growth. Population dynamics thus is understood as the study of how these components vary and interact to affect changes in populations (i.e., growth). Fuller et al. (2003) provided a comprehensive review and meta-analysis of studies of wolf populations, most of which analyzed telemetry data using a demographic or mechanistic paradigm.

An alternate paradigm considers the relationship between population growth and population density as the defining feature for understanding population dynamics (Sibly and Hone 2003). Under this paradigm, the form of the relationship between population density and population growth rate integrates the complexities of interacting age- and sex-specific vital rates, trophic interactions, and ecological effects. Analysis requires relatively rigorous information on population trends over a meaningful range of population densities during a time frame that is meaningful with respect to a given species' capacity for growth. Density, demographic, and mechanistic paradigms are, of course, complimentary approaches for gaining a fuller understanding of population dynamics. Wolves are wide-ranging and exist at relatively low densities on the landscape and hence are difficult to count. Thus, for generating population-level data, it may be more efficient to study samples of radiocollared individuals than to accumulate annual censuses. The trade-off is that research using radio telemetry is expensive and labor intensive. Consequently, most (but not all: see Vucetich and Peterson this volume) radiotelemetry studies tend to be of short duration and involve relatively small samples of radiocollared individuals thereby raising issues of inference to dynamics occurring over longer time periods and larger spatial scales. For example, Fuller et al. (2003:182) presented a table showing the relationship between growth rate and annual mortality in 19 studies. Duration of these studies ranged from two to nine years with a mean of five years – roughly equivalent to the lifespan of a single adult wolf.

Understanding the growth of the SLS wolf population is not merely an academic exercise. Wolf depredations have increased steadily with increasing population size (Ruid et al. this volume). In addition, hunting for white-tailed deer has very high economic and cultural importance for residents of Wisconsin and Michigan, and support for wolf recovery is likely to decline if a growing wolf population is seen as a threat to deer hunting (Lohr et al. 1996; Meadow et al. 2005). At current population levels, wolf depredations (Ruid et al. this volume) and Federal delisting are forcing state managers to re-evaluate the population goals and the degree to which management activities such as regulated hunting, trapping, and culling may be needed to achieve these goals now or after additional increases. Understanding population growth and its relationship to ecological factors is the basis of harvest management theory (McCullough 1979). According to the most widely appealed-to theory of harvest management (Sustained-Yield theory) sustainability is defined primarily by the relationship between population growth rate and population density (McCullough 1979; Ludwig 2001; Skalski et al. 2005). With respect to population goals, empirically derived information on population growth is the most important single source of information for determining the relationship between harvest and resultant equilibrium population sizes. My goal in this chapter is to empirically estimate the population growth rate and form of density dependence displayed by the SLS wolves. Using these estimates, I will then predict the carrying capacity or the equilibrium number of wolves likely if growth were to continue under current levels of population management.

9.2   Study Area

The study area for this analysis includes parts of northern Wisconsin and the Upper Peninsula of Michigan that were occupied by wolves as of 2007 (Beyer et al. this volume; Wydeven et al. this volume). Wolves began re-colonizing this SLS region during the mid-1970s following termination of extermination campaigns that had resulted in effective extirpation by about 1960 (Wydeven et al. 1995). The qualifier "effective" here refers to a population-level understanding. Sightings of individual wolves (likely dispersers) were reported throughout the 1960s and 1970s in both states (Baker 1983; Thiel 2001). Despite the fact that re-colonization almost certainly occurred through dispersal from wolf populations in Minnesota and Ontario (Mech et al. 1995; Jensen et al. 1986; Thiel 2001), the SLS region is spatially distinct from areas continuously occupied by wolves in Minnesota and Ontario because of important though incomplete barriers to wolf movement imposed by the Great Lakes and their connecting waterways and by discontinuities in favorable habitat (Mladenoff et al. 1995; Haight et al. 1998).

The part of the SLS region currently occupied by wolves is diverse, containing examples of 25 unique regional landscape ecosystems according to a classification based on variation in climate, bedrock geology, glacial landform, and soils (Albert 1995). In general terms, the landscape of Wisconsin and Upper Michigan grades from intense agriculture in central and western Wisconsin to extensive forest in the north and northeast. Within this gradient, relief varies from level to rugged, and extensive wetlands, lakes, human population centers, and pockets of agriculture provide variation.

Rigorous classification of this habitat with respect to quality for wolves is controversial (Mech 2006; Mladenoff et al. 2006). Nonetheless, relative to other parts of midwestern North America, northern Wisconsin and Upper Michigan have low road densities, low human densities, and high deer populations (Haight et al. 1998; Mladenoff et al. 1995, 1999). Collectively, these habitat features (Mladenoff et al. 1995, 1999) and legal protection since 1974 (Mech 1995) indicate ecological conditions favorable for re-colonization. Robust population growth since 1980 (Beyer et al. this volume; Wydeven et al. this volume) supports this contention.

9.3   Methods

I estimated the growth rate and predicted potential equilibrium population levels for the SLS wolf population by fitting population growth models to a time series of yearly population estimates. The time series represents the sum of the yearly winter counts done in both Michigan and Wisconsin from 1980 to 2007 (Fig. 9.1). Managers in both states used a combination of aerial monitoring of radiocollared wolves and intensive snow tracking to estimate the total number of wolves alive during winter. These estimates are considered a complete minimum count because lone wolves are probably undercounted, indicating the potential for a small negative bias (Wydeven et al. this volume). Year-to-year estimates also reflect an increasing number of wolves captured in response to depredation complaints. This number has increased from 7 in 1998 to 25 in 2006; through 2002 these captured wolves were translocated, but since 2003 most have been euthanized (Ruid et al. this volume).

I evaluated the growth characteristics of the SLS wolf population by fitting a set of simple growth models (Skalski et al. 2005) to the population trend from 1982 to 2005. Skalski et al. (2005: 22) listed discrete-time formulations of seven growth models commonly used in wildlife and fishery science (Table 9.1). Except for the exponential model, each model is a different representation of the relationship between

Table 9.1 Discrete-time growth models (Skalski et al. 2005: 22) used to describe the growth of the Southern Lake Superior (SLS) wolf population (1982–2005)

Model[a]	Form
Exponential	$N_{t+1} = N_t \lambda_{max}$
Logistic	$N_{t+1} = N_t + N_t (\lambda_{max} - 1) \left(1 - \dfrac{N_t}{K} \right)$
Generalized logistic	$N_{t+1} = N_t + N_t (\lambda_{max} - 1) \left(1 - \left(\dfrac{N_t}{K} \right)^z \right)$
Ricker	$N_{t+1} = N_t \lambda_{max}^{\left(1 - \frac{N_t}{K} \right)}$
Beverton and Holt	$N_{t+1} = \dfrac{N_t \lambda_{max}}{(1 + (N_t / K'))}$
Hessell	$N_{t+1} = \dfrac{N_t \lambda_{max}}{(1 + (N_t / K'))^z}$
Shepard	$N_{t+1} = \dfrac{N_t \lambda_{max}}{1 + (N_t / K')^z}$

Parameters are λ_{max} = discrete growth rate; K, K' = constants associated with asymptotic growth; z = shape parameter
[a]References in Skalski et al. (2005)

density or abundance, growth rate, and an asymptotic population size (defined as carrying capacity). Differences among models reflect differences in forms of the relationship between realized growth rate and density and the flexibility with which non-linear forms (e.g., Ellis and Post 2004) can be fit to empirical data. In this context, inclusion of the exponential model (no density dependence, no asymptotic growth level) allows for evaluation of a null condition wherein the SLS wolf population is growing without limit (Dennis and Taper 1994; Taper and Gogan 2002).

Significantly, all seven models depend on one to three parameters whose inter-pretations have biological significance (Table 9.1). Each model includes a param-eter representing a characteristic average yearly growth rate in the absence of density-dependent effects, λ_{max}. The six density-dependent models each have a parameter (K or K') that determines asymptotic population growth. In the case of the logistic and generalized logistic, K is the value of the asymptote. The parameter z in the generalized logistic, Hessell, and Shepard models controls the shape of the growth curve as it approaches the asymptote.

Because of serial dependence in time series of population sizes, parameter esti-mates generated from model-fitting routines in statistical software are valid but variance calculations for parameter estimates are suspect (Dennis and Taper 1994; Ives and Zhu 2006). I used parametric jackknifing to generate meaningful estimates of parameter variances. Parametric jackknifing is a re-sampling technique wherein $N - 1$ pseudo-samples are created by sequentially removing a single observation. A model is then fit to each of the pseudo-samples and the mean and variance of the $N - 1$ parameter estimates are taken to be rigorous estimates of the model parameters fit to the full data set (Efron 1981).

I fit growth models to the SLS wolf population data using PROC NLIN (SAS version 9.1; SAS Inc., Cary, North Carolina). I evaluated the relative value of each model for describing the data on the basis of Akaike's Information Criterion for small sample sizes (AICc, Burnham and Anderson 1998). Akaike weights are calculated from a model's AICc value and are interpreted as the probability or confidence that a given model is the optimal model given a suite of models under consideration and a common data set. Thus, to incorporate uncertainty in model selection, my inferences are based on a ≥95% confidence set of models (i.e., the subset of models that jointly represents a ≥95% probability of containing the optimal model, Burnham and Anderson 1998).

Because each candidate model contained a discrete-time growth rate parameter (λ_{max}), I used Akaike weights and model averaging (Burnham and Anderson 1989) to calculate an unconditional estimate of λ_{max} and its variance. These estimates are unconditional in the sense that they are not conditioned on a single best model that may be only marginally better than alternate models but rather are conditioned on a ≥95% confidence set of models. The suite of models does not contain a single parameter equivalent to K (of the logistic models) whose value represents asymptotic carrying capacity. Moreover, logistic growth becomes asymptotic at values progressively lower than K with increasing variance in K. Thus, to estimate a predicted asymptotic growth level that incorporates both model selection uncertainty and the variance associated with estimated parameters for the fitted models, I used jackknifed estimates of variance and 1,000 Monte Carlo simulations to derive mean asymptotic growth values and their variances for each model in the ≥95% confidence set. I then computed an AICc-based unconditional estimate of asymptotic growth using model averaging (Burnham and Anderson 1989) across simulated mean asymptotic growth levels and their variances.

9.4   Results

All seven models were highly significant ($P < 0.0001$) when evaluated as alternatives to the standard null hypothesis of $N_{t+1} = N_t$. Evaluation of the suite of models in terms of which was the best approximating model given the data (i.e., which had the lowest AICc value, Burnham and Anderson 1998) suggested that the Beverton–Holt model was optimal ($w_{Beverton–Holt} = 0.27$, Table 9.2) although it was only marginally better than the Ricker ($w_{Ricker} = 0.24$) and logistic ($w_{logistic} = 0.21$) models. In general, models with three estimable parameters (λ_{max}, K or K', and an error term) performed better than models with an additional shape parameter (z, Table 9.2) whose w_is ranged from 0.9 to 0.11. In the interest of parsimony, AICc comparisons impose a penalty for additional variables that do not improve the trade-off between bias associated with few parameters and variance associated with additional parameters. In this context, it is notable that the exponential model had essentially no support ($w_{exponential} = 0.00$, Table 9.2) despite having the fewest parameters.

Jackknifed estimates of annual growth rates of wolf populations not including density effects (λ_{max}) varied from 1.27 to 1.41 and standard deviations were small (range: 0.01–0.04) for the six models in the ≥95% confidence set (Table 9.3). The model-averaged estimate of growth rate was 1.31 (95% confidence interval

Table 9.2   Model selection statistics (Burnham and Anderson 1998) for seven growth models fit to a time series of minimum population counts for the Southern Lake Superior (SLS) wolf population (1982–2005)

Rank	Model	k^a	AICc[b]	w_i^c	Cumulative w_i
1	Beverton–Holt	3	144.2	0.27	0.27
2	Ricker	3	144.4	0.24	0.51
3	Logistic	3	144.7	0.21	0.72
4	Hessell	4	146.0	0.11	0.83
5	Shepard	4	146.4	0.09	0.92
6	Generalized logistic	4	146.5	0.09	1.01[d]
7	Exponential	2	159.4	0.00	1.01[d]

[a]Number of estimable parameters ($k - 1$ structural parameters + an error term)
[b]Akaike's information criterion for small samples
[c]Akaike weights
[d]Cumulative $w_i > 1.00$ due to rounding errors

Table 9.3 Parameter estimates for models in a ≥95% confidence set of growth models fit to a time series of minimum population counts for the Southern Lake Superior (SLS) wolf population (1982–2005)

			Parameters					
			λ_{max}		K or K′		z	
Rank	Model	wi	$\bar{x}$	SD $\bar{x}$	$\bar{x}$	SD $\bar{x}$	$\bar{x}$	SD $\bar{x}$
1	Beverton – Holt	0.27	1.29	0.01	4,429	196		
2	Ricker	0.24	1.28	0.01	1,228	24		
3	Logistic	0.21	1.27	0.01	1,194	23		
4	Hessell	0.11	1.36	0.02	285	156	0.16	0.04
5	Generalized logistic	0.09	1.41	0.04	1,435	76	0.44	0.09
6	Shepard	0.09	1.38	0.03	7,459	1,160	0.61	0.09
7	Exponential[a]	0.00	1.13					

Means and standard deviations were estimated through parametric jackknifing (Efron 1981)aThe exponential model was not part of the 95% confidence set but is included in this table for comparison

[CI]: 1.28–1.34). Estimates of K, K′, and z were more variable and are not directly comparable across models.

For the sake of comparison to the data, I scaled growth simulations to the growth phase of the SLS wolf dataset by beginning each simulation with the 18 wolves counted in 1987 (Fig. 9.2). Simulated growth became asymptotic roughly 30 years later, between 2015 and 2020. Simulated growth integrated the variances in the input parameters; thus, not surprisingly, simulated asymptotic growth levels were more variable across models (range 1,194–1,674 wolves) and exhibited relatively wide CIs within models (Fig. 9.2). Similarly, models with three parameters had wider CIs for asymptotic growth. The model-averaged estimate of the asymptotic annual growth levels attained during the Monte Carlo simulations was 1,321 wolves (95% CI: 1,215–1,427).

9.5   Discussion

The wolf population in the SLS region has been growing at a rate of about 30% per year since the early 1990s, although the growth rate has decreased as the number of wolves has increased. Lack of support for the exponential growth model relative to the suite of density-dependent growth models would support an ecological interpretation of increasingly important density-dependent factors acting to restrict growth (Hayes and Harestad 2000).

The estimated growth rate is high relative to growth rates from 30 wolf populations in studies reviewed by Fuller et al. (2003) and equivalent to the median rate of six years of growth in a wolf population in the Yukon (λ_{Yukon} = 1.3)

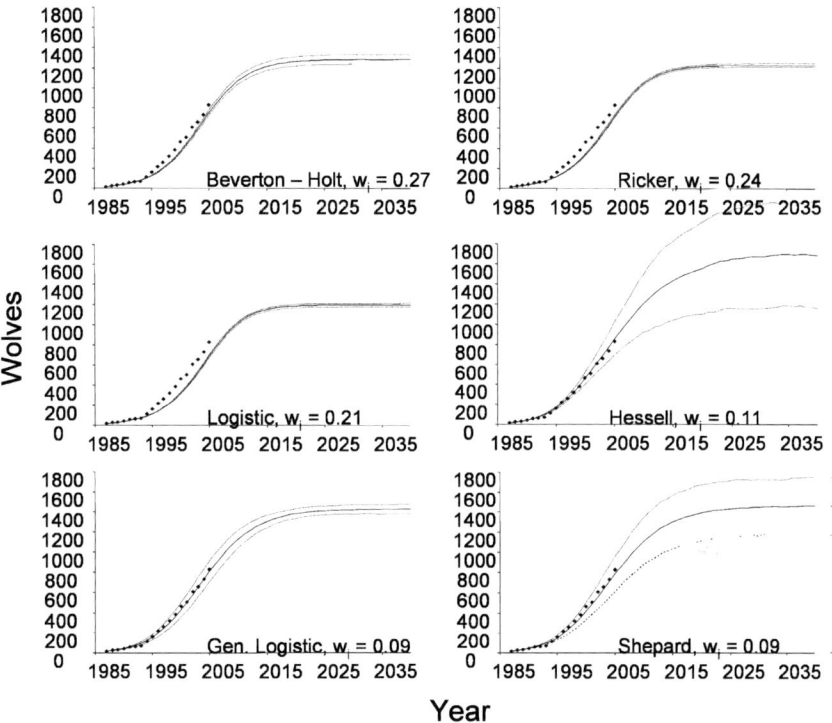

Fig. 9.2 Simulated growth of the Southern Lake Superior (SLS) wolf population using six growth models. Ninety-five percent CIs were estimated using Monte-Carlo simulation ($N = 1,000$) with parameter variance estimates derived from parametric jackknifing (see text) of a time series of population estimates (1982–2005). Models were scaled to a growth phase beginning in 1987 (points) with a beginning population of 18 wolves. Akaike weights (w_i) indicate the probability of each model being the optimal model given a common data set (Burnham and Anderson 1989)

recovering from an intense reduction (Hayes and Harestad 2000) and the growth phase of a recovering population of Scandinavian wolves ($\lambda_{Scandinavia} = 1.29$, Wabakkan et al. 2000). Fuller et al. (2003) concluded that high growth rates are features of re-colonizing populations or populations released from severe exploitation. Notably, the growth rate for the SLS population is for a 27-year time period whereas most other studies reporting growth rate are for shorter time periods (Fuller et al. 2003).

Estimates of wolf population growth in Minnesota since 1970 are a bit speculative because of reliance on varying population estimates and indices (Erb and DonCarlos this volume). Growth following protection during the 1970s after enactment of U. S. Endangered Species laws was considered rapid at 5% per year ($\lambda = 1.05$, Minnesota Department of Natural Resources 2001). Statewide population estimates using standard methods were infrequent for Minnesota, occurring in 1979, 1989, 1998, and 2004.

Erb and DonCarlos (this volume) estimated annual growth rates of 3–6% from 1970 to 2004. More modest growth in Minnesota is attributed to lower prey biomass (Minnesota Department of Natural Resources 2001; Erb and DonCarlos this volume).

The pattern of higher support for the Beverton–Holt, Ricker, and logistic models relative to the generalized logistic, Hessell, and Shepard models suggests that there was little advantage to having an additional shape parameter. Consequently, given the model structures (e.g., logistic) and their estimated parameter values, this analysis suggests that the form of density dependence is growth as a simple negative linear function of density resulting in symmetric sigmoid curves in the relationship of population size over time (Fig. 9.2). Ellis and Post (2004) found that that the relationship between density and growth for wolves on Isle Royale was non-linear (the strength of density dependence was itself density dependant) and interacted with climate effects. They reported that the consequences of mischaracterizing the form of density dependence are poorly understood. To the extent that there is any departure from linearity in this analysis, the $z < 1.0$ values for the generalized logistic, Hessell, and Shepard models and the parameter values of the Beverton–Holt and Ricker models suggest that the form would be concave (when viewed from above, Sibly et al. 2005) rather than convex as suggested by Fowler (1981) for long-lived mammals with complex life histories. Sibly et al. (2005) analyzed 1,780 time series of population growth for 674 vertebrate species and found that the concave non-linearity was most common.

The mechanism responsible for regulating density-dependent population growth is usually attributed to increasing intra-specific competition for a limiting resource – usually food (Sibly and Hone 2003). Thus, one interpretation of evidence of density dependence in growth models may be that growth of the wolf population is declining in response to increased competition (Hayes and Harestad 2000) for its primary prey in the SLS region, white-tailed deer. That abundance of white-tailed deer may ultimately be limiting population growth of SLS wolves seems unlikely since abundance of white-tailed deer was at or near historic high levels throughout the region and the population trend for deer has been generally increasing during the period of rapid growth in the population of SLS wolves. Estimates of density of deer in winter for Wisconsin's northern forest region (38,850 km^2, roughly the northern third of Wisconsin) ranged from 5 to13 deer/km^2 during the 1980s and 1990s (Wisconsin Department of Natural Resources 2001). Using data from 2000 as an example, if one assumed that the entire maximum number of wolves estimated (248, Wydeven et al. this volume) were confined to the northern forest region (they were not), wolf density would be about 6 wolves/1,000 km^2. There would have been roughly 900 deer per wolf and an ungulate biomass index (Fuller 1989) of 6 deer-equivalents/km^2, enough to support a wolf population density of 21 wolves/1,000 km^2 according to the regression relationship in Fuller et al. (2003:171). Thus, the recovering wolf population was <1/3 of the level expected given the food supply in 2000.

Since 2003, an increasing number of wolves have been euthanized in response to depredation complaints, thus one component of a density-dependent reduction in annual growth rate may be a density-dependent increase in culling. The magnitude of these removals was small (maximum of 25 in 2006, Ruid et al. this volume). Moreover, compensatory responses typically accompany human exploitation (Boyce et al. 1999), thus it is unlikely that culling, by itself, is beginning to regulate the SLS wolf population.

Fuller et al. (2003) concluded that the question of regulatory mechanisms for wolf populations was partially one of earlier citations arguing for intrinsic (e.g., social) regulation whereas later citations argued for regulation by prey abundance, although recent analyses suggested that social factors can influence the form of density dependence (Hayes and Harestad 2000; Ellis and Post 2004). Fuller et al. (2003) concluded that this controversy had been resolved in favor of the later view that prey regulates abundance of wolves. In particular, it was the relative biomass of vulnerable prey, defined as "vulnerable young, old, and weak individuals" (Mech 1970: 318) that was a primary driver of population dynamics along with wolf density and human exploitation. In this context, the wolf population in the SLS region may be prey-limited despite high abundance of deer if a relatively small fraction of the deer herd is vulnerable (Mech 1970; Fuller et al. 2003; Vucetich and Peterson 2004).

Discussion of density dependence and equilibrium population growth is compli-cated by the fact that the southwestern boundary of the area occupied by wolves in the SLS region, currently in central Wisconsin, has not yet stabilized. The expand-ing wave-front of wolves is moving in a southwesterly direction against a gradient of generally increasing agricultural land use and human presence on the landscape (Haight et al. 1998). Under this scenario, density effects could occur if the population was increasing at a faster rate than the increase in area or quality of the range. The increased mortality risks associated with humans (Fuller et al. 2003) may be mediated somewhat by increasing densities of deer associated with agriculture and milder winters in southern and central Wisconsin. Ahead of this front, dispersing wolves are penetrating well into the intensive agricultural regions of the Midwest as evidenced by dispersing wolves showing up in Indiana, Illinois, and Missouri (Haight et al. 1998; Treves et al. this volume). These anecdotes contrast with char-acterizations of colonizing wolves as strongly philopatric with dispersers tending to colonize the edges of their parents' territories to minimize the risk of mortality associated with long-distance dispersal (Mech 1987; Hayes and Harestad 2000).

It is difficult to predict where the southern boundary of the SLS wolf population will stabilize or what combination of prey biomass and human tolerance will enable that stabilization. The spatial pattern of wolf recovery in the SLS region correlates well with the predicted spatial pattern of high-quality wolf range (Haight et al. 1998; Mladenoff et al. 1999). Given this, evidence of a density-dependent decline in growth rate may indicate that the high-quality wolf range is becoming saturated (Haight et al. 1998) and that colonists in marginal areas are contributing less to population growth. Dispersal is highly variable (Mech 1987; Wabbakken et al. 2000) and may help regulate growing wolf populations that are expanding from a core area outward (Haight et al. 1998; Hayes and Harestad 2000). Higher road density and human activity in marginal habitat of the agricultural regions (Mladenoff et al. 1995) could subject colonizing wolves to higher mortality, and reduced or delayed pack formation leading to reduced reproduction.

The suite of models predicts that the SLS wolf population will, if allowed, grow to a carrying capacity of roughly 1,300 wolves. This value is equivalent to Mladenoff et al.'s (1997) predictions of 1,131–1,424 for the carrying capacity of the SLS population. Mladenoff et al. (1997) predicted that about twice as many wolves would occur in Michigan relative to Wisconsin, but surveys since the mid-1990s suggest that proportions were closer to 50% in each state. My estimated carrying capacity of 1,300 for both states is three times the minimum goals set by both states of 450 wolves, and the portion living in Wisconsin (~650) would be nearly twice the current management goal of 350. This prediction must be interpreted with appropriate caution and skepticism. To begin with, the closure assumption of density-dependent models was not met. Also, I ignored the impact of wolves that were euthanized because of depredations, thus my characterization of both growth rate and estimates of carrying capacity as those of a completely unexploited population may be biased low. More importantly, most of the data points reflected growth at low density, thus levels of asymptotic growth may be largely artifacts of how the various model structures fit yearly growth increments at low density. That said, there is little reason from theory or experience (Hayes and Harestad 2000; Wabakkan et al. 2000; US Fish and Wildlife Service et al. 2007) to believe the SLS wolf population will abruptly level off. Thus, conservationists should expect additional growth and plan accordingly.

Despite the uncertainty in predicting asymptotic growth, the close association between increasing wolf populations and depredation complaints (Ruid et al. this volume) suggests that one can be certain that the absolute number of depredation events and their spatial distribution will increase as wolf numbers grow and colonize more agricultural regions (Haight et al. 1998). Expanding wolf populations will force managers to confront the question of population reduction to maintain wolf populations at goals below carrying capacity.

Maintaining a population at a level below carrying capacity requires determining the number of individuals that need to be removed to offset the annual recruitment of the population at the desired density. The relationship between population density and the off-take needed to sustain that density is a function primarily of growth rate and carrying capacity (Ludwig 2001). For example, given a growth rate of 1.31 and the K' estimated for the Beverton–Holt model, an additive maximum sustained yield of 92 wolves (7%) would maintain the SLS population at 60% of estimated carrying capacity (770 wolves). This is much lower than the maximum sustained harvest of 23–28% estimated by Keith (1983) or the 35% suggested by Fuller et al. (2003). Keith (1983) assumed a higher growth rate and neither author accounted explicitly for density-dependent growth. Moreover, Keith's (1983) and Fuller et al.'s (2003) estimates were made from empirical data on the population trends of exploited wolf populations thereby implying that compensatory mechanism were operating (Fuller et al. 2003). Estimates of off-take may be modified to account for different forms of density dependence and system variability (Ludwig 2001). In practice, population management using sustainable harvest must cope with incomplete or imprecise data and population and environmental stochasticity. Population management, whether deliberately or not, therefore often converges on an adaptive cycle of prediction, harvest, and evaluation. This analysis could be a useful first step.

With delisting, wolf management is likely to become even more controversial because those who are experiencing wolf depredation or who see wolf recovery as a threat to their deer hunting (Lohr et al. 1996; Williams et al. 2002) will see delisting as a timely removal of Federal pre-emption of Wisconsin's and Michigan's ability to manage the wolf population at levels below current population sizes. Wisconsin's management goal of 350 wolves (Michigan currently does not have a management goal) was established before there were empirical data on how the recovering wolf population would respond to the unique ecological and human sociological landscapes of the SLS region. Hence, re-evaluation or re-validation of state goals with respect to population growth and estimates of carrying capacity of wolves, as well as the management effort needed to stabilize a wolf population below carrying capacity, may be needed.

References

Albert, D. A. 1995. Regional landscape ecosystems of Michigan, Minnesota, and Wisconsin: a working map and classification (fourth revision: July 1994). U.S.D.A. Forest Service, North Central Forest Experiment Station, General Technical Report NC-178.

Baker, R. H. 1983. Michigan mammals. East Lansing: Michigan State University Press.

Boyce, M. S., Sinclair, A. R. E., and White, G. C. 1999. Seasonal compensation of predation and harvesting. Oikos 87:419–426.

Burnham, K. P., and Anderson, D. R. 1998. Model selection and inference: a practical information-theoretic approach. New York: Springer-Verlag.

Dennis, B., and Taper, M. L. 1994. Density dependence in time series observations of natural populations: estimation and testing. Ecological Monographs 64:205–224.

Ellis, A. M., and Post, E. 2004. Population response to climate change: linear vs. non-linear modeling approaches. BMC Ecology 4:2 (http://www.biomedcentral.com/1472-6785/4/2).

Efron, B. 1981. Nonparametric estimates of standard error: the jackknife, the bootstrap, and other methods. Biometrika 68:589–599.

Fuller, T. K. 1989. Population dynamics of wolves in north-central Minnesota. Wildlife Monographs 105.

Fuller, T. K., Mech, L. D., and Cochrane, J. F. 2003. Wolf population dynamics. In Wolves: behavior, ecology, and conservation, eds. L. D. Mech and L. Boitani, pp. 161–191. Chicago: University of Chicago Press.

Fowler, C. W. 1981. Density dependence as related to life history strategy. Ecology 62:602–610.

Ives, A. R., and Zhu, J. 2006. Statistics for correlated data: phylogenies, space, and time. Ecological Applications 16:20–32.

Haight, R. G., Mladenoff, D. J., and Wydeven, A. P. 1998. Modeling disjunct gray wolf populations in semi-wild landscapes. Conservation Biology 12:879–888.

Hayes, R. D., and Harestad, A. S. 2000. Demography of a recovering wolf population in the Yukon. Canadian Journal of Zoology 78:36–48.

Jensen, W. F., Fuller, T. K., and Robinson, W. L. 1986. Wolf, *Canis lupus*, distribution on the Ontario-Michigan border near Sault Ste. Marie. Canadian Field-Naturalist 100:363–366.

Keith, L. B. 1983. Population dynamics of wolves. In Wolves in Canada and Alaska: their status, biology, and management, ed. L. N. Carbyn, pp. 66–77. Edmonton, Alberta: Canadian Wildlife Service, Report Series Number 45.

Lohr, C., Ballard, W. B., and Bath, A. 1996. Attitudes toward gray wolf reintroduction in New Brunswick. Wildlife Society Bulletin 24:414–420.

Ludwig, D. 2001. Can we exploit sustainably? In Conservation of exploited species, eds. J. D. Reynolds, G. M. Mace, K. H. Redford, and J. G. Robinson, pp. 16–38. Cambridge: Cambridge University Press.

McCullough, D. R. 1979. The George Reserve deer herd: population ecology of a K- selected species. Ann Arbor: University of Michigan Press.

Meadow, R., Reading, R. P., Phillips, M., Mehringer, M., and Miller, B. J. 2005. The influence of persuasive arguments on public attitudes toward a proposed wolf restoration in the southern Rockies. Wildlife Society Bulletin 33:154–163.

Mech, L. D. 1970. The wolf: the ecology and behavior of an endangered species. Garden City, New York: The Natural History Press.

Mech, L. D. 1987. Wolf dispersal from a Minnesota pack. In Mammalian dispersal patterns, eds. B. D. Chepko-Sade and Z. T. Halpin, pp. 55–74. Chicago: University of Chicago Press.

Mech, L. D. 1995. The challenge and opportunity of recovering wolf populations. Conservation Biology 9:270–278.

Mech, L. D. 2006. Prediction failure of a wolf landscape model. Wildlife Society Bulletin 34:874–877.

Mech, L. D., Fritts, S. H., and Wagner, D. 1995. Minnesota wolf dispersal to Wisconsin and Michigan. American Midland Naturalist 133:368–370.

Minnesota Department of Natural Resources. 2001. Minnesota wolf management plan. St. Paul, MN.

Mladenoff, D. J., Cayton, M. K., Sickley, T. A., and Wydeven, A. P. 2006. L. D. Mech critique of our work lacks scientific validity. Wildlife Society Bulletin 34:878–881.

Mladenoff, D. J., Haight, R. G., Sickley, T. A., and Wydeven, A. P. 1997. Cause and implications of species restoration in altered ecosystems: a spatial landscape projection of wolf population recovery. BioScience 47:21–31.

Mladenoff, D. J., Sickley, T. A., Haight, R. G., and Wydeven, A. P. 1995. A regional landscape analysis and prediction of favorable gray wolf habitat in the northern Great Lakes Region. Conservation Biology 9:279–294.

Mladenoff, D. J., Sickley, T. A., and Wydeven, A. P. 1999. Predicting gray wolf landscape colonization: logistic regression models vs. new field data. Ecological Applications 9:37–44.

Potvin, M. J., Drummer, T. D., Vucetich, J. A., Beyer, D. E. Jr., Peterson, R. O., and Hammill, J. H. 2005. Monitoring and habitat analysis for wolves in Upper Michigan. Journal of Wildlife Management 69:1660–1669.

Sibly, R. M., and Hone, J. 2003. Population growth rate and its determinants: an overview. In Wildlife population growth rates, eds. R. M. Sibley, J. Hone, and T. H. Clutton-Brock, pp. 11–40. Cambridge: Cambridge University Press.

Sibly, R. M., Barker, D., Denham, M. C., Hone, J., and Pagel, M. 2005. On the regulation of populations of mammals, birds, fish, and insects. Science 309:607–608.

Skalski, J. R., Ryding, K. E., and Millspaugh, J. J. 2005. Wildlife Demography. Burlington, MA: Elsevier Academic Press.

Taper, M. L., and Gogan, P. J. P. 2002. The northern Yellowstone elk: density dependence and climatic conditions. Journal of Wildlife Management 66:106–122.

Thiel, R. P. 2001. Keepers of the wolves: the early years of wolf recovery in Wisconsin. Madison: University of Wisconsin Press.

US Fish and Wildlife Service, Nez Perce Tribe, National Park Service, Montana Fish Wildlife and Parks, Idaho Fish and Game, and USDA Wildlife Services. 2007. Rocky Mountain wolf recovery 2006 interagency annual report, eds. C. A. Sime and E. E. Bangs, USFWS, Ecological Services, Helena, MT, USA.

Vucetich, J. A., and Peterson, R. O. 2004. The influence of prey consumption and demographic stochasticity on population growth of Isle Royale wolves. Oikos 107:309–320.

Wabakkan, P., Sand, H., Liberg, O., and Bjärvall, A. 2000. The recovery, distribution, and population dynamics of wolves on the Scandinavian Peninsula 1978–1998. Canadian Journal of Zoology 79:710–725.

Wisconsin Department of Natural Resources. 2001. Management workbook for white-tailed deer, second edition. Bureaus of Wildlife Management and Integrated Science Services, Madison, WI, USA.

Williams, C. K., Ericsson, G., and Heberlein, T. A. 2002. A quantitative summary of attitudes towards wolves and their reintroduction (1972–2000). Wildlife Society Bulletin 30:575–584.

Wydeven, A. P., Schultz, R. N., and Thiel, R. N. 1995. Gray wolf (*Canis lupus*) population monitoring in Wisconsin 1979–1991. In Ecology and conservation of wolves in a changing world, eds. L. N. Carbyn, S. H. Fritts, and D. R. Seip, pp. 147–156. Edmonton, Alberta: Canadian Circumpolar Institute.

Chapter 10
Prey of Wolves in the Great Lakes Region

Glenn D. DelGiudice, Keith R. McCaffery,
Dean E. Beyer Jr., and Michael E. Nelson

10.1 Introduction

Wolves (*Canis lupus*) were abundant in the Great Lakes region just prior to early
European settlement (1800–1850). The subsequent extirpation of wolves and most
of their large prey from much of this region is just one of the many threats humans
have posed to North American wildlife by exploitation and indifference. To understand and learn from the ongoing recovery of wolves in the Great Lakes region (for
our purposes here, Minnesota, Wisconsin, and Michigan) and the historic declining
trend that preceded it (Erb and DonCarlos this volume; Wydeven et al. this volume;
Beyer et al. this volume) requires consideration of their primary prey, both historically and today. Because the nutrition afforded by food is essential to survival,
reproductive success, and population persistence, both for wild animals and
humans, it is not difficult to comprehend how during those early times, and still
today for some, ungulate prey species are often at the center of the conflict between
wolves and humans. History has taught us that it was not predation by wolves that
diminished the diversity and abundance of ungulate species in the Great Lakes
region, but rather the human appetite for ungulates, and unfortunately for wolves
and their prey, the unprecedented drive to satisfy it.

On the other hand, humans have a great capacity for conservation when that is
their true intention. However, the success of conservation efforts also relies largely
on species-specific biology, in this case, of wolves and their prey. Wolves are adaptable,
opportunistic predators, but the species that serve as prey for them in the Great
Lakes region and elsewhere has depended largely on their size, abundance, and how
"catchable" they are (Peterson and Ciucci 2003). Consequently, the relative contributions of primary and secondary prey to the diets of wolves of the Great Lakes
region, to their individual health and welfare, and the long-term persistence of their
populations have changed historically, and continue to vary seasonally, annually,
and across the landscape.

We begin this chapter with a brief description of the historic (1800s) trends in
distribution and relative abundance of the ungulates that were likely most important
in the multi-prey system of the wolves of the Great Lakes region. Our major focus,
however, is the more recent trends of white-tailed deer (*Odocoileus virginianus*),

A.P. Wydeven et al. (eds.), *Recovery of Gray Wolves in the Great Lakes*
Region of the United States,
DOI: 10.1007/978-0-387-85952-1_10, © Springer Science+Business Media, LLC 2009

the primary prey of wolves in a single-ungulate prey system that has persisted throughout the twentieth century and during their recent ongoing recovery. We discuss specific aspects of the ecology of deer that have enabled their populations to thrive despite relatively heavy exploitation by humans, increasing numbers of wolves, and a concomitant expansion of the geographic range of wolves. Our discussion is based upon management efforts and data generated from studies of coexisting white-tailed deer and wolves in the Great Lakes region.

10.2   Historic Trends of Wolf Prey

Although the diet of wolves includes a wide range of items, from grass, berries, carrion, and garbage to prey varying in size from mice to bison (*Bison bison*), their greatest dependence has been on ungulates (Peterson and Ciucci 2003). In the early 1800s, prior to the influx of Euro-American settlers, a greater diversity of ungulates inhabited the pristine vegetative communities of the Great Lakes region, including elk (*Cervus elaphus*), bison, woodland caribou (*Rangifer tarandus caribou*), white-tailed deer, moose (*Alces alces*), mule deer (*Odocoileus hemionus*), and pronghorn (*Antilocapra americana*; Dahlberg and Guettinger 1956; Hall and Kelson 1959; Fashingbauer 1965a, b, c, d, e; Idstrom 1965; Petraborg and Burcalow 1965; Baker 1983). Undoubtedly, the historic food economy of wolves of the Great Lakes region was a multi-prey system with varying contributions made by these ungulates, but we may never know the relative importance of each species or how it may have changed with time. Understanding the complexity of these systems has been facilitated by recent studies of wolf foraging behavior (Mech and Peterson 2003; Peterson and Ciucci 2003). However, predominance of ungulate species in the diet of wolves with access to multiple prey has varied across North America (e.g., Ballard et al. 1997; Dale et al. 1994; Weaver 1994; Bergerud and Elliot 1998; Kunkel et al. 1999).

Modern "guesstimates" of historic abundance and distribution of ungulates have relied largely upon commercial records of hides and meat; documented observations of early pioneers and "take" of trappers; calculated extrapolations based on densities of native Americans, their subsistence needs, and known use of ungulate parts for food, clothing, tools, and weapons; and current knowledge of these ungulates, their densities and ecology (e.g., O.J. Murie 1951; McCabe 1982; McCabe and McCabe 1984; Reeves and McCabe 1997; Mech and Peterson 2003). These estimates facilitated an understanding of how the historic ranges of these ungulates varied in size, distribution, and overlap, and how these prey collectively supported their historic coexistence with wolves throughout the Great Lakes region. Although mule deer and pronghorn ranged along the western edge of Minnesota in the 1900s (Hall and Kelson 1959; Fashingbauer 1965d,e), it is unlikely that self-sustaining populations were supported within the Great Lakes region. Consequently, these species would have been of only incidental importance to wolves as prey, and we do not discuss them further.

10.2.1 Elk, Bison, and Caribou

Elk (Manitoban and Eastern), plains bison, and woodland caribou were among the most abundant ungulates collectively, but their existence was relatively brief once exploitation by settlers began (Schorger 1954; Fashingbauer 1965a,b,c; Moran 1973) (Fig. 10.1). Elk ranged throughout Michigan's Lower Peninsula (LP), southern and western Wisconsin, and Minnesota's southern prairies and much of its transition zone (mixed hardwoods–northern conifers), but evidence supporting their occupation of northern Wisconsin and Michigan's Upper Peninsula (UP) is somewhat equivocal (O.J. Murie 1951; Schorger 1954; Fashingbauer 1965a; O'Gara and Dundas 2002). Bison were largely associated with the prairies and oak savannas of the Great Lakes region, their range being most extensive in Minnesota but extending to southeastern Wisconsin and extreme southern Michigan (Swanson 1940; Schorger 1942; Skinner and Kaisen 1947). Observations of herds ranging from hundreds to tens of thousands were recorded (Fashingbauer 1965b). Caribou were less abundant than elk or bison, but at one time they probably numbered in the thousands in the far northern boreal coniferous forests, muskegs, and bogs of the Great Lakes region and on Isle Royale (Dahlberg and Guettinger 1956; Fashingbauer 1965c; Bergerud 1978). The natural abundance of caribou declined from the northern to the southern reaches of their historic range in Minnesota, and although occasionally present near Lake Superior, the occurrence of viable populations in northern Wisconsin during recent centuries

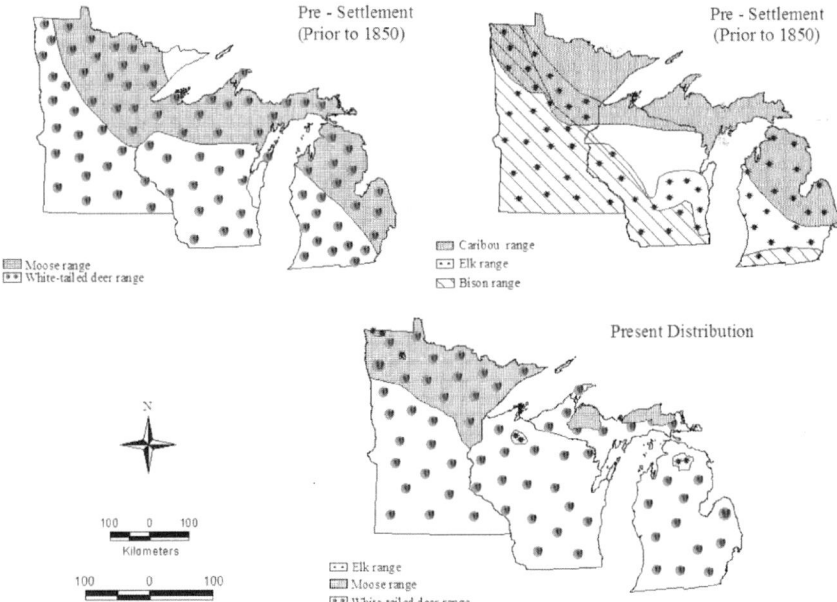

Fig. 10.1 Historic and present distributions of ungulate prey of wolves in the Great Lakes region (Minnesota, Wisconsin, and Michigan). Distributions were adapted from several sources (see text)

is doubtful (Schorger 1942). They may have occurred as far south as the southeastern LP of Michigan until the mid-1800s (Baker 1983).

Uncontrolled hunting, vegetative changes and habitat alterations by logging, uncontrolled fires, and clearing of land for cultivation and settlements were major factors in decimating these species. By the 1870s, elk were extirpated from Michigan and Wisconsin and rare in Minnesota (Caton 1877; O.J. Murie 1951; Jackson 1961). Subsequent reintroductions of elk have met with varied levels of success (Schorger 1942; Jackson 1961; Fashingbauer 1965a; Moran 1973; Lizotte 1998). Currently, populations of about 110 and 800–900 elk inhabit northern Wisconsin and Michigan's northern LP, respectively (Fig. 10.1; L. R. Stowell, Wisconsin DNR, pers. comm.; Walsh 2007). In Wisconsin, wolf predation in 2006 surpassed vehicle-kills for the first time in 10 years as the primary cause of mortality for elk (Stowell et al. 2007). Wolves have not had a significant impact on the productivity of elk herds, and between 1995 and 2006 had killed 20 elk, including 16 bulls, 2 female calves, and 2 old, barren cows (Stowell and McKay 2006). In northwestern Minnesota, a population of about 55 elk still inhabits a 177-km^2 (45-mi^2) area between Marshall and Beltrami counties, and 100–125 elk still range on the Minnesota–Manitoba border in Kittson County (Fig. 10.1; Minnesota DNR, in litt.).

By 1800, few bison remained east of the Mississippi River, and by 1830, they no longer resided regularly in Minnesota (Fashingbauer 1965b), having been extirpated from this portion of the wolf's range by uncontrolled harvest before the actual surge of settlers had begun. Caribou persisted somewhat longer than elk and bison, due largely to their remoteness on Isle Royale and in more northern portions of Minnesota. However, as settlers increasingly moved north, caribou also declined (Swanson 1940). It has also been speculated that as wolf populations increased in response to increasing numbers of deer in the north, predation by wolves also may have contributed to the loss of caribou (Bergerud 1974). Bergerud and Mercer (1989) also present evidence that meningeal worms (*Parelaphostrongylus tenuis*), a common parasite of deer that is lethal to caribou, may have contributed to their decline. Bergerud and Mercer (1989) noted that the decline of caribou and the increase in deer coincided with the end of the 500-year "Little Ice Age," which probably created climatic conditions unsuitable for caribou but favorable for deer. By the 1920s, scattered sightings of small groups of caribou, probably not even year-round residents of the region, were all that remained (Swanson 1940; Cochrane 1996).

10.2.2   White-Tailed Deer and Moose

Prior to settlement of the Great Lakes region by Europeans, white-tailed deer were most commonly associated with the scattered hardwood forests, marshes, prairies, and grassland openings of southern Minnesota, Wisconsin, and Michigan, where local densities have been estimated to have been as high as 8–19 deer per km^2 (20–50/mi^2; Fig. 10.1; Swanson 1940; Dahlberg and Guettinger 1956; Bohley 1964; Baker 1983; McCabe and McCabe 1984). Petraborg and Burcalow (1965) reported that much of northern Minnesota was uninhabited by white-tailed deer, although archaeological evidence indicates that deer were a source of food for indigenous people there during 200 BC to 1750 AD (Lukens 1973; Le Vasseur 2000; Mulholland 2000; Valppu and Rapp 2000; Arzigian and Stevenson 2003). Deer also were observed and eaten on expeditions to Minnesota in 1805 (Pike 1895) and 1831 (Schoolcraft 1834), as well as by Ojibwe living in north-central Minnesota in 1860 (Hannes 1994). In the northern forests of Wisconsin and Michigan, natural disturbances may have created openings and stimulated early successional growth frequently enough to supplement existing pine-barrens. Together, these favorable habitats may have helped to support an average of up to 6 deer per km^2 (16 deer/mi^2; Doepker et al. 1995; McCaffery 1995). Similarly, disturbance of forests in northern Minnesota also would have been conducive to habitation by deer, albeit likely at low densities.

By the late 1800s, the more accessible populations of deer in the southern portions of the Great Lakes region were nearly decimated by over-harvesting and habitat alteration. At this time, the little there was in the way of wildlife and habitat management, harvest regulations, or law enforcement had minimal impact towards conserving populations (Swanson 1940; Dahlberg and Guettinger 1956; McNeil 1962; Petraborg and Burcalow 1965; Ludwig and Isley 1983). Conversely, logging and forest fires in the central and northern forests were enhancing habitat quality for deer (Schorger 1953; Petraborg and Burcalow 1965). Deer were increasing in numbers and expanding their range northward in Minnesota. This trend of increasing abundance was short-lived in some areas such as Michigan's UP, where market-hunting, excessive cutting of hardwoods, catastrophic fires, and depletion of forest cover began to dramatically reverse the trend (Barlett 1938). By the early 1900s, the scarcity of deer in portions of the Great Lakes region prompted an era of wildlife protection, which included increased hunter education and hunting regulations, restricted and periodically closed seasons, bag limits, refuges, and greater enforcement (Dahlberg and Guettinger 1956; McNeil 1962; Petraborg and Burcalow 1965; Bersing 1966; Ludwig and Isley 1983). Predator control and increased suppression of forest fires also contributed to the increasing trend in populations of deer in the Great Lakes region that was apparent by the 1920s to 1930s (Dahlberg and Guettinger 1956).

As populations of wolves steadily declined in the northern forests throughout much of the 1900s (Fuller et al. 1992; Thiel 1993), populations of deer fluctuated, sometimes considerably, in response to varying winter severities and hunter harvest (Barlett 1938; Bersing 1966; Ludwig and Isley 1983; Langenau 1994; McCown 1994; Grund et al. 2004). In the mid-1930s, the population of deer in the UP was estimated to be 300,000, while in the forest zone of northern Minnesota the population was estimated at 473,000 deer (Barlett 1938; Ludwig and Isley 1983). In fall 1943, the population of deer in Wisconsin's northern forest may have peaked at >700,000 based on the harvest reported by Bersing (1966). Wolves were federally protected in 1974 and their numbers steadily increased from about 750 in 1970 to 3,020 in 2004 in Minnesota (Erb and DonCarlos this volume), from about 25 in 1980 to 467 in 2006 in Wisconsin (Wydeven et al. this volume), and from about 3 in 1989 to 434 in 2006 in Michigan's UP (Beyer et al. this volume). Populations of deer in the northern Great Lakes region, where recovery of wolves has been strongest, have continued to fluctuate in abundance, but have been managed consistently at relatively high densities (Fig. 10.2; M. S. Lenarz, Minnesota DNR, pers. comm.; R. E. Rolley, Wisconsin DNR, pers. comm.; R. Clute, Michigan DNR, pers. comm.). Average deer populations between 1989 and 2006 were 375,000 in northern Minnesota, 559,000 in Michigan's UP, and 444,000 in northern and central Wisconsin, and numbers were often above local management goals.

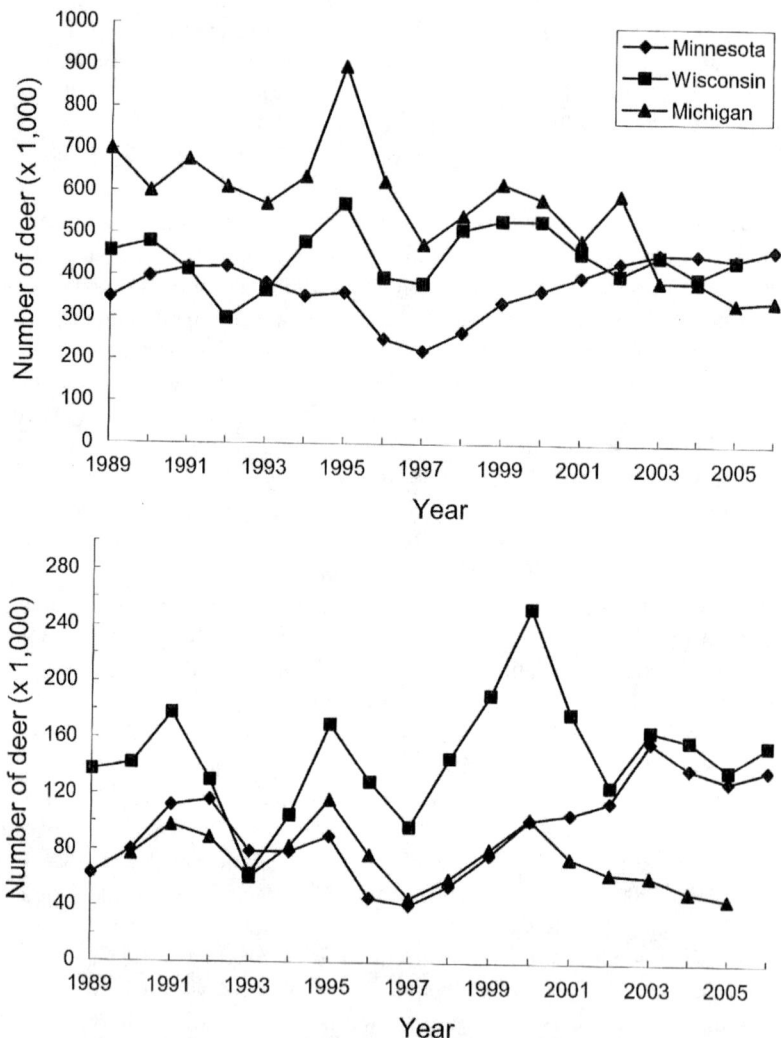

Fig. 10.2 Annual estimates of populations of white-tailed deer (*top*) primarily in wolf ranges in northern Minnesota's forest zone (pre-fawning), the northern and central forests of Wisconsin (post-harvest), and the Upper Peninsula (UP) of Michigan (pre-harvest) during wolf recovery, and annual total deer harvests (bottom, rounded to the nearest 1,000) for the same regions, 1989–2006

Presently, moose are the only other ungulates since Euro-American settlement that have continued to serve as prey of wolves in the Great Lakes region, but their importance as a primary food source is restricted to Isle Royale, where they are the only ungulate, to the Boundary Waters Canoe Area (BWCA), and to a small portion of the Superior National Forest (northeastern Minnesota), where very few white-tailed deer presently reside in winter (Mech 1966; Peterson et al. 1998; Nelson and Mech 2006). Although present in Michigan's UP, moose are rarely preyed upon by wolves, because their range overlaps little with wolf pack territories and their density is low compared to deer.

Into the mid-1800s, moose were perhaps the most common ungulates in the boreal and coniferous-deciduous forests of the northern Great Lakes region, but they also occurred from the northern to southeastern portions of Michigan's LP (Fig. 10.1; Peterson 1955; Hall and Kelson 1959; Idstrom 1965; Karns et al. 1974; Krefting 1974; Karns 1997). Moose were considered rare in extreme northeastern Minnesota while woodland caribou were common (Idstrom 1965). From the late 1800s to early 1900s, the same logging operations and wild fires that enhanced habitat for deer degraded habitat for moose, and along with increased subsistence hunting, caused moose to decline in portions of central Minnesota, northern Wisconsin, and Michigan's LP (Wood and Dice 1923; Idstrom 1965). Throughout the 1900s, Minnesota's population of moose fluctuated, but their numbers probably never exceeded 8,000–10,000 (M. S. Lenarz, Minnesota DNR, pers. comm.). In 2006, the population of northeastern Minnesota was estimated at about 8,400 moose, but declined to 6,460 moose in 2007. In northwestern Minnesota, the population of moose was estimated at around 1,000 individuals in 2000–2001, but a declining population was estimated at just under 100 moose in 2007 and this trend was expected to continue (Murray et al. 2006; Minnesota DNR, in litt.). Densities of moose are lower in the more southern portions of its range. In northern Wisconsin, moose have been rare since 1900 with no confirmed observations from 1921 to 1960, but with an estimated 30–40 animals throughout the state as of 1987 (Karns 1997). Similarly, only 25–50 moose inhabited Michigan's UP since the mid-1900s (Karns 1997; Verme 1984), that is, until translocations in 1985 and 1987 induced an increase in the population in the western UP to its current 300–400 moose (D. E. Beyer, Jr., Michigan DNR, in litt.). Moose range in the eastern UP is rather fragmented (Fig. 10.1) and currently supports a population of <100 animals. The 1,748-km^2 core range (area of high moose density) of Michigan's mainland population occurs in the western UP and is surrounded by a peripheral range of low moose density (D. E. Beyer, Jr., Michigan DNR, in litt.).

10.3   Important Prey in the Diet of Wolves in the Great Lakes Region

To begin to determine the importance of white-tailed deer, moose, and smaller prey to the existence of wolves in this region requires study of food habits. Analysis of wolf scats has been a common method used to assess the seasonal importance of specific prey in their diet (Floyd et al. 1978), and has included estimating relative frequency of occurrence (%) of prey in scat, number of individuals and biomass consumed, and their variation over time and the landscape. Diets of wolves may vary by location and by month, season, or year and are influenced, in part, by abundance of specific prey (Thompson 1952; Theberge et al. 1978; Fritts and Mech 1981; Fuller 1989). Before wolves were extirpated in Wisconsin, white-tailed deer occurred in >90% of scats annually (Thompson 1952). In early re-colonizing wolves in Wisconsin, about 55% of scat volume was deer, but samples were biased toward warmer months (Mandernack 1983). During the early to mid-1970s, deer and moose comprised at least 94% of animal biomass consumed by wolves in the Beltrami State Forest (Minnesota), located in the western portion of primary wolf range (Fritts and Mech 1981), and occurred in 70% (60% deer, 14% moose) of wolf scats in northeastern Minnesota (Van Ballenberghe et al. 1975). By all measures, white-tailed deer were the major prey of wolves during winter and summer. Occurrence of deer in scats and as percent of total biomass consumed were higher in winter (86% and 75%) than in summer (75% and 57%) in the Beltrami. Moose were the second most important prey, but biomass of moose and number of individuals consumed was greater in summer than winter (Fritts and Mech 1981). In north-central Minnesota, where moose were scarce, deer accounted for 79–98% of animal biomass consumed by wolves each month (Fuller 1989). Similarly, but derived by a different approach, a recent study of wolf predation in Michigan's UP showed that white-tailed deer were the primary prey of wolves during winter and comprised 91% of all of their kills (Huntzinger 2006).

Juvenile ungulates are important prey of wolves in most areas of the Great Lakes region. Thompson (1952) documented this by analysis of wolf scats in summers of the late 1940s in Wisconsin. During April to May in Minnesota, deer fetuses or fawns comprised only 3% of the total biomass of deer eaten, but the ratio of individuals in the diet was 1 fawn:1.6 adults. During June to July, however, fawns accounted for 80% of deer occurrences in wolf scats, their estimated biomass ingested by wolves was 1.4 times that of adults, and the ratio of the number of fawns eaten per adult was 9:1 (Fritts and Mech 1981). The increase from early spring to summer likely occurs because parturition typically occurs in late May (Kunkel and Mech 1994; Carstensen Powell and DelGiudice 2005). Percentage of biomass consisting of moose in wolf scats was highest in April-May, peaking in May at 32%, and this appeared to be the only season when biomass of moose consumed by wolves was greater than biomass of deer eaten, even though the ratio of individuals consumed was 4.5 deer per moose (Fritts and Mech 1981). The peak in consumption of biomass of moose in spring was attributable more to consumption of adult moose than consumption of calves. Consumption of moose biomass was generally higher during March to June than during the rest of the year, with calves accounting for about 40% of the remains of moose in wolf scats during May to July. The calf:adult ratio of moose was 1:0.7, but the corresponding biomass ratio was 1:7.8.

In northeastern Minnesota, percent occurrence of deer in wolf scats varied from 42% to 81% during mid-May to late September, with lowest apparent consumption during late summer (Van Ballenberghe et al. 1975). Fawns contributed most (48% of occurrences of deer in scats) to the diets of wolves during mid-June to mid-July, but reliance on fawns continued to be higher during late summer than prior to mid-June. Likewise, in north-central Minnesota, the number of fawns consumed by wolves relative to the total number of deer eaten was greatest in June and July, but fawns only accounted for an estimated 15–37% of total biomass of prey consumed (Fuller 1989).

Small prey contribute to the diversity of the diets of wolves, particularly in summer. Snowshoe hares (*Lepus americanus*), however, were part of the diet of wolves in northwestern Minnesota throughout the year. In terms of the number of individuals consumed, snowshoe hares ranked second only to deer (Fritts and Mech 1981), but they accounted for <1% of the total biomass consumed by wolves. Beavers (*Castor canadensis*) were eaten by wolves in northwestern Minnesota during April to July, but were relatively unimportant as food items due to the low number of individuals and small percent of total biomass consumed. In northeastern Minnesota, overall relative occurrence of snowshoe hares and beavers in wolf scats was 3% and 9%, respectively, but the sampling was biased toward summer (Van Ballenberghe et al. 1975). In north-central Minnesota, beavers comprised 20–47% of all items in wolf scats and an estimated 12% and 19% of biomass consumed in April and May, but <7% of the biomass consumed during the rest of the year (Fuller 1989). Snowshoe hares were killed and consumed as often as deer during most of the year, but their biomass contributed only 2–3% of the annual consumption by wolves (Fuller 1989). In Wisconsin during re-colonization, beavers comprised 17% and snowshoe hares 10% of relative volume in scats annually (Mandernack 1983).

Analyses of scats indicate that small prey are far less important than ungulates in the diets of wolves in the Great Lakes region, particularly when assessed by percent of total biomass consumed. Primarily, wolves supplement their diet with snowshoe hares and beavers to "fill the gaps" nutritionally as necessary.

10.4   Survival and Causes of Mortality of White-Tailed Deer

Our understanding of the persistence of white-tailed deer as the primary prey of wolves throughout most of the wolf range in the Great Lakes region for the past century comes from several long-term studies of deer ecology and life history in this region. Changes in deer populations are dictated by population performance, which is determined by the counteracting influences of mortality (inverse of survival) and reproduction. In turn, the dynamics of these populations of deer over time and space depend largely on the direct and indirect effects of humans (e.g., population and habitat management) and natural forces (e.g., winter severity, predation) on survival and reproductive success. Ongoing research has demonstrated that key aspects of the biology and life history of white-tailed deer have enabled them to exist at high densities in much of this region despite relatively high annual hunter harvests, the vagaries of winter weather, and predation by increasing numbers of wolves.

Our understanding of the components of population performance of white-tailed deer was limited through the mid-1900s when wolves were extirpated from Wisconsin, Michigan, and much of Minnesota. Prior to the use of radio-collars in the 1960s our knowledge of how long they lived, and when and why they died, was incomplete. An accurate technique for ageing deer by evaluating their teeth (Gilbert 1966) enabled researchers to show that wolves in northeastern Minnesota killed disproportionately more fawns and older deer compared to deer killed by hunters (Erickson et al. 1961; Mech and Frenzel 1971). On average, deer killed by wolves and hunters were 4.7 and 2.6 years old, respectively. These age-specific impacts on deer were later confirmed relative to the recovering population of wolves in north-central Minnesota (DelGiudice et al. 2002, 2006).

Since wolves were federally protected in 1974, three studies in Minnesota have examined survival and mortality factors of radio-collared white-tailed deer within the range of wolves during a combined 30 years. The first study ("Ely," 1973–1984) monitored deer in the Superior National Forest of northeastern Minnesota, which has been continuously occupied by wolves (Hoskinson and Mech 1976; Nelson and Mech 1986a,b). The second ("Bearville," 1981–1986) and third ("Grand Rapids," 1991–2003) studies monitored deer in different areas of north-central Minnesota (Fuller 1990; DelGiudice et al. 2002, 2006). The Bearville study was conducted on the edge of primary wolf range, whereas the Grand Rapids study was conducted in an area where wolves were extirpated by the 1950s to 1960s, but had become re-established just five years before that study began. Most of the deer in all the studies were radio-collared when ≥ 7 months old, but two of the studies also radio-collared newborn fawns (Kunkel and Mech 1994; Carstensen et al. 2009). Survival of radio-collared deer (adults and fawns) was assessed during 1986–1989 in northwestern Wisconsin before wolves re-colonized (Lewis and Rongstad 1998), and in Michigan's UP, deer ≥ 7 months old were radio-collared and monitored during January 1992–1995, shortly after wolf re-colonization (Van Deelen et al. 1997).

For females, the reproductive component of the deer population with the greatest impact on its dynamics, the lifetime risk of death (hazard) due to all causes is represented by a U-shaped curve (Fig. 10.3). For females in the Grand Rapids study, the greatest risk occurred within the first year of life, when 50% of the radio-collared females died by 0.8 years of age (Fig. 10.3). This risk decreased through two years of age, and thereafter remained at a relatively low constant level until increasing at six or nine years of age, depending upon the occurrence of deep snow and hunting pressure on antlerless deer (DelGiudice et al. 2006). With deep snow, risk of mortality by wolf predation increases, as the increased energy demands of moving through deep snow lowers body conditions of deer over time and impedes escape (Mech et al. 1971; Nelson and Mech 1986b; DelGiudice 1998; DelGiudice et al. 2002). Deep snow enables wolves to kill not only more females but also younger females of prime reproductive ages than during years of normal snow depths (DelGiudice et al. 2006).

Despite a young median age of survival (0.8 years old), annual survival of females after that first year in the northern Great Lakes typically was quite high.

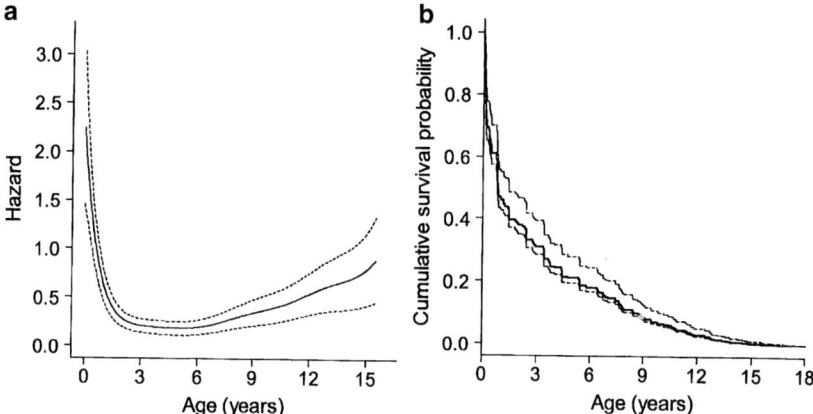

Fig. 10.3 The hazard curve (**a**) depicts the age-specific instantaneous risk of death from all causes of mortality and the survival curve (**b**) shows the probability of survival relative to age of female white-tailed deer during their life cycle in north-central Minnesota (from DelGiudice et al. 2006). The hazard curve shows that the greatest risk of death occurs during the deer's first year of life, and the survival curve shows that the median age of survival is 0.8 years old (50% probability of surviving to this age). The risk of death decreases through age two years old, remains relatively low and stable through six years of age, then steadily increases with old age (90% confidence limits are shown by *dashed lines*)

Variations in survival of deer relative to sex, age, or overall at the population level occur across the landscape and among years due to local fluctuations in seasonal mortality, winter severity, and hunting pressure. Due to the high reproductive potential of white-tailed deer (e.g., see McCafferey et al. 1998; DelGiudice et al. 2007), populations, with few exceptions, are consistently productive. In the Ely study, where hunting was mainly for antlered males, 79% of adult (≥1 year old) females survived annually compared to 69% at Bearville where females were vulnerable to antlerless permit hunting. In Michigan's UP, where hunting was also mainly for antlered males, and in northwestern Wisconsin, where hunting pressure was relatively light, annual survival of adult females has been high (77–89%). As one would expect, annual survival of adult males (~45%) was lower than for females in the Ely, Bearville, and Wisconsin studies, where each hunter was permitted to shoot one male. In the UP, heavy hunting pressure contributed to an even lower annual survival of males (22–25%). Fewer fawns than older deer survived each year (31% at Ely and 22% at Bearville).

For white-tailed deer in the northern part of their geographic range, parturition (fawning) occurs predominantly in late May to early June (Kunkel and Mech 1994; Carstensen Powell and DelGiudice 2005), and the majority of mortality for fawns in their first year occurred within the first six months when black bears (*Ursus americanus*), bobcats (*Lynx rufus*), wolves, and "unknown" predators (insufficient evidence to identify) killed at least 51% of fawns (Kunkel and Mech 1994, Carstensen et al. 2009). Conversely, adult deer rarely died in the period from June to start of the November firearm season; survival rates were 95–97% for females and 88–95% for males in the Ely and Bearville studies in northern Minnesota, and 90–100% for females and 83–100% for males in the UP.

The first major period of mortality for adult white-tailed deer occurs during the firearms hunting season in November, which allows unlimited shooting of males and a variable, limited kill of antlerless deer. Hunters shot about 30% of adult males and up to 10% of adult females during the Ely and Bearville studies; however, including other causes of death, survival rates during this period were 82–98% for females and 53–68% for males. The Grand Rapids study did not include fawns, and hunters there took between 0% and 16% of adult females annually, although in a year of increased antlerless permits, 41% were shot. In the UP, hunters harvested 51% and 72% of the yearling and adult males, respectively, but only 5% and 4% of the yearling and adult females. Throughout the range of wolves in the Great Lakes region, total annual hunter-harvest of white-tailed deer by all methods (i.e., archery, firearm, muzzleloader) has fluctuated but remained high during wolf recovery, averaging 96,000 deer (1989–2006) in Minnesota, 148,000 deer (1989–2006) in Wisconsin, and 74,300 deer (1990–2005) in Michigan (Fig. 10.2).

The next major period of mortality for adults occurred during December to May; snow covered the ground during this period, except for May. Wolves killed 7% (0–13%) of all fawn and adult females during most winters of the Grand Rapids study, but during a historically severe winter, mortality by wolf predation was 23%. Miscellaneous causes of mortality, including bobcat predation, vehicle collisions, and poor nutrition accounted for up to 14% of winter mortality. Overall, annual changes in winter severity accounted for 52% of the variability in winter mortality of adult (≥1 year old) female deer over a 16-year period (DelGiudice et al. 2006). In the Ely study, wolves killed 51% of fawns during January to May, and wolves killed 9–14% of yearling and adult females, and 23–32% of yearling and adult males during December to May. Poor nutrition killed only 2% of fawns during January to May in the Ely study. Wolves killed up to 30% of all deer during severe winters and up to 9% during mild winters. Evidence from the Ely study also indicated that adult deer were particularly vulnerable to wolf predation during fall migration (Nelson and Mech 1991). Killing of prey with little to no consumption of the carcass (surplus killing) is atypical for wolves, but has been documented in the northern Great Lakes region during unusually severe winters (Mech et al. 1971; DelGiudice 1998). Extreme deterioration of body condition in deer was implicated as the major link between snow conditions (depth and penetrability) and excessive killing by wolves (DelGiudice 1998). In the Michigan UP, where wolf densities were still relatively low, wolf predation accounted for only one death of radio-collared deer. Despite relatively mild winters in the Michigan study, losses of 5–12% of fawns due to poor nutrition were documented. In northwestern Wisconsin, survival of deer was high overall (85%) during mild to moderate winters, and no wolf predation occurred on the sample of radio-collared deer. The mortality recorded was attributable to coyote predation, starvation, and accidents (e.g., breaking through ice). In the northern forests of the Great Lakes region, hunter-harvests of antlered deer have been inversely related to snow depths or a combination of snow depths and cold ambient temperatures (i.e., winter severity index) during the previous winter (Creed et al. 1984; Nelson and Mech 2006). This was indicative of the greater losses of deer from the populations during winters of increased severity (Lenarz 2003), although it is not known whether those losses were attributable to wolf predation or other mortality factors.

Considering the variety of mortality factors impacting populations of deer in the northern parts of their range, the reported longevity of some deer in the Great Lakes region, and the number that reach old age, can seem surprising. Indeed, up to 15% of live-captured and radio-monitored females in the studies in Minnesota discussed above were ≥10 years old, and ages up to 19 years old were recorded for recovered carcasses (Nelson and Mech 1990; DelGiudice et al. 2007). Similar observations of longevity have been documented in Michigan's UP (Ozoga 1969; Van Deelen et al. 1996).

10.5   The Future

Decades of research and management focused on the ecology of white-tailed deer and other prey of wolves within the Great Lakes region and beyond continue to contribute to the information base necessary for their effective and ecologically sound management as wolves increase in number and expand their range. The primary focus of wolves on deer in most of the Great Lakes region makes this predator–prey system relatively simple compared to more complex multi-ungulate prey systems involving wolves in other regions of North America. Yet, a critical component of our current understanding of the relationship between white-tailed deer and wolves is the inherent variation in their numerous interactions over space and time. Although many of the studies conducted during the past 30 years in different areas of Minnesota revealed a similar influence of wolves on mortality of deer, the magnitude of that influence varied with demographic (e.g., health, sex, and age structure of deer populations) and environmental factors (e.g., winter severity) and their effect on prey vulnerability (Mech and Peterson 2003). Importantly, relative numbers of predators and prey can play a pivotal role in the dynamics of prey populations (Mech 1970; Van Ballenberghe 1987; Eberhardt 1997). Assuming an annual kill-rate of 18 deer per wolf (Fuller 1989; Mech and Peterson 2003), the most recently estimated numbers of wolves in Minnesota (3,000 wolves), Wisconsin (540 wolves), and Michigan (509 wolves) are estimated to have reduced the pre-hunter-harvest deer populations in those three states during 1989–2006 by <15%, <1.8%, and about 1.3%, respectively. Such factors are not directly or fully controllable by management, consequently, as wolves become established in local areas and predation on deer increases, their impact, particularly relative to winter severity, should be considered in local management strategies (e.g., harvest) for deer (Nelson and Mech 1986a, b; Kunkel and Mech 1994; DelGiudice et al. 2002, 2006). Although wolves may have their greatest impact on the demographics of deer and other prey through

Fig. 10.4 Typical sparse remains within 48 h of wolves preying on an adult white-tailed doe (*left*) and a white-tailed doe with her surviving fawn (*right*) in late winter in the northern Great Lakes. Even though white-tailed deer are the primary prey for wolves and primary game for hunters in the Great Lakes region, the high reproductive potential and success of these deer and their survival capabilities have enabled their populations to thrive as wolves recover. *Left photo by Glenn D. DelGiudice and right photo by William E. Berg*

predation on young-of-the-year (Pimlott 1967; Mech 1970), recent studies have shown that mortality of neonatal deer by specific predator species also varies over time and across a landscape of relatively high densities of wolves (Kunkel and Mech 1994; Carstensen et al. 2009). The thriving deer populations in the parts of the Great Lakes region currently occupied by wolves are the result of high fertility and fecundity (McCaffery et al. 1998; DelGiudice et al. 2007) and frequently occurring mild winters (Fig. 10.4), in spite of predation of deer by wolves. Indeed, the high reproductive output of local populations may be in part an indirect effect of wolf predation on densities of deer relative to carrying capacity (Mech and Peterson 2003). The increase of wolves to numbers beyond recovery goals set for the Great Lakes states, along with concomitant record high populations of deer, provides conclusive evidence that wolves and deer can fulfill their natural relationship as predator and prey in this region.

Acknowledgments We would like to thank P. R. Krausman and M. S. Lenarz for their critical reviews of an early draft of this chapter and their constructive comments. We also appreciate the technical assistance of B. A. Sampson in developing the prey distribution figure. Finally, we would like to acknowledge the many researchers and managers whose efforts over the decades in the Great Lakes region contributed to our current understanding of wolves and their prey and their increasingly sound management.

References

Arzigian, C. M., and Stevenson, K. P. 2003. Minnesota's Indian mounds and burial sites: a synthesis of prehistoric and early historic archaeological data. Publication no. 1. St. Paul: The Minnesota Office of the State Archaeologist.
Baker, R. H. 1983. Michigan mammals. East Lansing: Michigan State University Press.

Ballard, W. B., Ayres, L. A., Krausman, P. R., Reed, D. J., and Fancy, S. G. 1997. Ecology of wolves in relation to a migratory caribou herd in northwest Alaska. Wildlife Monographs 135:1–47.

Barlett, I. H. 1938. Whitetails, presenting Michigan's deer problem. Lansing: Michigan Department of Conservation Game Division. 64pp.

Bergerud, A. T. 1974. The decline of caribou in North America following settlement. Journal of Wildlife Management 38:757–770.

Bergerud, A. T. 1978. Caribou. In Big game of North America: Ecology and management, eds. J. L. Schmidt and D. L. Gilbert, pp. 83–101. Harrisburg: Stackpole Books.

Bergerud, A. T., and Elliott, J. P. 1998. Wolf predation in a multiple-ungulate system in northern British Columbia. Canadian Journal of Zoology 76:1551–1569.

Bergerud, A. T., and Mercer, W. E. 1989. Caribou introductions in eastern North America. Wildlife Society Bulletin 17:111–120.

Bersing, O. S. 1966. A century of Wisconsin deer. Madison: Wisconsin Conservation Department. 272pp.

Bohley, L. 1964. The great Hinkley hunt. Ohio Conservation Bulletin 18:11, 31–32.

Carstensen Powell, M., and DelGiudice, G. D. 2005. Birth, morphological, and blood characteristics of free-ranging white-tailed deer neonates. Journal of Wildlife Diseases 41:171–183.

Carstensen, M., DelGiudice, G. D., Sampson, B. A., and Kuehn, D. W. 2009. Understanding survival, birth characteristics, and cause-specific mortality of northern white-tailed deer neonates. Journal of Wildlife Management 73: In Press.

Caton, J. D. 1877. The antelope and deer and America. New York: Forest and Stream Publication Company.

Cochrane, J. F. 1996. Woodland caribou restoration at Isle Royale National Park. United States National Park Service Technical Report No. 96-03. 85pp.

Creed, W. A., Haberland, F., Kohn, B. E., and McCaffery, K. R. 1984. Harvest management: the Wisconsin experience. In White-tailed deer: ecology and management, ed. L. K. Halls, pp. 243–260. Harrisburg: Stackpole Books.

Dahlberg, B. L., and Guettinger, R. C. 1956. The white-tailed deer in Wisconsin. Madison: Wisconsin Conservation Department. 282pp.

Dale, B. W., Adams, L. G., and Bowyer, R. T. 1994. Functional response of wolves preying on barren-ground caribou in a multiple prey ecosystem. Journal of Animal Ecology 63:644–652.

DelGiudice, G. D. 1998. Surplus killing of white-tailed deer by wolves in northcentral Minnesota. Journal of Mammalogy 79:227–235.

DelGiudice, G. D., Fieberg, J., Riggs, M. R., Carstensen Powell, M., and Pan, W. 2006. A long-term age-specific survival analysis of female white-tailed deer. Journal of Wildlife Management 70:1556–1568.

DelGiudice, G. D., Lenarz, M. S., and Carstensen Powell, M. 2007. Age-specific fertility and fecundity in northern free-ranging white-tailed deer: evidence for reproductive senescence? Journal of Mammalogy 88:427–435.

DelGiudice, G. D., Riggs, M. R., Joly, P., and Pan, W. 2002. Winter severity, survival, and cause-specific mortality of white-tailed deer in north-central Minnesota. Journal of Wildlife Management 66:698–717.

Doepker, R. V., Beyer, D. E., Jr., and Donovan, M. 1995. Deer population trends in Michigan's Upper Peninsula. Michigan Department of Natural Resources Wildlife Division Report No. 3254. Lansing. 11pp.

Eberhardt, L. L. 1997. Is wolf predation ratio-dependent? Canadian Journal of Zoology 75:1940–1944.

Erickson, A. W., Gunvalson, V. E., Stenlund, M. H., Burcalow, D. W., and Blankenship, L. H. 1961. The white-tailed deer of Minnesota. St. Paul: Minnesota Department of Conservation Technical Bulletin 5. 64pp.

Fashingbauer, B. A. 1965a. The elk in Minnesota. In Big game in Minnesota, ed. J. B. Moyle, pp. 99–132. St. Paul: Minnesota Department of Conservation.

Fashingbauer, B. A. 1965b. The bison in Minnesota. In Big game in Minnesota, ed. J. B. Moyle, pp. 167–173. St. Paul: Minnesota Department of Conservation.

Fashingbauer, B. A. 1965c. The caribou in Minnesota. In Big game in Minnesota, ed. J. B. Moyle, pp. 133–166. St. Paul: Minnesota Department of Conservation.

Fashingbauer, B. A. 1965d. The mule deer in Minnesota. In Big game in Minnesota, ed. J. B. Moyle, pp. 49–56. St. Paul: Minnesota Department of Conservation.

Fashingbauer, B. A. 1965e. The pronghorn antelope in Minnesota. In Big game in Minnesota, ed. J. B. Moyle, pp. 175–178. St. Paul: Minnesota Department of Conservation.

Floyd, T. J., Mech, L. D., and Jordan, P. A. 1978. Relating wolf scat content to prey consumed. Journal of Wildlife Management 42:528–532.

Fritts, S. H., and Mech, L. D. 1981. Dynamics, movements, and feeding ecology of a newly protected wolf population in northwestern Minnesota. Wildlife Monographs 80: 1–79.

Fuller, T. K. 1989. Population dynamics of wolves in north-central Minnesota. Wildlife Monographs 105:1–41.

Fuller, T. K. 1990. Dynamics of a declining white-tailed deer population in north-central Minnesota. Wildlife Monographs 110: 1–37.

Fuller, T. K., Berg, W. E., Radde, G. L., Lenarz, M. S., and Joselyn, G. B. 1992. A history and current estimate of wolf distribution and numbers in Minnesota. Wildlife Society Bulletin 20:42–55.

Gilbert, F. F. 1966. Aging white-tailed deer by annuli in the cementum of the first incisor. Journal of Wildlife Management 30:200–202.

Grund, M. D., Cornicelli, L., and Osborn, R. G. 2004. History of Minnesota deer hunting and management. Internal report. St. Paul: Minnesota Department of Natural Resources. 27pp.

Hall, E. R., and Kelson, K. R. 1959. The mammals of North America. Volume 2. New York: Ronald Press Company.

Hannes, S. M. 1994. The faunal analysis of the Horseshoe Bay Site: a subsistence study of a nineteenth century fur trading post. M.S. Thesis, University of Iowa, Iowa City, Iowa. 153pp.

Hoskinson, R. L., and Mech, L. D. 1976. White-tailed deer migration and its role in wolf predation. Journal of Wildlife Management 40:429–441.

Huntzinger, B. A. 2006. Sources of variation in wolf kill rates of white-tailed deer during winter in the Upper Peninsula, Michigan. MS Thesis, Michigan Technological University, Houghton.

Idstrom, J. M. 1965. The moose in Minnesota. In Big game in Minnesota, ed. J. B. Moyle, pp. 57–98. St. Paul: Minnesota Department of Conservation.

Jackson, H. H. T. 1961. Mammals of Wisconsin. Madison: University of Wisconsin Press.

Karns, P. D. 1997. Population distribution, density and trends. In Ecology and management of the North American moose, eds. A. W. Franzmann and C. C. Schwartz, pp. 125–139. Washington D. C.: Smithsonian Institution Press.

Karns, P. D., Haswell, H., Gilbert, F. F., and Patton, A. E. 1974. Moose management in the coniferous-deciduous ecotone of North America. Nat. Can. (Que.) 101:643–656.

Krefting, L. W. 1974. Moose distribution and habitat selection in North America. Nat. Can. (Que.) 101:81–100.

Kunkel, K. E., and Mech, L. D. 1994. Wolf and bear predation on white-tailed deer fawns in northeastern Minnesota. Canadian Journal of Zoology 72:1557–1565.

Kunkel, K. E., Ruth, T. K., Pletcher, D. H., and Hornocker, M. G. 1999. Winter prey selection by wolves and cougars in and near Glacier National Park, Montana. Journal of Wildlife Management 63:901–910.

Langenau, E. E., Jr. 1994. 100 years of deer management in Michigan. Wildlife Division Report No. 3213. Lansing: Michigan Department of Natural Resources. 15pp.

Lenarz, M. S. 2003. White-tailed deer of Minnesota's forested zone: harvest, population trends, and modeling. St. Paul: Minnesota Department of Natural Resources.

Le Vasseur, A. K. 2000. 10,000 years in the headwaters: archaeology on the Chippewa National Forest. The Minnesota Archaeologist 59:11–21.

Lewis, T. L., and Rongstad, O. J. 1998. Effects of supplemental feeding on white-tailed deer, Odocoileus virginianus, migration and survival in northern Wisconsin. Canadian Field-Naturalist 112:75–81.

Lizotte, T. E. 1998. Productivity, survivorship, and winter feeding ecology of an experimentally reintroduced elk herd in northern Wisconsin. MS Thesis, University of Wisconsin, Stevens Point.

Ludwig, J., and T. Isley. 1983. Minnesota's rich deer and deer hunting history. In Minnesota deer classic record book, ed. M. LaBarbara, pp. 9–21. Minneapolis: Minnesota Wildlife Heritage Foundation.

Lukens, P. W., Jr. 1973. The vertebrate fauna from Pike Bay Mound, Smith Mound 4, and McKinstry Mound. In The laurel culture in Minnesota, ed. J. B. Stoltman, pp. 37–45. Minneapolis: Minnesota Historical Society.

Mandernack, B. A. 1983. Food habits of Wisconsin wolves. MS Thesis, University of Wisconsin, Eau Claire.

McCabe, R. E. 1982. Elk and Indians: historical values and perspectives. In Elk of North America: ecology and management, eds. J. W. Thomas, and D. E. Toweill, pp. 61–123. Harrisburg: Stackpole Books.

McCabe, R. E., and T. R. McCabe. 1984. Of slings and arrows: an historical retrospection. In White-tailed deer: ecology and management, ed. L. W. Hall, pp. 19–72. Harrisburg: Stackpole Books.

McCaffery, K. R. 1995. History of deer populations in northern Wisconsin. In Hemlock ecology and management: proceedings of a regional conference on ecology and management of eastern hemlock, eds. G. Mroz and J. Martin, pp. 109–114. Houghton: Michigan Technological University.

McCaffery, K. R. Ashbrenner, J. E., and Rolley, R. E. 1998. Deer reproduction in Wisconsin. Transactions of the Wisconsin Academy of Sciences, Arts, and Letters. 86:249–261.

McCown, W. ed. 1994. Wisconsin's deer management program: the issues involved in decision-making. Publication RS-911–94. Madison: Wisconsin Department of Natural Resources. 31pp.

McNeil, R. J. 1962. Population dynamics and economic impact of deer in southern Michigan. Game Division Report No. 2395. Lansing: Michigan Department of Conservation. 143pp.

Mech, L. D. 1966. The wolves of Isle Royale. Washington, D. C.: U. S. National Park Service Fauna Series No. 7.

Mech, L. D. 1970. The wolf: the ecology and behavior of an endangered species. Minneapolis: University of Minnesota Press.

Mech, L. D., and Frenzel, L. D., Jr. 1971. An analysis of the age, sex, and condition of deer killed by wolves in northeastern Minnesota. In Ecological studies of the timber wolf in northeastern Minnesota, eds. L. D. Mech and L. D. Frenzel, pp. 35–51. St. Paul, Minnesota: U. S. Forest Service Research Paper NC-52.

Mech, L. D., Frenzel, L. D., Jr., and Karns, P. D. 1971. The effect of snow conditions on the vulnerability of white-tailed deer to wolf predation. In Ecological studies of the timber wolf in northeastern Minnesota, eds. L. D. Mech and L. D. Frenzel, pp. 51–59. St. Paul, Minnesota: U. S. Forest Service Research Paper NC-52.

Mech, L. D., and Peterson, R. O. 2003. Wolf-prey relations. In Wolves: behavior, ecology, and conservation, eds. L. D. Mech and L. Boitani, pp. 131–160. Chicago: The University of Chicago Press.

Moran, R. J. 1973. The Rocky Mountain elk in Michigan. Wildlife Division of Research and Development Report no. 267. East Lansing: Michigan Department of Natural Resources. 93pp.

Mulholland, S. C. 2000. The Arrowhead since the glaciers: the prehistory of northeastern Minnesota. The Minnesota Archaeologist 59:1–10.

Murie, O. J. 1951. The Elk of North America. Harrisburg: Stackpole Company.

Murray, D. L., Cox, E. W., Ballard, W. B., Whitlaw, H. A., Lenarz, M. S., Custer, T. W., Barnett, T., and Fuller, T. K. 2006. Pathogens, nutritional deficiency, and climate influences on a declining moose population. Wildlife Monographs 166:1–30.

Nelson, M. E., and Mech, L. D. 1986a. Mortality of white-tailed deer in northeastern Minnesota. Journal of Wildlife Management 50:691–698.

Nelson, M. E., and Mech, L. D. 1986b. Relationship between snow depth and gray wolf predation on white-tailed deer. Journal of Wildlife Management 50:471–474.

Nelson, M. E., and Mech, L. D. 1990. Weights, productivity, and mortality of old white-tailed deer. Journal of Mammalogy 71:689–691.

Nelson, M. E., and Mech, L. D. 1991. Wolf predation risk associated with white-tailed deer movements. Canadian Journal of Zoology 69:2696–2699.

Nelson, M. E., and Mech, L. D. 2006. A 3-decade dearth of deer (*Odocoileus virginianus*) in a wolf (*Canis lupus*)-dominated ecosystem. American Midland Naturalist 155:373–382.

O'Gara, B. W., and Dundas, R. G. 2002. Distribution: past and present. In North American elk: ecology and management, eds. D. E. Toweill and J. W. Thomas, pp. 67–119. Washington, D. C.: Smithsonian Institution Press.

Ozoga, J. J. 1969. Some longevity records for female white-tailed deer in northern Michigan. Journal of Wildlife Management 33:1027–1028.

Peterson, R. L. 1955. North American moose. Toronto: University of Toronto Press.

Peterson, R. O., and Ciucci, P. 2003. The wolf as a carnivore. In Wolves: behavior, ecology, and conservation, eds. L. D. Mech and L. Boitani, pp. 104–130. Chicago: The University of Chicago Press.

Peterson, R. O., Thomas, N. J., Thurber, J. M., Vucetich, J. A., and Waite, T. A. 1998. Population limitation and the wolves of Isle Royale. Journal of Mammalogy 79:828–841.

Petraborg, W. H., and Burcalow, D. W. 1965. The white-tailed deer in Minnesota. In Big game in Minnesota, ed. J. B. Moyle, pp. 11–48. St. Paul: Minnesota Department of Conservation.

Pike, Z. M. 1895. The expeditions of Zebulon Pike to headwaters of the Mississippi River, through Louisiana Territory, and in New Spain, during the years 1805–6–7. New York: F. P. Harper.

Pimlott, D. H. 1967. Wolf predation and ungulate populations. American Zoology 7:267–278.

Reeves, H. M., and McCabe, R. E. 1997. Of moose and man. In Ecology and management of the North American moose, eds. A. W. Franzmann, and C. C. Schwartz, pp. 1–75. Washington, D. C.: Smithsonian Institution Press.

Schoolcraft, H. R. (1834). 1858. Schoolcraft's expedition to Lake Itasca: the discovery of the source of the Mississippi, ed. P. P. Mason. East Lansing: Michigan State University Press.

Schorger, A. W. 1942. Extinct and endangered mammals and birds of the Upper Great Lakes Region. Transactions of the Wisconsin Academy of Science, Arts, and Letters 34:23–44.

Schorger, A. W. 1953. The white-tailed deer in early Wisconsin. Transactions of the Wisconsin Academy of Science, Arts, and Letters. 42:197–247.

Schorger, A. W. 1954. The elk in early Wisconsin. Transactions of the Wisconsin Academy of Science and Arts, Letters 43:5–23.

Skinner, M. F., and Kaisen, O. C. 1947. The fossil Bison of Alaska and preliminary revision of the genus. Bulletin of the American Museum of Natural History 89:155–163.

Stowell, L. R., and McKay, M. 2006. Eleven years of elk mortality characteristics for Wisconsin elk restoration project and their management implications. In 11th Annual Eastern Elk Workshop. Higgins Lake: Michigan Department of Natural Resources.

Stowell, L. R., McKay, M., and Jonas, K. W. 2007. Elk considerations regarding trail use in the Great Divide District of the Chequamegon-Nicolet National Forest. Hayward, Wisconsin: Wisconsin Department of Natural Resources Report to the United States Forest Service. 31pp.

Swanson, E. B. 1940. The use and conservation of Minnesota game, 1850–1900. Ph.D. Dissertation, University of Minnesota, St. Paul.

Theberge, J. B., Oosenbrug, S. M., and Pimlott, D. H. 1978. Site and seasonal variation in foods of wolves, Algonquin Park, Ontario. Canadian Field-Naturalist 92:91–94.

Thiel, R. P. 1993. The timber wolf in Wisconsin: the death and life of a majestic predator. Madison: The University of Wisconsin Press.

Thompson, D. Q. 1952. Travel, range, and food habits of timber wolves in Wisconsin. Journal of Mammalogy 33:429–442.

Valppu, S. H., and Rapp, G. 2000. Paleoenthnobotanical context and dating of the Laurel use of wild rice: the Big Rice site. The Minnesota Archaeologist 59:81–87.

Van Ballenberghe, V. A. 1987. Effects of predation on moose numbers: a review of recent North American studies. Swedish Wildlife Research 1 (Supplement):431–460.

Van Ballenberghe, V., Erickson, A. W., and Byman, D. 1975. Ecology of the timber wolf in northeastern Minnesota. Wildlife Monographs 43: 1–44.

Van Deelen, T. R., Campa, H., III, Hamady, M., and Haufler, J. B. 1996. Longevity of white-tailed deer, *Odocoileus virginianus*, does in Michigan. Canadian Field-Naturalist 110:630–633.

Van Deelen, T. R., Campa, H., III, Haufler, J. B., and Thompson, P. D. 1997. Mortality patterns of white-tailed deer in Michigan's Upper Peninsula. Journal of Wildlife Management 61:903–910.

Verme, L. J. 1984. Some background on moose in Upper Michigan. Michigan Department of Natural Resources, Wildlife Division Report No. 2973. 6pp.

Walsh, D. P. 2007. Population estimation and fixed kernel analyses of elk in Michigan. Ph.D. Dissertation, Michigan State University, East Lansing.

Weaver, J. L. 1994. Ecology of wolf predation amidst high ungulate diversity in Jasper National Park, Alberta. Ph.D. Dissertation, University of Montana, Missoula.

Wood, N. A., and Dice, L. R. 1923. Records of the distribution of Michigan mammals. Michigan Academy of Science, Arts, and Letters 3:425–469.

Chapter 11
Factors Influencing Homesite Selection by Gray Wolves in Northwestern Wisconsin and East-Central Minnesota

David E. Unger, Paul W. Keenlance, Bruce E. Kohn, and Eric M. Anderson

Preface One of the most critical aspects of population dynamics in any animal species is the birth and successful rearing of young. Therefore, understanding the characteristics of areas where wolves give birth and rear pups (den and rendezvous sites) is important for proper management. In the Great Lakes region, the gray wolf has made a remarkable recovery, from a small remnant population in northeastern Minnesota to the recolonization of most of northern Wisconsin and the Upper Peninsula of Michigan. In this chapter, we review relevant literature on wolf dens and rendezvous sites and attempt to determine those factors most critical in the selection of homesites in the upper Great Lakes region.

11.1 Introduction

Much research has been conducted on den and rendezvous sites (collectively, "homesites") of gray wolves (*Canis lupus*). However, with few exceptions (e.g., Norris et al. 2002; Theuerkauf et al. 2003; Capitani et al. 2006), this research has been essentially descriptive in nature, with little or no attempt to quantify those characteristics selected by wolves.

Den sites are often burrows in the ground, but wolves have also been known to den in beaver lodges, hollow logs, beaver dams, caves, or open pits (Joslin 1967; Mech 1970, 1993; Peterson 1977). It has been suggested that the den area is selected for its slope, aspect, sandy soil, and adequate drainage. Norris et al. (2002) found that wolves selected areas of pine and suggested that dens be protected at a relatively large scale. Theuerkauf et al. (2003) found that wolves selected dry conifer forests for both den and rendezvous sites. Rendezvous sites have been described as grassy areas, ~0.5 ha in size, with semiopen canopy. With few exceptions (Van Ballenberghe et al. 1975; Ballard and Dau 1983), rendezvous sites were found in lowland areas bordering bogs, beaver ponds, or wetlands with open water, a large system of trails, and beds or play areas where pups trampled extensive areas of grass (Joslin 1966).

A.P. Wydeven et al. (eds.), *Recovery of Gray Wolves in the Great Lakes Region of the United States*,
DOI: 10.1007/978-0-387-85952-1_11, © Springer Science+Business Media, LLC 2009

In addition to these habitat characteristics, there is evidence that homesites may also be affected by spatial factors. Ballard and Dau (1983) and Gehring (1995) noted that den sites tended to be located in roughly the center of a wolf's territory. By contrast, Ciucci and Mech (1992) suggested that dens were randomly distributed within the territorial boundaries. Peterson et al. (1984) found that the distance from the natal den to the first rendezvous site was < 2 km, with successive rendezvous sites being located farther and farther from the natal den. Groebner (1991) noted that rendezvous sites fell in the center of a male wolf's territory, but along the edges of a female wolf's territory. Theuerkauf et al. (2003) found that wolves selected areas away from villages, forest edges, and intensively used roads for both den and rendezvous sites.

These findings suggest that multiple spatial and habitat factors affect homesite placement. Characteristics such as location within a wolf's territory and proximity to features such as water, roads, and a particular habitat type may affect the spatial placement of a homesite. Variables such as habitat type, level of fragmentation, human disturbance, and prey density may affect the gross placement of the site within the greater landscape. Microhabitat variables such as vegetation, visibility, availability of water, and other features may determine the specific physical location of the homesite. Others have suggested that it is important to investigate both spatial and habitat factors affecting resource use and homesite placement (Clark et al. 1993; Mladenoff et al. 1995; Arjo and Pletscher 2004).

As homesites may determine the reproductive success of a wolf pack (Harrington and Mech 1982), understanding the factors affecting their placement may prove critical to managers as wolves continue to expand numerically and geographically in the Great Lakes region. Our objectives were to characterize gray wolf den and rendezvous sites in northwestern Wisconsin and east-central Minnesota and suggest what features most strongly affect site selection.

11.2   Study Area

Research was conducted in northwestern Wisconsin and east-central Minnesota (Fig. 11.1). The habitat in the study area (21,591 km^2) is primarily a patchwork of second growth northern deciduous (aspen-birch, *Populus tremuloides-Betula papyrifera*; sugar maple, *Acer saccharum*) and coniferous (white pine, *Pinus strobus*; balsam fir, *Abies balsamea*) forest, wetland deciduous shrubs (*Alnus rugosa, Salix, Fraxinus pennsylvanica*), wetland forest (*Thuja occidentalis, Picea mariana, Fraxinus nigra, Ulmus rubra*), emergent wet meadow (*Carex, Calamagrostis*), bogs (Ericaceae shrubs, *Sphagnum*), and agricultural lands (Curtis 1959). The topography in the area is a rolling plain with elevations mostly from 250 to 500 m above sea level. Land ownership includes private land, county and state forests, private industrial forest land, tribal lands, and federal land including the St. Croix National Riverway. Road densities within the study area range from 0 to 1.5 km/km^2 and human density is low, with an average of seven people/km^2 (Mladenoff et al. 1995).

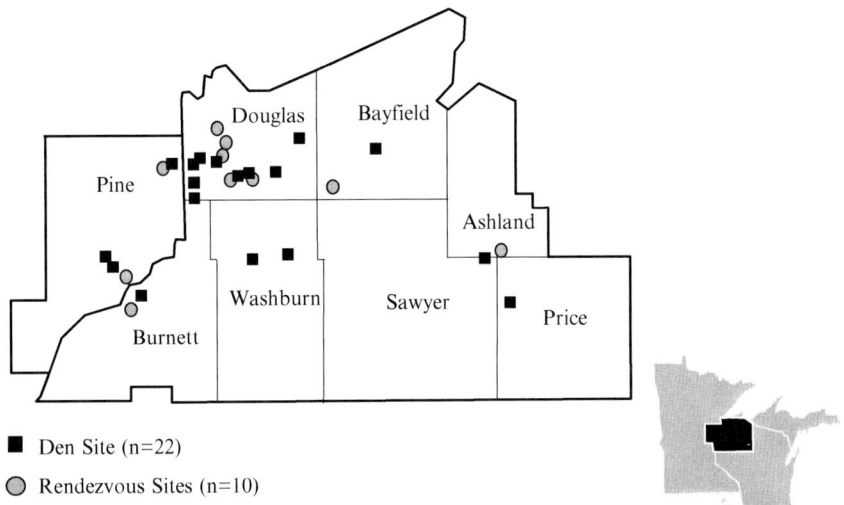

Fig. 11.1 Study area including the locations of timber wolf den and rendezvous sites investigated from July 1994 to August 2001 in counties (*thin lines*) of northwestern Wisconsin and east-central Minnesota (*bold lines*)

Available prey species include white-tailed deer (*Odocoileus virginianus*) and beavers (*Castor canadensis*), both of which have been shown to be primary food sources for wolves in the Great Lakes region (Mandernack 1983; Fuller 1989a; DelGiudice et al., this volume). Estimates of deer densities within the study area have ranged from 9.7 to 13.4/km² ($\overline{X}$ = 11.7/km²) since 1995 (Wisconsin DNR, unpublished data). Beavers are common in the study area, with helicopter surveys in 1998 estimating 0.61 active beaver colonies per kilometer square in northwest Wisconsin (Wisconsin DNR, unpublished data).

11.3   Methods

Wolf trapping and radio-collaring was conducted as part of the Wisconsin wolf monitoring program (Wydeven et al. 1995, this volume), and a study on impacts of highway development on wolves (Kohn et al. 2000, this volume). Wolves were trapped using modified Newhouse #14 foothold traps (Kuehn et al. 1986) and fitted with VHF radio collars (Telonics, Mesa, Arizona). Wolves were located by radio telemetry from both ground and fixed-wing aircraft two to five times weekly. Those areas that showed a tight cluster of locations (≥3 locations in a 1-km² area) within a 3–5 week period were investigated on foot after wolves had abandoned the area.

Homesites were confirmed by the presence of a burrow and excavated soils, scat, tracks, kills, activity areas, and heavily used trails (Joslin 1966; G.B. Kolenosky and Johnston 1967; Mech 1970; Ballard and Dau 1983). Location of the homesite was recorded using a handheld global positioning system (GPS; Garmin™ GPS 45, GPS III, Garmin International, Lenexa, KS) and mapped using Arcview 3.2 and ArcGIS 9.0 (hereafter, GIS; ESRI, Redlands, CA). Homesites were studied at three scales: (1) microhabitat (biotic and abiotic variables measured within 50 m of a site); (2) macrohabitat (location relative to roadways and spatial/structural analysis of habitat cover types located within 1.2 km of site); and (3) location within annual territory relative to an inner 25% core area. Field research was conducted from July 1994 through August 2001.

11.3.1 Location Within Annual Territory

Annual territories were based on locations from May 20 of the pre-den year to May 19 of the denning year. This time period was selected to begin after abandonment of the previous year's den and include only one parturition event. Territory boundaries were determined using the 95% minimum convex polygon method (MCP; Mohr 1947) within the animal movement extension (Hooge and Eichenlaub 1997) in GIS. We used the 95% MCP to maximize the distribution of wolf locations while eliminating outliers. In addition, MCP avoids bias in territory boundary estimation due to concentrations of telemetry locations. Territories were required to have ≥30 locations obtained in at least six different months of the year to be included in core analysis (Fuller and Snow 1988).

We used GIS to create a central core that was the same shape as the annual territory but only 25% of its area. The core area was created to determine if wolves were selecting the center of their territory when placing homesites. We recorded the occurrence of homesite and corresponding random sites in relation to this core. To our knowledge, no studies have analyzed specifically whether wolf territorial boundaries are dictated by den site placement or vice versa. Therefore, we also analyzed den site location in territories created from locations obtained between June and December of the previous year (pre-denning), January–May of the same year (denning, including construction and utilization), and June–December of the same year (post-denning).

11.3.2 Macroscale Analyses

Macroscale analyses were conducted within a 1.2-km radius buffer (4.2 km^2) around homesites. This buffer was based on the average distance of radio locations for the only radio-collared female from her pup-occupied den (April 1–July1). While this technique relied on data from only one individual, we believed it was more tightly tied to the biology of the species than arbitrarily chosen distances used in previous studies (Unger 1999; Norris et al. 2002). Random sites corresponding to each homesite were also selected and buffered to allow comparisons between areas used as homesites and unused sites. The number of random sites in each annual territory varied with the size of the territory, based on one random site per 40 km^2. This area was chosen because it appeared to give adequate coverage of the home range while minimizing overlap between random sites.

To examine the effect of roads on homesite selection, presence or absence of roads within the buffers around homesites or random sites was recorded. Distance to closest road and density of roads within buffers were also calculated. Road types included in the analysis were highways, other paved roads, and improved (graded) unpaved roads passable by two-wheel drive auto but did not include unimproved forest roads and trails.

Vegetation cover type layers provided by the Wisconsin DNR (http://dnr.wi.gov/maps/gis/datalandcover.html#overview) and Minnesota Department of Natural Resources Data Deli (www.deli.dnr.state.mn.us) were used to analyze the habitat surrounding homesites and corresponding random sites. Habitats were grouped into 12 cover types, including 8 forests types (oak, pine, maple, aspen, mixed deciduous/coniferous, mixed other conifers, mixed other deciduous, and forested wetlands), grasslands, upland shrub, lowland shrub, and emergent wetlands. Minor habitat areas (<0.1% of home range) were combined with other habitats, and open water and urban areas were excluded from analysis. The proportion of the area within buffers around each homesite and random site classified as each of the above cover types was calculated and then analyzed to examine the effect of vegetation cover type on the selection of homesites. The GIS-based landscape structure and spatial analysis extension Patch Analyst (Elkie et al. 1999) was used to explore relationships and test for differences in landscape structure and pattern between homesites and randomly selected sites.

11.3.3 Microscale Analyses

Microhabitat variables at homesites were analyzed using a nested sampling technique (Higgins et al. 1994). Variables measured at each site included percent canopy cover, percent visual obstruction, tree species composition, slope, aspect, and whether or not a homesite was within 50 m of a permanent water source. These variables were selected as attributes likely to be biologically important in the selection of a home site (Joslin 1966, 1967; G.B. Kolenosky and Johnston 1967; Mech 1970; Stephenson 1974; Ryon 1977; Ballard and Dau 1983; Fuller 1989b). The type of den structure (burrow, hollow log, beaver lodge, and cave) and den dimensions were also recorded. Den sites were generally analyzed after deciduous trees had shed their leaves in autumn to more closely resemble the vegetative conditions of late winter/early spring when wolves likely selected sites (Thiel et al. 1997). Rendezvous sites were generally investigated within 2 weeks of abandonment.

11.3.4 Statistics

11.3.4.1 Macroscale

The proportions of each cover type within buffers around homesites were compared to the proportions in buffers around random sites within the boundaries of annual territories using a Wilcoxon matched-pairs signed-rank test (SPSS, Inc. 1994). The mean distance to roads from homesites and random sites and the density of roads in buffers were analyzed using this same test.

For homesite location within a territory, we used the binomial probability analysis (SPSS, Inc. 1994) to determine if homesites were more likely to be located within the center 25% core of the territory. For location in relation to roadways, we used Fisher's exact test (Zar 1984) to examine whether buffers around homesites were less likely to contain a road. Data obtained from Patch Analyst were analyzed using the Wilcoxon matched-pairs signed-rank test (SPSS, Inc. 1994) to test for differences in measures of landscape pattern and structure around homesites versus around random sites. Homesites from a particular territory were statistically compared to random sites analyzed in that same territory. A paired analysis protocol was used because at our study scale (1.2-km radius circle) the habitat in one wolf pack territory could differ widely from that of another.

All macroscale variables that demonstrated significant differences ($P < 0.05$) between homesites and randomly selected sites were retained for further multivariable analysis. Spearman rank correlation analysis was used to identify significantly correlated macroscale variables. When two variables were correlated ($|r| > 0.5$), only one was kept, based on its ecological relevance. Noncorrelated macroscale variables were subjected to forward stepwise logistic regression (SPSS, Inc. 1994). Logistic regression was chosen because it is generally robust to violations of normality and can be used with discrete and continuous variables (Gorenzel and Salmon 1995).

11.3.4.2 Microscale

For microscale analysis, two sites located randomly within the annual territory were selected. Data for each variable were then collected in a similar fashion at all sites for statistical analysis. Mann Whitney U tests, Fisher's exact tests, and Spearman's rank correlation analysis were used to test for differences between data collected at homesites and that collected at random sites.

The data presented in this chapter represent combined information from two separate studies conducted by Unger (1999) and Keenlance (2002). Because of methodological differences, much of the data needed to be reanalyzed. Multivariate analysis was not performed on microhabitat scale variables, and in some cases analyses were restricted to only those homesites examined in one of these studies.

11.4 Results

11.4.1 Den Sites

Twenty-two dens in 15 individual packs were identified from July 1, 1994 to August 31, 2001 (Fig. 11.1). Based on territory size, random sites between 1 and 8 were selected within each territory ($n = 63$). We were able to calculate an annual territory ($\bar{X} = 202$ km^2, SD = 115) for 18 den sites, and 14 fell within the 25% central core ($X^2 = 26.7$, $p < 0.001$). We calculated a pre-denning ($\bar{X} = 133$ km^2, SD = 82), denning ($\bar{X} = 120$ km^2, SD = 72.), and post-denning ($\bar{X} = 177$ km^2, SD = 69) territory for 9, 9, and 6 den sites, respectively. Wolves also selected the 25% central core during each of these periods ($p = 0.001$, 0.001, and 0.005, respectively). Buffers (1.2-km radius) around dens were less likely to contain a road (7 of 22) than buffers around random sites (40 of 63, $P < 0.001$). Mean distance to roads from a den was more (1,562 m, SE = 207) than from a random site (821 m, SE = 117, $P = 0.006$). Mean road density within buffers around dens (1.08 km/km^2, SE = 1.03) was not significantly different than in buffers around random sites (1.31 km/km^2, SE = 0.86, $P = 0.37$).

Habitat analysis revealed that buffers around dens contained significantly less jack pine ($P = 0.005$), grassland ($P = 0.009$), emergent wet meadow ($P = 0.043$), and mixed/other coniferous forest ($P = 0.043$), and greater amounts of lowland shrub ($P = 0.047$) than buffers around random sites. There was no statistical difference in the average value of any index of landscape structure in buffers around homesites compared to buffers around random sites.

We entered location within territory, presence/absence of roads, proportion of lowland shrub, open grassland, and jack pine habitats within buffers around dens into a forward stepwise selection procedure within logistic regression to further determine which variables most influence den site selection. Our analysis revealed selection of the central core to be the only significant and most useful predictor of den site placement (Wald Statistic, $P = 0.0045$). The model had $R^2 = 0.29$ and correctly classified 80% of all sites.

We statistically analyzed microscale data from 12 den sites. Wolves selected areas of steeper slope ($p = 0.016$) for den placement. Spearman's rank correlation indicated that the relative percentages of individual tree species between den and randomly selected sites were not correlated, and therefore different ($r = 0.215$, $P = 0.551$). Den sites had higher percentages of upland tree species such as aspen, sugar maple, and balsam fir, while random sites had high percentages of more hydric species such as tag alder, black ash, and tamarack. Of 22 dens investigated, 20 were burrows and two were located under uprooted trees. The entrances of nine of these burrows averaged 50 cm (SD = 11.2 cm) high by 47 cm (SD = 10.7 cm) wide. The tunnels generally sloped downward from the entrance into the den. The burrows averaged 230 cm (SD = 51 cm) in length, 68 (SD = 19 cm) in width, 47 cm

(SD = 15 cm) in height, with an average volume of 0.79 m³ (SD = 0.52 m³). Dens were found to be clean and dry with no evidence of debris, leaves, or wolf scat. Some dens did have small amounts of porcupine (*Erethizon dorsatum*) scat deposited after wolves had vacated the den. Of nine dens that were entered, seven were found to be simple tubes or round chambers without a separate birthing chamber. Wolves showed no distinct selection for the orientation of the entrance of the burrow or hill aspect. Of 12 burrows, 4 were on a northerly aspect, 5 were on a southerly aspect, and 3 were in flat terrain with no noticeable aspect.

11.4.2 Rendezvous Sites

Ten rendezvous sites in nine pack territories were located (Fig. 11.1). One had a den associated with it, in the form of an uprooted white cedar (*Thuja occidentalis*). Mean distance to roads (1,296 m, SE = 290 vs 1,166 m, SE = 149, P = 0.68), and road density in buffers (1.12 km/km², SE = 0.31 vs 1.37 km/km², SE = 0.26, P = 0.37) were not significantly different between rendezvous and random sites, respectively.

We calculated an annual territory for nine rendezvous sites. Wolves did not show selection of the 25% central core area of their territory, with four of nine being located within the inner core (X^2 = 1.81, P = 0.337). Buffers around rendezvous sites were not less likely to contain a road than those around random sites (P = 1.000). Wolves selected buffer areas with significantly more aspen (P = 0.046). No indices of landscape structure and pattern were found to be significantly different in buffers around rendezvous sites compared to buffers around random sites. Because only one variable (aspen habitat) was shown to be significant, we were unable to perform multivariate analysis on rendezvous sites.

At the microhabitat level, wolves selected rendezvous sites more often associated with water (P = 0.007) and higher visual obscurity (P = 0.050). Spearman's rank correlation indicated that the tree species between rendezvous sites and randomly selected sites were not correlated and, therefore, different (r = 0.525, P = 0.119). Rendezvous sites had higher percentages of wetland species such as tag alder, red maple, and black ash while random sites had higher percentages of upland species such as aspen and sugar maple.

11.5 Discussion

11.5.1 Den Sites

Spatial location appears to be crucial in the selection of den sites by wolves in northwestern Wisconsin and east-central Minnesota. Wolves selected the inner core of their annual territory when placing a den. This supports previous assumptions

(Ballard and Dau 1983; Gehring 1995) and findings (Theuerkauf et al. 2003). Dens also were located in the central core of the territory regardless of time of year (pre-den, denning, and post-denning periods) when the territory boundaries were determined. Although wolves will form territories in the absence of a den (Rothman and Mech 1979), our data, particularly our pre-denning results, suggest that territorial boundaries have a strong influence in the placement of a den site.

Possible reasons for placing a den in the central core of an annual territory are optimal foraging and avoidance of interpack strife. Prey abundance may also influence the selection of a den site (Jordan et al. 1967; Lawhead 1983). In the Great Lakes region, wolf pups are generally born in March and April (Fuller 1989b). This coincides roughly with spring deer migration and the abandonment of winter yard areas (Nelson and Mech 1981; Messier and Barrette 1985), resulting in prey that are more widely dispersed (Nelson and Mech 1981). Wolves have been shown to travel long straight-line distances to bring prey to the den (Young and Goldman 1944; Mech 1970; Mech et al. 1999), and must travel repeatedly to food sources to maintain pups (Mech 1970; Groebner 1991). In placing a den near the center of their territory, wolves may be showing a central foraging tendency (Stephens and Krebs 1986) to minimize travel distance to prey, and thus reduce handling time before returning to the den.

Previous research has shown that wolf territories often overlap (Van Ballenberghe et al. 1975; Peterson 1977; Fritts and Mech 1981; Nelson and Mech 1981; Peterson et al. 1984; Jędrzejewski et al. 2007). This tendency also occurred in northwestern Wisconsin and east-central Minnesota (Gehring 1995; Shelley and Anderson 1995; Unger 1999). In these "buffer zones" (Mech 1977), aggression between packs can occur, sometimes resulting in death (Mech 1994). Kohn et al. (2000) reported that aggression between packs accounted for 2 of 18 mortalities of collared wolves in northwestern Wisconsin. Locating the den in the central part of the territory should minimize intrusion on the den area by neighboring packs. In addition, placing the den in the central core would minimize the distance required for reaching all edges of the territory when marking and patrolling these boundaries (Mech 1970; Briscoe et al. 2002).

Theuerkauf et al. (2003) suggested that human disturbance and persecution in and around the Białowieża Forest of Poland resulted in wolves choosing the center of their territory for homesite placement. Lower human population densities and legal protection likely lessened the impact of human disturbance in our study area. Wolves in our study area did choose areas farther from roads. However, we did not detect differences in road density between den and random sites, and road occurrence was not shown to be a significant variable in multivariate analysis. As wolves expand into more human-dominated range, anthropogenic disturbance may become a more important factor in den site selection, necessitating further research.

Twenty of 22 dens studied were burrows in the ground, which have been suggested to be the preferred den structures for wolves (Joslin 1967). Also, 13 of 15 packs used different den sites in subsequent years. These findings suggest that suitable den sites were not limited in our study area.

Ciucci and Mech (1992) determined that wolves in the Superior National Forest in northeastern Minnesota placed dens randomly within their territories. They used a probability distribution of wolf dens within 60% of the mean radius from the approximate center of the territory to the edge using MCP. Our differences could be due to our methodology. Ciucci and Mech (1992) found that den location was related to territory size, with larger territories having dens more centrally located. Our mean annual territory was comparable to previous studies in Minnesota (Van Ballenberghe et al. 1975; Fritts and Mech 1981; Berg and Kuehn 1982; Fuller 1989a), but it appeared wolves in our study selected for the central core of their territory, regardless of size.

More random distribution of dens and traditional den use (used ≥2 years) in northern Minnesota may be due to continuous occupation by wolves, and some influence of artificial feeding sources in the form of garbage dumps (Ciucci and Mech 1992). We identified only two packs using traditional dens. The recolonizing nature of the population during our research (1994–2001) may have caused more territorial fluidity than in the established population studied by Ciucci and Mech (1992). We were unaware of any open garbage dumps available in our study area.

Our analysis of habitat showed dens were found in areas with less emergent wet meadow, grassland, jack pine, and mixed conifer/deciduous forest and more low-land shrub. We believe these results were influenced by ease of digging, escape and thermal cover, and the selection of dry upland areas for den sites. Emergent wet meadows precluded digging during den construction, while grasslands and jack pine (mostly plantations <10 years old) lacked adequate cover.

The greater amount of lowland shrub habitat in buffers around dens was due to placement of dens on slightly elevated areas surrounded by dense alder wetlands. Such areas would provide security and escape cover while also placing dens near a water source for the nursing female.

At the microscale level, wolves selected for areas of steeper slope and drier habitats. Den sites were placed in tree communities indicative of upland, well-drained soils (trembling aspen, sugar maple, balsam fir). This finding agrees with previous den site descriptions (Murie 1944; Stenlund 1955; Jordan et al. 1967; Joslin 1967; Mech 1970; Stephenson 1974; Peterson 1977; Ballard and Dau 1983; Fuller 1989b). Steep slope and sandy (dry) soil conditions have been suggested as important for ease of digging and drainage purposes (Jordan et al. 1967; Stephenson 1974). We did not find that visual obscurity was an important attribute for selection in the immediate vicinity of the dens, in contrast to Joslin (1967) and Stephenson (1974). While wolves did select more dense lowland shrub nearby, the area immediately around dens (<20 m) tended to be relatively open. We did not detect selection of a specific aspect for den placement, but as our research was conducted at lower latitudes than most studies, southern exposures may have been less important (Stephenson 1974; Ballard and Dau 1983; Fuller 1989b). Our den dimensions were similar to those reported by Murie (1944), Joslin (1967), Mech (1970), Stephenson (1974), and Ryon (1977), but most dens did not end in an enlarged chamber as described by some (Murie 1944; Joslin 1967; Mech 1970; Stephenson 1974; Ryon 1977; Trapp 2004).

11.5.2   Rendezvous Sites

Rendezvous site selection appeared to be determined mainly by habitat factors as opposed to spatial factors such as territory boundaries and roads. In contrast to our findings on den sites, wolves did not select for the center of their territory or areas devoid of roads for rendezvous site. The increased mobility of maturing pups at rendezvous sites, and therefore their greater ability to avoid danger (predators, humans), may explain this trend. Location of rendezvous sites within aspen habitats may reflect prey availability. Deer, the main food of wolves, feed heavily on forage species found in aspen habitat during spring and summer (McCaffery et al. 1974).

Rendezvous sites were located in wetland community types. Within this wetland, habitat sites were in close proximity to open water, suggesting that available water may play an important role in the site selection process. Wolves usually move from the den to the first rendezvous site when pups are 8–10 weeks of age (Mech 1970; Peterson et al. 1984), which coincides roughly with when they are weaned (Mech 1970; Harrington and Mech 1982). Water is very important to adult wolves for digestion after gorging (Mech 1970). Pups are relatively sedentary at the rendezvous site (Joslin 1967; Theberge and Pimlott 1969). Thus, a permanent water source nearby would be beneficial to pups for digestion, hydration, and evaporative cooling during the warm summer months.

Wolves also selected for higher visual obscurity within this wetland habitat. Higher visual obstruction might be selected to minimize possible conflicts between pups and intruders (Theberge and Pimlott 1969; Harrington and Mech 1982; Veitch et al. 1993; Thiel et al. 1998).

11.6   Summary and Management Recommendations

We studied gray wolf den and rendezvous site selection in northwestern Wisconsin and east-central Minnesota. Spatial location within the wolves' territories appeared the most important factor in the selection of den sites, whereas rendezvous site location was determined primarily by habitat. In the first several weeks after birth, the resource needs of the pups are provided for at the den by the mother. With these needs met, protection of the pups from intruders or competing packs may become the dominant concern in the placement of the den. The increased mobility of pups by 8 weeks of age probably makes variables such as location within the territory and lower road density less important than habitat variables for rendezvous site selection. In addition, as pups mature and are weaned, an increased need for resources may dictate the move to a rendezvous site in wetland habitat where water and higher visual cover are available.

Hierarchy theory is important for studies of habitat selection and foraging behavior (Allen and Starr 1982; Pribil and Picman 1997). Wolves likely select homesites in this manner with the selection of a territory on the landscape followed by selection of an area within this territory, and finally the selection of a specific microscale location to birth and raise the pups. In our study, wolves selected for the central core of their territory when placing a den site. Within this central core area selected areas of lower road density, upland habitat, and steep slope. Wolves selected areas of aspen habitat for rendezvous sites and within this habitat chose wetland areas in close proximity to water and high visual obscurity.

The data and conclusions presented here provide a further understanding of a factor heavily influencing wolf population dynamics, but must be viewed in the context of larger scale habitat selection by wolves. Without suitable habitat in which to establish a territory, the selection of high quality homesites may become more difficult with a consequent potential reduction in pup production.

Currently, homesite availability does not seem to be a limiting factor in northwest Wisconsin and east-central Minnesota. While areas around dens were less likely to contain a road than areas around random sites, the avoidance of human disturbance does not seem to be driving homesite selection. Based on these findings, the healthy growth of the wolf population in the region during the last 15 years (Erb and DonCarlos, this volume; Beyer et al., this volume; Wydeven et al., this volume), and given current levels of protection and public attitudes toward wolves, broadscale protection of homesites seems unnecessary. Situations involving vulnerable population segments or areas with a high potential for human–wolf conflict (e.g., the edge of the recolonization front), however, may require protection of homesites. Based on our data, the best indicator of den site location is the identification of the 25% inner core of the annual territory as estimated by the 95% MCP of home range locations of radio-collared wolves.

All discussions of wolf resource selection and population dynamics must be viewed in the context of current levels of legal protection and public acceptance. Changes in either could quickly lead to changes in wolf population levels and resource selection, including homesite selection.

Acknowledgments Funding and support was provided by the Wisconsin Department of Transportation, Wisconsin Department of Natural Resources (WDNR), University of Wisconsin – Stevens Point (UWSP), Michigan State University Department of Fish and Wildlife (MSU), and Milwaukee Zoological Society. Additional assistance was provided by the Wisconsin Bureau of Land Management, Wildlife Science Center, Douglas County Forestry, Minnesota Department of Natural Resources, and Mosinee Paper Company. We thank from WDNR, A. Wydeven, R. Shultz, H. Reese, F. Krueger, D. Kallenbach, and P. Miller, as well as K. Millenbah (MSU), E. Merrill (UWSP), K. Rice (UWSP), Alexa Spivey (volunteer), and Michelle Lassige (volunteer).

References

Arjo, W. M., and Pletscher, D. H. 2004. Coyote and wolf habitat use in northwestern Montana. Northwest Science 78:24–32.

Allen, T. F. H., and Starr, T. B. 1982. Hierarchy: perspectives for ecological complexity. Chicago: University of Chicago Press.

Ballard, W. B., and Dau, J. R. 1983. Characteristics of Gray Wolf, *Canis lupus*, den and rendezvous sites in south-central Alaska. Canadian Field-Naturalist 97:299–302.

Berg, W. E., and Kuehn, D. W. 1982. Ecology of wolves in north-central Minnesota. In Wolves of the World, eds. F. H. Harrington and P. C. Paquet, pp. 4–11. Park Ridge: Noyes Publishing.

Briscoe, B. K., Lewis, M. A., and Parrish, S. E. 2002. Home range formation in wolves due to scent marking. Bulletin of Mathematical Biology 64:261–284.

Capitani, C., Mattioli, L., Avanzinelli, E., Gazzola, A., Lamberti, P., Mauri, L., Scandura, M., Viviani, A., and Apollonio, M. 2006. Selection of rendezvous sites and reuse of pup raising areas among wolves *Canis lupus* of north-eastern Apennines, Italy. Acta Theriologica 51:395–404.

Ciucci, P., and Mech, L. D. 1992. Selection of wolf dens in relation to winter territories of northeastern Minnesota. Journal of Mammalogy 73:899–905.

Clark, J. D., Dunn, J. E., and Smith, K. G. 1993. A multivariate model of female black bear habitat use for a geographic information system. Journal of Wildlife Management 57:519–526.

Curtis, J. T. 1959. The Vegetation of Wisconsin: an Ordination of Plant Communities. Madison: University of Wisconsin Press.

Elkie, P. C., Rempel, R. S., and Carr, A. P. 1999. Patch Analyst User's Manual: a tool for quantifying landscape structure. Thunder Bay: NWST Technical Manual TM-002, Ontario Ministry of Natural Resources.

Fritts, S. H., and Mech, L. D. 1981. Dynamics, movements, and feeding ecology of a newly protected wolf population in northwestern Minnesota. Wildlife Monographs 80:1–70.

Fuller, T. K. 1989a. Population dynamics of wolves in north-central Minnesota. Wildlife Monographs 105:1–41.

Fuller, T. K. 1989b. Denning behaviors of wolves in north-central Minnesota. American Midland Naturalist 121:188–193.

Fuller, T. K., and Snow, W. J. 1988. Estimating winter wolf densities using radio-telemetry data. Wildlife Society Bulletin 16:367–370.

Gehring, T. M. 1995. Winter wolf movements in northwestern Wisconsin and east-central Minnesota: a quantitative approach. Stevens Point: Thesis, University of Wisconsin Stevens Point.

Gorenzel, W. P., and Salmon, T. P. 1995. Characteristics of American crow urban roosts in California. Journal of Wildlife Management 59:638–645.

Groebner, D. J. 1991. Summer movement rates of a pair of northeastern Minnesota timber wolves. MS Thesis, Northern Michigan University, Marquette, Michigan.

Harrington, F. H., and Mech, L. D. 1982. Patterns of homesite attendance in two Minnesota wolf packs. In Wolves of the World, eds. F. H. Harrington and P. C. Paquet, pp. 81–105. Park Ridge: Noyes Publishing.

Higgins, K. F., Oldenmeyer, J. L., Jenkins, K. J., Clambey, G. K., and Harlow, R. F. 1994. Vegetation sampling and measurement. In Research and Management Techniques for Wildlife and Habitats, Fifth Edition, ed. T.A. Bookhout, pp. 567–591. Bethesda: The Wildlife Society.

Hooge, P. N., and Eichenlaub, B. 1997. Animal movement extension to Arcview. Version 1.1. Anchorage: Alaska Biological Science Center, U.S. Geological Survey.

Jędrzejewski, W., Schmidt, K., Theuerkauf, J., Jędrzejewska, B., and Kowalczyk, R. 2007. Territory size of wolves *Canis lupus*: linking local (Białowieża Primeval Forest, Poland) and Holarctic-scale patterns. Ecography 30:66–76.

Jordan, P. A., Shelton, P. C., and Allen, D. L. 1967. Numbers, turnover, and social structure of the Isle Royale wolf population. American Zoology 7:233–252.

Joslin, P. W. B. 1966. Summer activities of two timber wolf packs in Algonquin Park. MS Thesis, University of Toronto.

Joslin, P. W. B. 1967. Movements and home sites of timber wolves in Algonquin Park. American Zoology 7:279–288.

Keenlance, P. W. 2002. Resource selection of recolonizing gray wolves in northwest Wisconsin. PhD Dissertation, Michigan State University, East Lansing.

Kohn, B. E., Frair, J. L., Unger, D. E., Gehring, T. M., Shelley, D. P., Anderson, E. M., and Keenlance, P. W. 2000. Impact of the US Highway 53 expansion project on wolves in northwestern Wisconsin. Madison: Wisconsin Department of Natural Resources.

Kolenosky, G. B., and Johnston, D. H. 1967. Radio-tracking timber wolves in Ontario. American Zoology 7:289–303.

Kuehn, D. W., Fuller, T. K., Mech, L. D., Paul, W. J., Fritts, S. H., and Berg, W. E. 1986. Trap related injuries in gray wolves in Minnesota. Journal of Wildlife Management 50:90–91.

Lawhead, B. E. 1983. Wolf den site characteristics in the Nelchina Basin, Alaska. MS Thesis, University of Alaska, Fairbanks.

Mandernack, B.A. 1983. Food habits of Wisconsin timber wolves. MS Thesis, University of Wisconsin-Eau Claire.

McCaffery, K. R., Tranetzki, J., and Piechura, Jr., J. 1974. Summer foods of deer in northern Wisconsin. Journal of Wildlife Management 38:215–219.

Mech, L. D. 1970. The wolf: the ecology and behavior of an endangered species. Garden City: Natural History Press.

Mech L. D. 1977. Wolf pack buffer zones as prey reservoirs. Science 198:320–321.

Mech, L. D. 1993. Resistance of young wolf pups to inclement weather. Journal of Mammalogy 74:485–486.

Mech, L. D. 1994. Buffer zones of territories of gray wolves as regions of intraspecific strife. Journal of Mammalogy 75:199–202.

Mech, L. D., Wolf, P. C., and Packard, J. M. 1999. Regurgitative food transfer among wild wolves. Canadian Journal of Zoology 77:1192–1195.

Messier, F., and Barrette, C. 1985. The efficiency of yarding behavior by white-tailed deer as an anti-predator strategy. Canadian Journal of Zoology 63:785–789.

Mladenoff, D. J., Sickley, T. A., Haight, R. G., and Wydeven, A. P. 1995. A regional landscape analysis and prediction of favorable gray wolf habitat in the northern Great Lakes region. Conservation Biology 9:279–294.

Mohr, C. O. 1947. Table of equivalent populations of North American mammals. American Midland Naturalist 37:223–249.

Murie, A. 1944. The wolves of Mount McKinley. U.S. National Park Service Fauna Series 5.

Nelson, M. E., and Mech, L. D. 1981. Deer social organization and wolf predation in northeastern Minnesota. Wildlife Monographs 77:1–53.

Norris, D. R, Theberge, M. T., and Theberge, J. B. 2002. Forest composition around wolf (*Canis lupus*) dens in eastern Algonquin Provincial Park, Ontario. Canadian Journal of Zoology 80:866–872.

Peterson, R. O. 1977. Wolf ecology and prey relationships on Isle Royale. U.S. National Park Service Scientific Monograph Series 11.

Peterson, R. O., Woolington, J. D., and Bailey, T. N. 1984. Wolves of the Kenai Peninsula, Alaska. Wildlife Monographs 88:1–52.

Pribil, S., and Picman, J. 1997. The importance of using the proper methodology and spatial scale in the study of habitat selection by birds. Canadian Journal of Zoology 75:1835–1844.

Rothman, R. J., and Mech, L. D. 1979. Scent-marking in lone wolves and newly formed pairs. Animal Behavior 27:750–760.

Ryon, J. C. 1977. Den digging and related behavior in a captive timber wolf pack. Journal of Mammalogy 58:87–89.

Shelley, D. P., and Anderson, E. M. 1995. Final report: impacts of US Highway 53 expansion on timber wolves – baseline data. Stevens Point: University of Wisconsin Stevens Point.

SPSS Inc. 1994. SPSS user's guide. Version 6.1. M.J. Norusis, ed. SPSS, Inc., Chicago, Illinois.

Stenlund, M. H. 1955. A field study of the timber wolf (*Canis lupus*) on the Superior National Forest, Minnesota. Minnesota Department of Conservation Technical Bulletin 4:1–55.

Stephenson, R. O. 1974. Characteristics of wolf den sites. Alaska Department of Fish and Game. Federal Aid to Wildlife Restoration Final Report, W-17–2 through W-17–6.

Stephens, D. W., and Krebs, J. R. 1986. Foraging theory. Princeton: Princeton University Press.

Theberge, J. B., and Pimlott, D. H. 1969. Observations of wolves at a rendezvous site in Algonquin Park. Canadian Field-Naturalist 83:122–128.

Theuerkauf, J., Rouys, S., and Jedrzejewski, W. 2003. Selection of den, rendezvous, and resting sites wolves in the Bialowieza Forest, Poland. Canadian Journal of Zoology 81:163–167.

Thiel, R. P., Hall, W. H., and Schultz, R. N. 1997. Early den digging by wolves, *Canis lupus*, in Wisconsin. Canadian Field-Naturalist 111:481–482.

Thiel, R. P., Merrill, S., and Mech, L. D. 1998. Tolerance by denning wolves, *Canis lupus*, to human disturbance. Canadian Field-Naturalist 112:340–342.

Trapp, J. T. 2004. Wolf den site selection and characteristics in the northern Rocky Mountains: a multiscale approach. Prescott: Prescott College.

Unger, D. E. 1999. Timber wolf den and rendezvous site selection in northwestern Wisconsin and east-central Minnesota. MS Thesis, University of Wisconsin – Stevens Point.

Van Ballenberghe, V., Erickson, A. W., and Byman, D. 1975. The ecology of the timber wolf in northeastern Minnesota. Wildlife Monographs 43:1–43.

Veitch, A. M., Clark, W. E., and Harrington, F. H. 1993. Observations of an interaction between a barren-ground black bear, *Ursus americanus* and a wolf, *Canis lupus*, at a wolf den in northern Labrador. Canadian Field-Naturalist 107:93–97.

Wydeven, A. P., Schultz, R. N., and Thiel, R. P. 1995. Monitoring a recovering gray wolf population in Wisconsin, 1979–1991. In Ecology and Conservation of Wolves in a Changing World, eds. L. N. Carbyn, S. H. Fritts, and D. R. Seip, pp. 147–156. Edmonton: Canadian Circumpolar Institute.

Young, S. P., and Goldman, E. A. 1944. The wolves of North America. Washington, D. C.: Wildlife Institute.

Zar, J. H. 1984. Biostatistical analysis. Englewood Cliffs: Prentice-Hall.

Chapter 12
Dispersal of Gray Wolves in the Great Lakes Region

Adrian Treves, Kerry A. Martin, Jane E. Wiedenhoeft,
and Adrian P. Wydeven

12.1   Introduction

In less than 40 years, gray wolves (*Canis lupus*) rebounded from a population of
<700 wolves restricted to northeastern Minnesota to >4,000 wolves across northern
Minnesota, Wisconsin, and Michigan (Chaps. 4–6, this volume). This recovery is
due in part to changing human attitudes toward wolves, protection by the U.S. Fish
and Wildlife Service (USFWS) under the Endangered Species Act (ESA), and
favorable ecological conditions (Mladenoff et al. 1997; USFWS 2007; Schanning,
this volume). Furthermore, two features intrinsic to wolf life history facilitated
rapid recovery: long-range movements and broad habitat tolerance. The wolf is a
habitat generalist, using all habitat types of the northern hemisphere except tropical
rain forests and deserts (Mech 1970). Equally important is their tremendous capacity
for rapid, long-distance movement, allowing them to colonize distant, suitable areas
rapidly. Mech and Boitani (2003) describe wolf packs as "dispersal pumps" that
convert prey into wolves and scatter these dispersers across the landscape, "pumping
out" half of their pack each year.

Long-range movements of wolves deserve scrutiny by managers and scientists
because such movements are driving recolonization of their historical range. Wolf
populations in the western Great Lakes and the northern Rocky Mountains have
been assigned to distinct population segments proposed for removal of federal protections under the ESA (USFWS 2007, 2008), but would continue to produce dispersers that reach states were wolves still would receive ESA protection. Predicting
where such long-range movements take wolves requires an understanding of dispersal and habitat selection during long-range movements.

Dispersing gray wolves can travel vast distances, and have moved as far as
1,092 km in a straight-line distance from their original pack territories (Wabakken
et al. 2007). One wolf radio-collared in Wisconsin dispersed >689 km into eastern
Indiana, a trip that likely entailed skirting the greater Chicago metropolitan area
(Thiel et al., this volume). Certainly, short-range movements are more common
among wolves, but long-range movements are disproportionately important from a
management perspective because they create the possibility of recolonizing historic

range and new regions. Indeed, it took Minnesota wolves <30 years to recolonize the northern third of Wisconsin and Upper Peninsula (UP) of Michigan (Wydeven et al., this volume; Beyer et al., this volume).

With few exceptions, unless a wolf attains alpha status, it will eventually leave its pack (Mech 1970, 1999). However, the appealing story of wolves setting out on lengthy voyages of discovery is misleading. Most extraterritorial movements (ETMs), short or long, do not result in new packs or immigration into an existing pack. Indeed, when a wolf leaves its natal pack, it may return with or without encountering other wolves. Wolves that do not return to the natal pack may spend years as loners before joining packs or being joined by other wolves. Some long-range movements do result in residence in a new pack (dispersal). Because intentions are inscrutable and initial ETMs do not seem to predict the eventual outcomes, wolf movements defy easy classification. Furthermore, efforts to predict future dispersal and colonization are hindered by methods available for monitoring wolves. Radio telemetry remains the current best option, but transmitters fail, batteries weaken, and people cannot search everywhere when a wolf goes missing (Mech 1974, 1983; Coffey et al. 2006). All these factors make it challenging to predict future sites of colonization or dispersal.

In this chapter, we review findings on wolf dispersal, and examine a subset of long-range movements from Wisconsin that illuminates habitat selection by wolves when they disperse long distances. Understanding habitat use by dispersing wolves in this human-dominated ecosystem is important for planning wolf conservation and habitat protection, particularly for regions outside of vast wilderness areas.

12.2   Review of Wolf Dispersal in the Great Lakes Region

Long-range wolf movements have been critical to recolonization of the Great Lakes region. Before breeding packs were detected in Wisconsin and Michigan, individual wolves were detected, probably dispersing from Minnesota or Ontario (Hendrickson et al. 1975; Thiel 1978; Thiel and Hammill 1988). Extensive data on wolf dispersal in Minnesota, Michigan, and Wisconsin have been collected (Table 12.1), although definitions of dispersal, starting and ending points, and methods for estimating movement have varied across studies. Nevertheless, comparisons between studies illuminate consistent patterns that help us to infer specific regional characteristics relating to sex and age of dispersers and travel patterns.

Studies conducted in the Great Lakes region (Table 12.1) reveal no clear pattern in the sex ratio of dispersers. Fuller (1989) and Gese and Mech (1991) detected no difference in the numbers of males and females dispersing in Minnesota. This corroborates research in Montana and Alaska where male and female wolves were equally likely to disperse (Boyd and Pletscher 1999; Peterson et al. 1984). In central Alaska, males dispersed at higher rates than did females (Ballard et al. 1987). Peterson et al. (1984) found males from the Kenai Peninsula dispersed farther than females, whereas Ballard et al. (1987) found that females from central Alaska

Table 12.1 Summary of gray wolf dispersal patterns in the Great Lakes region

Location	n	Mean distance (km)	Mean age (years)	Sex ratio ($\male$:$\female$)	Percentage of yearling	Percentage of success	Reference
Northwestern Minnesota	9	NA[a]	NA	0.5:1.0	78	44	Fritts and Mech 1981
North-central Minnesota	15	148	1.7	2.5:1.0	60	NA	Berg and Kuehn 1982
North-central Minnesota	28	29	NA	NA	39	42	Fuller 1989
Northeastern Minnesota	75	88 ($\male$) 65 ($\female$)	1.5	1.1:1.0	53	63	Gese and Mech 1991
Northern Wisconsin	16	114	1.7	0.8:1.0	50	31	Wydeven et al. 1995
Central Wisconsin	15	83 ($\male$) 67 ($\female$)	2.2 ($\male$)[b] 1.7 ($\female$)	0.9:1.0	NA	58	Thiel et al., this volume

[a]*NA* not available
[b]Excluding a 7.8-year-old male

dispersed farther. Costs and benefits of dispersal are likely to differ between males and females given differences in costs of reproduction and competition for mates, as well as local variation in breeding vacancies (Shields 1987).

In most studies, only small percentages of pups dispersed, and mostly at the end of winter when they neared 1 year of age. Percentages of yearlings among dispersers ranged from 38% to 78% across studies (Table 12.1). In Wisconsin, average age at dispersal was between 1.5 and 2.2 years, and appeared slightly higher for males in central Wisconsin (Thiel et al., this volume). The percentage of yearlings dispersing changed with population status in northeastern Minnesota. Seventy percent of yearlings dispersed during population declines, 47% when the population was stable, and 83% while the population was increasing (Gese and Mech 1991). Average age of dispersal was higher and percent of yearlings dispersing was lower in Quebec, Alaska, and Montana (Messier 1985; Ballard et al. 1987, 1997; Boyd and Pletscher 1999). Dispersal rates may be higher for pups of eastern wolves (*Canis lycaon,* see Nowak, this volume, for discussion of taxonomic status) in Ontario, where dispersal was suspected to have occurred in 20% of monitored pups and occurred as early as 4.5 months of age (Mills et al. 2008).

The oldest disperser detected by Gese and Mech (1991) in northeastern Minnesota was 4.5 years old. In contrast, one male in central Wisconsin joined a new pack at 7.8 years of age. In Minnesota and Wisconsin, older wolves generally were more successful at establishing new territories or joining other packs, and usually traveled shorter distances (Gese and Mech 1991; Wydeven et al. 1995). Individual success in establishing new home ranges and attaining breeding status varied from 31% to 63% across studies (Table 12.1).

Dispersal distances in Alaska and Montana were similar to those in north-central Minnesota in the 1970s (Berg and Kuehn 1982), and in Wisconsin during early colonization (Wydeven et al. 1995). Radio-collared wolves in the Great Lakes states have shown tremendous ability to disperse long distances (Table 12.2). The longest known dispersal in North America occurred when an adult male moved 886 km northwest from north-central Minnesota through Manitoba and eventually to eastern Saskatchewan (Fritts 1983). A dispersing Scandinavian wolf topped this, moving 1,092 km from southern Norway to the Finnish–Russian border (Wabakkan et al. 2007). Scandinavian wolves averaged 313 km per dispersal

Table 12.2 Long-distance movements of gray wolves in the Great Lakes region (1976–2004)

Wolf/sex/ age[a]	Origin	Final destination	Minimum distance (km)	Movement duration (years)	Reference
5167/M/P	Northwestern Minnesota	Western Ontario	390	2.8 (1976–1979)	Fritts and Mech 1981
555/M/Y	North-central Minnesota	Southern Manitoba	432	0.7 (1979)	Berg and Kuehn 1982
/M/Y	North-central Minnesota	Eastern Saskatchewan	886	0.82 (1981)	Fritts 1983
035/F/A	Northwestern Wisconsin	Central UP Michigan	227	1.0 (1985–1986)	Thiel 1988
177/F/A	Eastern Minnesota	Northwestern Wisconsin	304	0.13 (1993)	Wydeven 1994
113/F/Y	North-central Wisconsin	Western Ontario	480	1.0 (1988–1989)	Wydeven et al. 1995
/M/A[b]	Northern Minnesota?	Southern South Dakota	530+	1991[c]	Licht and Fritts 1994
/M/A[b]	Northern Minnesota?	Western North Dakota	343+	1992[c]	Licht and Fritts 1994
395/M/P	Northeastern Minnesota	Central UP Michigan	275	2.8 (1991–1994)	Mech et al. 1995
487/F/Y	Northeastern Minnesota	Southern Wisconsin	555	0.33 (1994)	Mech et al. 1995
7809/F/A	Central Minnesota	Central Minnesota	494[d]	0.49 (1999)	Merrill and Mech 2000
0071/F/P	Eastern UP Michigan	Southern Wisconsin	483	0.16 (2001)	WDNR/MI DNR files
0018/M/P	Western UP Michigan	North-central Missouri	720	1.6 (2000–2001)	Mech and Boitani 2003
409/M/P	Central Wisconsin	Eastern Indiana	689	0.43 (2003)	Thiel et al., this volume
4914/M/A	Eastern UP Michigan	Northwestern Wisconsin	435	0.12 (2004)	WDNR/MI DNR files
2061/M/A	Central UP Michigan	Eastern Minnesota	427	0.25 (2004)	WDNR/MI DNR files

[a]Age at start except as listed below; P <1 year, Y >1 & <2 years, A >2 years
[b]Age at time of death
[c]Year found dead
[d]Straight line from origin and furthest point; actual movements were > 4251 km before return to original territory

movement, longer than averages for North American wolves (Linnell et al. 2005). Wolves from newly established subpopulations, such as Wisconsin (1979–1992, Wydeven et al. 1995) and north-central Minnesota (1970–1980, Berg and Kuehn 1982), moved farther than dispersers from established subpopulations.

Total distances moved by wolves during dispersal significantly exceeded minimum straight-line distances between start and end points (Table 12.2). For example, a Minnesota wolf tracked by satellite telemetry for 179 days traveled at least 4,251 km from northwestern Wisconsin eastward to Green Bay, then west again to LaCrosse, then northwest to Grantsburg, before leaving the state and returning to her home territory in Camp Ripley, Minnesota. The straight-line distance from her original territory to her farthest destination was 494 km (Merrill and Mech 2000). She traveled through at least 27 of Wisconsin's 74 counties in <3 months. The above-mentioned Scandinavian yearling may have traveled >10,000 km with at least 3,471 km traversed over 271 days (Wabakkan et al. 2007). Clearly, end points of movements capture only portions of the extensive movements made by dispersing wolves.

In addition to traveling long distances, wolves readily cross human-altered landscapes and areas without resident wolves. Wolves from the Great Lakes states of the United States have moved into three Canadian provinces (Ontario, Manitoba, and Saskatchewan), and into at least five surrounding states (Illinois, Indiana, Missouri, North Dakota, and South Dakota). Before 1992, 43% of dispersing wolves from Wisconsin ($n = 14$) moved into Minnesota (Wydeven et al. 1995). In winter 2006–2007, 11 of 63 (17%) radio-collared wolves monitored in Wisconsin came from the UP of Michigan. Between 1994 and 2006, dead wolves were detected in 47 of 72 counties in Wisconsin, despite packs occurring in only 13 (1994)–30 (2006) counties. The likelihood that these long travels result in colonization of new areas depends on finding safety, a mate, food, and suitable habitat.

12.3 Habitat Used in Long-Range Movements by Wisconsin Wolves

Landscape features that characterize areas of wolf colonization are popular topics of research (e.g., Mech et al. 1988; Massolo and Meriggi 1998; Corsi et al. 1999; Jędrzejewski et al. 2004; Potvin et al. 2005), particularly in Wisconsin (e.g., Thiel 1985; Mladenoff et al. 1995; Wydeven et al. 2001). Despite keen interest from researchers, habitat selection by wolves during long-range movements remains poorly understood (Mladenoff et al. 1999). Early work suggested that dispersing wolves use areas previously thought unsuitable for establishment of wolf packs (Licht and Fritts 1994; Wydeven 1994; Mech et al. 1995). At least three studies have attempted to characterize suitable landscapes for dispersing wolves (Harrison and Chapin 1998; Wydeven et al. 1998; Oakleaf et al. 2006), and assumed that dispersing wolves tolerated poorer quality habitat than wolves in pack territories.

Underlying predictions about habitat selection by wolves are basic models of animal movement behavior. Fretwell and Lucas (1970) predicted that animals select the most suitable habitats first. As population density in those habitats increases, relative suitability decreases proportionally until habitats of originally lower suitability become equivalent in quality to those selected initially. Following saturation of more suitable habitats, animals occupy less suitable habitats. This idea of relaxing criteria for habitat selection has been tested and is useful in understanding habitat use by animals (Whitham 1980; Petit and Petit 1996). Because animals on the move must avoid danger, locate resources (food, water, shelter, etc.), and search for mates, intermediate steps likely reflect choices between available habitat patches. This conceptual model informed our analysis of long-distance movements by wolves in Wisconsin.

12.3.1 Methods

Managers and researchers have trapped and radio-collared wolves in Wisconsin since 1979 (Wydeven et al., this volume). These wolves have been tracked mainly by aerial radio telemetry. Radio-collared wolves were generally located once per week from the air by Wisconsin Department of Natural Resources (WDNR) pilots with fixed-wing airplanes, but wolves detected outside their packs and translocated wolves were sometimes located two to three times per week (Treves et al. 2002; Wydeven et al. 2004). Between 1981 and 2004, the WDNR recorded 20,006 locations from 202 wolves.

If a wolf left a known pack or territory to establish or join another pack or territory, we classified that wolf as an unambiguous disperser. Unfortunately, only 32 wolves provided such clear examples of individual wolf dispersal, including 18 with inter-mediate locations and 14 providing only start and end points within packs. To estimate dispersal distance in each of the 32 unambiguous cases, we measured the distance between the last location in the known territory and the first location in the destination territory (Gese and Mech 1991).

Most movements were ambiguous because we could not determine outcomes. To increase our sample of habitat used during long-term movements, we inspected all ETMs of Wisconsin wolves. We classified ETMs as movements that occurred >5 km beyond estimated territory boundaries (Messier 1985; Fuller 1989), and 63% of radio-collared wolves had ≥1 ETM (Martin 2007). The Wisconsin dataset contained 295 ETM segments from 127 wolves. Given that time required for confirmed dispersals is longer than for temporary ETMs (Gese and Mech 1991), we used 29 days as a threshold for defining "long-term movements" (see results for justification of this threshold). We removed briefer ETMs from further analysis, assuming these are less informative about habitat use by dispersing and colonizing wolves. Combined with the radio locations of unambiguous dispersers, our sample then comprised 60 movements by 49 wolves, with 609 radio locations (OBSERVED). Of these, 455 OBSERVED locations were from Wisconsin, 138 from Minnesota, and 16 from the UP of Michigan.

To test if habitat used during long-term movements was random relative to available habitat, we defined a comparison area for analysis around those 609 OBSERVED locations. The comparison area was derived by buffering the OBSERVED locations with a 29.6-km radius (dark gray region in Fig. 12.1). The buffer's radius was derived from sequential radio locations of the 60 movements of 49 wolves, and represented a mean straight-line distance of 14.3 km (SD = 15.3) between consecutive radio locations. We used this mean distance + 1 SD as a buffer around OBSERVED locations, which produced two discrete regions (Fig. 12.1). We randomly selected points in the comparison area where wolves had not been detected as our EXPECTED sample in proportion to the numbers observed in the two regions and each state.

We placed two restrictions on EXPECTED locations to avoid pseudoreplication. EXPECTED locations had to be >1.8 km apart (see below) and could not lie on any body of water ≥9.8 km², the largest body of water on which wolves were radio located. We then collected data from an area of 908 m-radius around each OBSERVED and EXPECTED location. This area equals one section in the Public Land Survey System (PLSS, 2.59 km²), and allowed for measurement error in aerial radio telemetry (Martin 2007). This scale also provided a simple way to map probabilities of long-range movement across the state, was visible on commercially available atlases, and was amenable to management decisions (Turner et al. 1995).

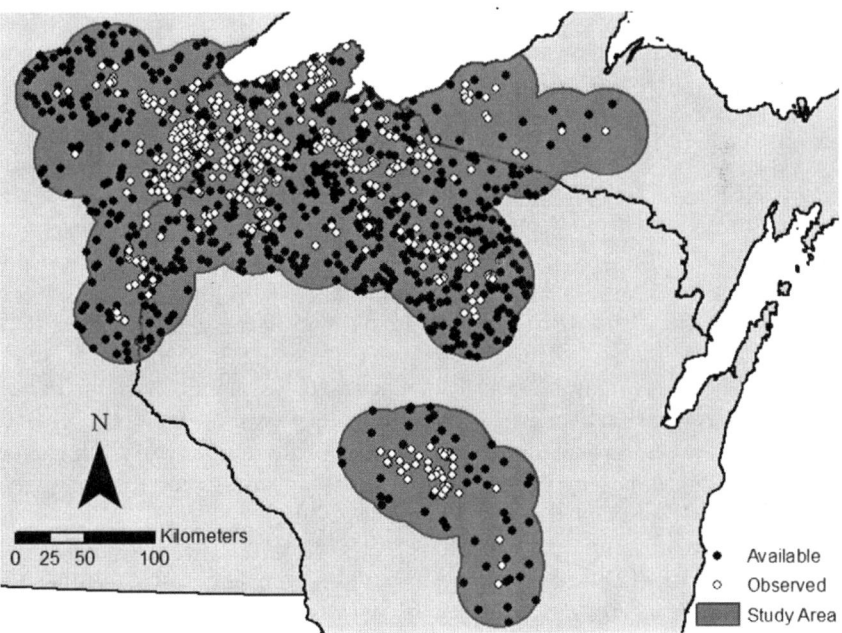

Fig. 12.1 The northern and southern sections of our analysis area, with *open circles* being OBSERVED locations of long-range movements and *closed circles* being an equal number of randomly placed EXPECTED locations

We measured spatial variation in landscape factors in four ways (Mladenoff et al. 1995; Wydeven et al. 1998; Treves et al. 2004; Potvin et al. 2005):

1. We estimated land cover composition using the 1992 National Land Cover Dataset (NLCD). NLCD divides cover into 21 classes at 30-m resolution (Vogelmann et al. 2001). We aggregated these 21 classes into 7 [water, urban, barren, forest, grasslands (generally pastures or hayfields), row crops, and wetlands] based on accuracy assessments for Wisconsin (Thogmartin et al. 2004; Wickham et al. 2004; Martin 2007).
2. We estimated white-tailed deer density as deer per km^2 averaged over the period 1995–2004 in discrete deer management units (DMUs) which generally cover 400–1,800 km^2 (WDNR 1998).
3. We estimated human population characteristics using the 2000 US Census Bureau TIGER/line files and included people and houses per km^2 at the scale of the census block (U.S. Census Bureau 2001), and road density (km/km^2).
4. We estimated agricultural characteristics using census data by county to quantify farms per km^2 and cattle per km^2 (U.S. Department of Agriculture 1997). We created and manipulated all data layers with ArcGIS 9.1 (ESRI 2005).

We used JMP 6.0.3 statistical software (SAS Institute Inc. Cary, NC) to compare habitat variables around OBSERVED and EXPECTED locations. We evaluated differences in movements with Student's t-tests, assuming unequal variances and evaluated differences in habitat variables with Welch ANOVA t-tests (Welch 1951).

12.3.2   Results and Discussion

Dispersing wolves from Wisconsin averaged 55.1 km (SD = 49.6 km) between original and final territories, with no difference between males and females in distances moved (Student's $t = 1.56$; $P = 0.13$). The 295 ETMs detected for Wisconsin wolves lasted 1–214 days ($[\bar{x}] = 19.3$, SD = 31.2), with a median of 7.5 days. The distribution of ETMs was bimodal, with none in the interval 29–35 days (Fig. 12.2). This pattern generated our operational definition of long-range movements >29 days.

OBSERVED locations differed significantly from EXPECTED for 10 of 12 landscape variables (Table 12.3). Only wetland cover did not differ and farm density showed only a slight tendency. All five variables associated with the presence of humans (houses, humans, roads, farms, and cattle densities) were higher in EXPECTED locations, consistent with humans causing most wolf mortality (Woodroffe and Ginsburg 1998; Wydeven et al. 2001). EXPECTED locations had fourfold higher human and house densities than OBSERVED locations. Mean road density of OBSERVED locations was 0.93 km/km^2, compared with a maximum of 0.88 km/km^2 found previously in wolf territories in Wisconsin (Wydeven et al. 2001). Five land cover variables differed significantly between OBSERVED and EXPECTED habitat. Wolves undertaking long-range movements selected forest but avoided grassland, row crops, water, and urban areas. Deer density was significantly

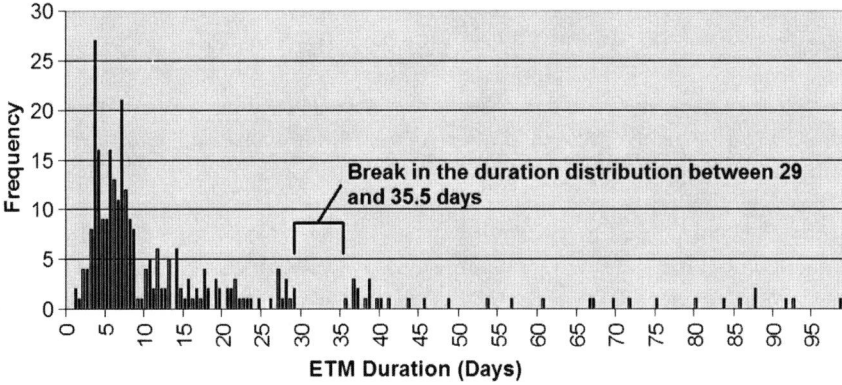

Fig. 12.2 The duration in days (*x-axis*) of extraterritorial movements (*ETMs*) by Wisconsin wolves, and the frequency (*y-axis*) with which these durations were observed

Table 12.3 Comparing landscape features of 609 OBSERVED and 609 EXPECTED locations for long-range movements

Land cover or use	EXPECTED Mean (SD)	OBSERVED Mean (SD)	Observed for females Mean (SD)	Observed for males Mean (SD)	Observed for adults Mean (SD)	Observed for sub-adults Mean (SD)
Water (% Cover)	3.9 (9.7)	1.9 (4.4)[a]	2.0 (4.6)	1.2 (3.0)	1.5 (3.9)	2.2 (4.7)
Urban (% Cover)	1.5 (8.3)	0.2 (0.8)[a]	0.2 (0.8)	0.2 (0.7)	0.2 (0.8)	0.2 (0.7)
Forest (% Cover)	54.7 (27.8)	68.9 (24.6)[a]	70.3 (24.0)	61.6 (25.9)[a]	70.7 (24.6)	67.3 (24.5)
Grassland (% Cover)	11.0 (15.0)	3.8 (7.8)[a]	4.0 (7.8)	3.2 (7.7)	2.8 (6.0)	4.7 (9.0)[a]
Row Crops (% Cover)	9.6 (14.8)	4.9 (9.0)[a]	5.0 (9.0)	4.4 (9.3)	3.2 (6.3)	6.3 (10.7)[a]
Wetlands (% Cover)	19.2 (22.5)	20.3 (22.5)	18.6 (21.5)	29.4 (25.2)[a]	21.6 (23.2)	19.3 (21.8)
Roads (km/ km²)	1.3 (1.2)	0.9 (0.8)[a]	1.0 (0.8)	0.7 (0.6)[a]	0.8 (0.7)	1.0 (0.9)[a]
Humans (per km²)	24.0 (157.5)	3.1 (7.9)[a]	3.4 (8.5)	1.2 (2.2)[a]	2.1 (6.5)	3.9 (8.8)[a]
Houses (per km²)	11.2 (70.9)[a]	2.8 (15.3)[a]	3.1 (16.8)	0.8 (1.0)[a]	1.3 (3.0)	4.0 (20.7)
Farms (per km²)	0.3 (0.2)	0.2 (0.2)	0.2 (0.2)	0.3 (0.2)[a]	0.2 (0.1)	0.3 (0.2)[a]
Cattle (per km²)	7.6 (9.0)	5.9 (5.6)[a]	5.1 (4.5)	10.0 (8.4)[a]	5.3 (5.6)	6.4 (5.7)
Deer (per km²)	10.7 (2.5)	9.9 (1.7)[a]	9.8 (1.5)	10.5 (2.5)	9.6 (1.9)	10.2 (1.5)[a]

See Fig. 12.1 for study area
[a]Indicates difference for Welch ANOVA, $P < 0.01$ values between paired means (comparisons are EXPECTED vs OBSERVED, male vs female, adult vs subadult)

lower in OBSERVED locations, in contrast with studies suggesting wolves follow deer (Poszig and Theberge 2000), and wolves prey on livestock in areas with higher deer density (Treves et al. 2004). However, the coarse resolution of information on deer density necessitates caution. Moreover, deer density increases in agricultural areas, hence apparent avoidance of areas with high density of deer may simply reflect avoidance of human-use areas.

Habitats used by female wolves during long-range movements differed significantly from those used by males in 7 of 12 variables (Table 12.3). Females used habitats with more forest cover, and higher densities of roads, humans, and houses, but lower densities of farms and cattle. Differences between adult and subadult (<2 year) wolves were clearer. Locations of adults differed significantly from those of subadults for 6 of the 12 variables (Table 12.3), with adults using less land cover associated with humans and less grassland cover.

12.4   Summary and Conclusions

Our analysis suggests that wolves undertaking long-range movements did not use habitat randomly. They selected wildland areas, while avoiding exposed habitats and areas modified by humans, including grasslands (pastures and hayfields), row crops, roads, houses, and farms. Preservation and restoration of forest and wetlands would help maintain suitable habitat for dispersal of wolves. Although dispersing wolves occasionally used more marginal habitat during movements, the most highly selected areas seemed similar to habitats selected by wolf packs (Mladenoff et al., this volume).

The observed differences between the sexes and ages are difficult to explain, although theory based on risk-taking, foraging strategies, and conspecific avoidance might prove useful (Linnell et al. 1999; Treves 2000). Females seemed to use habitats with a lower risk of encounter with agriculture and open habitats than did males, whereas males seemed to avoid people and houses. Further resolution of the timing or locations of long-range movements may resolve observed sex differences. Subadults seemed to use habitats with a higher risk than did adults. Subadults may have been avoiding adults or even pack territories because subordinate status may make them particularly vulnerable to attack from conspecifics. Alternatively, inexperienced subadults might be less aware of the hazards posed by people.

Wolves have successfully recolonized much of the forest areas of northern Minnesota, Wisconsin, and the UP of Michigan during the last 30 years. Dispersal movements of up to 886 km have been detected in this region, and have dispersed into at least five neighboring states and three Canadian provinces. Despite the extensive capacity of wolves to disperse over large areas, actual colonization and pack formation has occurred mainly in the northern portions of the region and more recently in central forest blocks and interstitial areas (Mladenoff et al. 1995; Mladenoff et al., this volume).

As wolves moved long distances, they used areas with relatively high forest cover, low road densities, wilder features, and lower human presence, similar to other habitat models for wolves. Ongoing multivariate analyses suggest that Wisconsin wolves have been relaxing their habitat criteria when moving and when establishing pack territories (Martin et al., unpublished data). Thus, current results about subadults relaxing criteria for long-range movements may indicate wolf populations in the western Great Lakes are saturating prime habitat and dispersers will now accept lower quality habitat when seeking new territories. Indeed, the behavior of subadults in our study suggests that young animals may choose or be forced into lower-quality habitat and thereby recolonize areas previously thought unsuitable. Although young dispersing wolves are willing to travel through more marginal habitat, it remains unclear whether survival will be high enough for persistence, territories can be established and maintained, and pups can be raised in these less suitable areas.

Given the long distances wolves can disperse and their broad habitat tolerances, we anticipate that regions and states beyond the three western Great Lakes states should expect occasional wolves to appear. The lessons learned in Michigan, Minnesota, and Wisconsin will undoubtedly help states manage recolonization by a federally protected species that can damage property and engender strong feelings in diverse stakeholders.

References

Ballard, W. B., Whitman, J. S., and Gardner, C. L. 1987. Ecology of an exploited wolf population in south-central Alaska. Wildlife Monographs 98:1–54.

Ballard, W. B., Ayres, L. A., Krausman, P. R., Reed, D. J., and Fancy, S. G. 1997. Ecology of wolves in relation to a migratory caribou herd in northwest Alaska. Wildlife Monographs 135:1–47.

Berg, W. E., and Kuehn, D. W. 1982. Ecology of wolves in north-central Minnesota. In Wolves of the World: Perspectives of Behavior, Ecology and Conservation, eds. F. H. Harrington and P. L. Paquet. pp. 4–11. Park Ridge: Noyes Publishing.

Boyd, D. K., and Pletscher, D. H. 1999. Characteristics of dispersal in a colonizing wolf population in the central Rocky Mountains. Journal of Wildlife Management 63:1094–1108.

Coffey, M. A., Cinnamon, S. K., and Thompson, J. L. 2006. A critique of wildlife radiotracking and its use in National Parks. Fort Collins: Biological Resources Management Division, U.S. National Park Services.

Corsi, F., Dupre, E., and Boitani, L. 1999. A large-scale model of wolf distribution in Italy for conservation planning. Conservation Biology 13:150–159.

ESRI. 2005. ArcGIS 9.1. ESRI, Inc., Redlands, California.

Fretwell, S. D., and Lucas, H. L. Jr. 1970. On territorial behavior and other factors influencing habitat distribution in birds. I. Theoretical development. Acta Biotheoretica 19:16–36.

Fritts, S. H. 1983. Record dispersal by a wolf in Minnesota. Journal of Mammalogy 64:166–167.

Fritts, S. H., and Mech, L. D. 1981. Dynamics, movements, and feeding ecology of a newly protected wolf population in northwestern Minnesota. Wildlife Monographs 80:1–79.

Fuller, T. K. 1989. Population dynamics of wolves in north-central Minnesota. Wildlife Monographs 105:1–41.

Gese, E. M., and Mech, L. D. 1991. Dispersal of wolves (Canis lupus) in northeastern Minnesota, 1969–1989. Canadian Journal of Zoology 69:2946–2955.

Harrison, D. J., and Chapin, T. G. 1998. Extent and connectivity of habitat for wolves in eastern North America. Wildlife Society Bulletin 26:767–775.

Hendrickson, J., Robinson, W. L., and Mech, L. D. 1975. Status of the wolf in Michigan, 1973. American Midland Naturalist 94:226–232.

Jędrzejewski, W., Niedziałkowska, M., Nowak, S., and J drzejewska, B. 2004. Habitat variables associated with wolf (*Canis lupus*) distribution and abundance in northern Poland. Diversity and Distributions 10:225–233.

Licht, D. S., and Fritts, S. H. 1994. Gray wolf (*Canis lupus*) occurrences in the Dakotas. American Midland Naturalist 132:74–81.

Linnell, J. D. C., Odden, J., Smith, M. E., Aanes, R., and Swenson, J. E. 1999. Large carnivores that kill livestock: do problem individuals really exist? Wildlife Society Bulletin 27:698–705.

Linnell, J. D. C., Brøseth, H., Solberg, E. J., and Brainerd, S. M. 2005. The origins of southern Scandinavian wolf *Canis lupus* population: potential for natural immigration in relation to dispersal distances, geography and Baltic ice. Wildlife Biology 11:383–391.

Martin, K. A. 2007. Long-range movements of wolves across the Wisconsin landscape. MS Thesis. Madison: University of Wisconsin-Madison.

Massolo, A., and Meriggi, A. 1998. Factors affecting habitat occupancy by wolves in northern Apennines (northern Italy): a model of habitat suitability. Ecography 21:97–107.

Mech, L. D. 1970. The wolf: the ecology and behavior of an endangered species. Garden City: Natural History Press.

Mech, L. D. 1974. Current techniques in the study of elusive wilderness carnivores. Proceedings of the 11th international Conference of Game Biology 11:315–322.

Mech, L. D. 1983. A Handbook of Animal Radio-tracking. Minneapolis: University of Minnesota Press.

Mech, L. D. 1999. Alpha status, dominance, and division of labor in wolf packs. Canadian Journal of Zoology 77:1196–1203.

Mech, L. D., and Boitani, L. 2003. Wolf social ecology. In Wolves: Behavior, Ecology, and Conservation, eds. L. D. Mech and L. Boitani, pp. 1–34. Chicago: The University of Chicago Press.

Mech, L. D., Fritts, S. H., Radde, G. L., and Paul, W. J. 1988. Wolf distribution and road density in Minnesota. Wildlife Society Bulletin 16:85–87.

Mech, L. D., Fritts, S. H., and Wagner, D. 1995. Minnesota wolf dispersal to Wisconsin and Michigan. American Midland Naturalist 133:368–370.

Merrill, S. B., and Mech, L. D. 2000. Details of extensive movements by Minnesota wolves (*Canis lupus*). American Midland Naturalist 144:428–433.

Messier, F. 1985. Solitary living and extraterritorial movements of wolves in relation to social status and prey abundance. Canadian Journal of Zoology 63:239–245.

Mills, K. J., Patterson, B. R., and Murray, D. L. 2008. Direct estimation of early survival and movements in eastern wolf pups. Journal of Wildlife Management 72:949–954.

Mladenoff, D. J., Haight, R. G., Sickley, T. A., and Wydeven, A. P. 1995. A regional landscape analysis and prediction of favorable gray wolf habitat in the northern Great Lakes region. Conservation Biology 9:279–294.

Mladenoff, D. J., Haight, R. G., Sickley, T. A., and Wydeven, A. P. 1997. Causes and implications of species restoration in altered ecosystems. BioScience 47:21–31.

Mladenoff, D. J., Sickley, T. A., and Wydeven, A. P. 1999. Predicting gray wolf landscape recolonization: logistic regression model vs. new field data. Ecological Applications 9:37–44.

Oakleaf, J. K., Murray, D. L., Oakleaf, J. R., Bangs, E. E., Mack, C. M., Smith, D. W., Fontaine, J. A., Jimenez, M. D., Meier, T. J., and Niemeyer, C. C. 2006. Habitat selection by recolonizing wolves in the northern rocky mountains of the United States. Journal of Wildlife Management 70:554–563.

Peterson, R. O., Woolington, J. D., and Bailey, T. N. 1984. Wolves of the Kenai Peninsula, Alaska. Wildlife Monographs 88:1–52.

Petit, L. J., and Petit, D. R. 1996. Factors governing habitat selection by Prothonotary Warblers: field tests of the Fretwell-Lucas models. Ecological Monographs 66:367–387.

Poszig, D., and Theberge, J. B. 2000. Gray wolf, *Canis lupus lycaon*, responses to shifts of white-tailed deer, *Odocoileus virginianus*, adjacent to Algonquin Provincial Park, Ontario. Canadian Field Naturalist 114:62–71.

Potvin, M. J., Drummer, T. D., Vucetich, J. A., Beyer, D. E., Peterson, R. O., and Hammill, J. H. 2005. Monitoring and habitat analysis for wolves in upper Michigan. Journal of Wildlife Management 69:1–10.

Shields, W. M. 1987. Dispersal and mating systems: investigating their causal connections. In Mammal Dispersal Patterns: The Effects of Social Structure on Population Genetics, eds. B. D. Chepko-Sade and Z. T. Halpin, pp. 3–26. Chicago: University of Chicago Press.

Thiel, R. P. 1978. The status of the timber wolf in Wisconsin – 1975. Transactions of the Wisconsin Academy of Science, Arts and Letters, 66:186–194.

Thiel, R. P. 1985. Relationship between road densities and wolf habitat suitability in Wisconsin. American Midland Naturalist 113:404–407.

Thiel, R. P. 1988. Dispersal of a Wisconsin wolf into Upper Michigan. The Jack-Pine Warbler 66:143–147.

Thiel, R. P., and Hammill, J. H. 1988. Wolf specimen records in upper Michigan, 1960s–1986. The Jack-Pine Warbler 66:149–153.

Thogmartin, W. E., Gallant, A. L., Knutson, M. G., Fox, T. J., and Suarez, M. J. 2004. Commentary: a cautionary tale regarding use of the National Landcover Dataset 1992. Wildlife Society Bulletin 32:970–978.

Treves, A. 2000. Theory and method in studies of vigilance and aggregation. Animal Behaviour 60:711–722.

Treves, A., Jurewicz, R. R., Naughton-Treves, L., Rose, R. A., Willging, R. C., and Wydeven, A. P. 2002. Wolf depredation on domestic animals: control and compensation in Wisconsin, 1976–2000. Wildlife Society Bulletin 30:231–241.

Treves, A., Naughton-Treves, L., Harper, E. K., Mladenoff, D. J., Rose, R. A., Sickley, T. A., and Wydeven, A. P. 2004. Predicting human-carnivore conflict: a spatial model derived from 25 years of data on wolf predation on livestock. Conservation Biology 18:114–125.

Turner, M. G., Arthaud, G. J., Engstrom, R. T., Hejl, S. T., Liu, J., Loeb, S., and McKelvey, K. 1995. Usefulness of spatially explicit population models in land management. Ecological Applications 5:12–16.

U.S. Census Bureau. 2001. Census 2000 TIGER/line files. Washington: U.S. Census Bureau, Washington.

U.S. Department of Agriculture. 1997. Census of Agriculture. Washington: USDA.

USFWS. 2007. Endangered and Threatened Wildlife and Plants; Final Rule Designating the Western Great Lakes Populations of Gray Wolves as a Distinct Population Segment; Removing the Western Great Lakes Distinct Population Segment of the Gray Wolf From the List of Endangered and Threatened Wildlife. Federal Register 72:6051–6103.

USFWS. 2008. Endangered and Threatened Wildlife and Plants; Final Rule Designating the Northern Rocky Mountain Population of Gray Wolf as a Distinct Population Segment and Removing This Distinct Population Segment From the Federal List of Endangered and Threatened Wildlife; Final Rule. Federal Register 73:10513–10560.

Vogelmann, J. E., Howard, S. M., Yang, L., Larson, C. R., Wylie, B. K., and Driel, N. V. 2001. Completion of the 1990s National Land Cover Data Set for the coterminous United States from Landsat Thematic Mapper data and ancillary data sources. Photogrammetric Engineering & Remote Sensing 67:650–652.

Wabakkan, P., Sand, H., Kojola, I., Zimmermann, B., Arnemo, J. M., Pedersen, H. C., and Liberg, O. 2007. Multistage, long-range natal dispersal by Global Positioning System-collared Scandinavian wolf. Journal of Wildlife Management 71:1631–1634.

Welch, B. L. 1951. On the comparison of several mean values: an alternative approach. Biometrika 38:330–336.

Whitham, T. G. 1980. The theory of habitat selection: examined and extended using Pemphigus aphids. American Naturalist 115:449–466.

Wickham, J. D., Stehman, S. V., Smith, J. H., and Yang, L. 2004. Thematic accuracy of the 1992 National Land-Cover Data for the western United States. Remote Sensing of Environment 91:452–468.

Wisconsin DNR. 1998. Wisconsin's deer management program; the issues involved in decision-making. Madison: Wisconsin Department of Natural Resources.

Woodroffe, R., and Ginsburg, J. R. 1998. Edge effects and the extinction of populations inside protected areas. Science 280:2126–2128.

Wydeven, A. P. 1994. Travels of a Midwestern disperser. International Wolf 4:20–22.

Wydeven, A. P., Schultz, R. N., and Thiel, R. P. 1995. Monitoring of a recovering gray wolf population in Wisconsin, 1979–1991. In Ecology and Conservation of Wolves in a Changing World, eds. L. N. Carbyn, S. H. Fritts, and D. R. Seip, pp.147–156. Edmonton: Canadian Circumpolar Institute.

Wydeven, A. P., Fuller, T. K., Weber, W., and MacDonald, K. 1998. The potential for wolf recovery in the northeastern United States via dispersal from southeastern Canada. Wildlife Society Bulletin 26:776–784.

Wydeven, A. P., Mladenoff, D. J., Sickley, T. A., Kohn, B. E., Thiel, R. P., and Hansen, J. L. 2001. Road density as a factor in habitat selection by wolves and other carnivores in the Great Lakes Region. Endangered Species Update 18:110–114.

Wydeven, A. P., Treves, A., Brost, B., and Wiedenhoeft, J. E. 2004. Characteristics of wolf packs in Wisconsin: identification of traits influencing depredation. In People and Predators: From Conflict to Coexistence, eds. N. Fascione, A. Delach, and M. E. Smith, pp. 28–50. Washington, D. C.: Island Press.

Chapter 13
Are Wolf-Mediated Trophic Cascades Boosting Biodiversity in the Great Lakes Region?

Thomas P. Rooney and Dean P. Anderson

13.1 Introduction

In recent years, conservation biologists broadened their efforts beyond genes, species, and ecosystems to include the conservation of species interactions such as mutualisms and predation (Kearns et al. 1998; Soulé et al. 2003, 2005). Ecologists generally agree that predators generate top-down effects in food webs, but the consensus ends there. Considerable disagreement remains over the strength of top-down effects, the relative importance of top-down versus bottom-up effects, and how the relative importance of these effects differs among systems, seasons and across scales (Polis 1999; Polis et al. 2000; Shurin et al. 2002; Schmitz et al. 2004). The question is further complicated because many top predators were significantly reduced in abundance or eliminated from temperate zone ecosystems decades or centuries before ecologists formally conceptualized trophic cascades (cf. Jackson 1997; Jackson et al. 2001). Still, many conservation biologists view the recovery of the gray wolf (*Canis lupus*) in the Great Lakes region as more than a conservation success story. This recovery carries with it the hope and expectation that the top-down effects generated by gray wolves will aid in the maintenance of regional biodiversity (McShea 2005; Ray 2005).

High densities of white-tailed deer (*Odocoileus virginianus*) throughout much of the upper Great Lakes region pose a challenge to conservation efforts. Densities are so great that deer harvests have set state records within the last 10 years (e.g., Michigan in 1998, Wisconsin in 2000, and Minnesota in 2003). High densities of deer come at an ecological cost: browsing contributed to the loss of plant diversity over the last few decades (Rooney et al. 2004), and, in turn, these losses might be generating additional indirect effects on insects, birds, and other species (Rooney and Waller 2003; McShea 2005). Several studies from western North America suggest that recovery of gray wolves generated strong top-down effects on vegetation and lateral effects on assemblages of scavengers (Ripple et al. 2001; Wilmers et al. 2003; Hebblewhite et al. 2005).

Are wolves having a similar effect in the Great Lakes states? While it might be tempting to simply extrapolate findings from western North America and apply them

A.P. Wydeven et al. (eds.), *Recovery of Gray Wolves in the Great Lakes Region of the United States*,
DOI: 10.1007/978-0-387-85952-1_13, © Springer Science+Business Media, LLC 2009

to this region, doing so would gloss over several important differences. Elk (*Cervus elaphus*) serves as the primary prey species in western North America, whereas white-tailed deer (and on Isle Royale, moose, *Alces alces*) is the primary prey of wolves in the Great Lakes states. White-tailed deer and moose are browsers, feeding primarily on forbs and woody plants. Elk rely on a mixed foraging strategy that includes both browsing and grazing, relying more heavily on grasses and other graminoid plants (Gordon 2003). Because their food resources are distributed differently in both space and time, the spatial distribution and behavior of these ungulates differs as well. Elk tends to be more social than white-tailed deer or moose (Kurta 1995). In western North America, high-quality browse is concentrated in riparian areas, but in the Great Lakes states, it is more evenly distributed throughout the landscape. Elk and white-tailed deer also differ in their seasonal movements. Elk migrate to their winter range each year, whereas migration of deer to winter yards varies annually in response to winter severity, and geographically, as the severity of winters is more pronounced toward the northern limits of the geographic range of deer.

The Great Lakes states lack the topographic relief and the extensive open grasslands of the Rocky Mountain region, and both could influence predator–prey dynamics. The more open, rugged landscapes of the west make it easier for wolves to locate prey. The food web structures also are very different between the regions. Western North American food webs contain more ungulates and predators than those in the Great Lakes states. Smith et al. (2003) showed greater complexity of food webs in Yellowstone National Park than on the smaller and more depauperate Isle Royale. While food web structure on the mainland in the Great Lakes states is more complex than on Isle Royale, it still lacks the complexity of Yellowstone (compare Smith et al. 2003 with Kurta 1995). The Great Lakes region also contains higher densities of people and roads. For these reasons, findings from the west might not be directly applicable to the Great Lakes states. However, insights from western North America do provide a starting point for trying to understand what, if any, trophic effects wolves may have in the Great Lakes region.

In this chapter, we explore the trophic effects that wolves have in Yellowstone National Park and the northern Rocky Mountains. We combine this body of research with studies conducted in the Great Lakes region, identifying trophic interactions that are common to both regions. We highlight some of the important factors that can modulate trophic effects in the Great Lakes region, and conclude with predictions how these trophic relationships will play out over time, and the research needed to test these predictions.

13.2 Trophic Interactions and Subsidies of Scavenger Food Webs

Soulé et al. (2003) consider the wolf to be a "strongly interactive" species, meaning that its removal or substantial reduction leads to significant changes in the ecosystems it inhabited. Their terminology is similar to Paine's (1966) keystone species concept, but it relaxes the requirement that the species has effects disproportionate to its population density. Evidence from western North America and the Great Lakes region support the idea that the wolf is strongly interactive and influencing biodiversity at some scales.

13.2.1   The Wolves of Yellowstone National Park and the Northern Rocky Mountains

Absent since 1926, wolves were reintroduced to Yellowstone National Park in 1995. Elk served as their primary prey, and White and Garrott (2005) reported that in their first decade of recovery, wolves had no appreciable effect on densities of mule deer (*Odocoileus hemionus*), bison (*Bison bison*), moose, bighorn sheep (*Ovis canadensis*), or pronghorn antelope (*Antilocapra americana*). They further noted a 50% decline in population density of elk between 1995 and 2004, suggesting a pronounced effect of wolves on elk. This precipitous decline, however, also may have been influenced by coincident extreme weather events and increased harvest rates by humans (Vucetich et al. 2005).

Studies of the age structure of trembling aspen (*Populus tremuloides*) and cottonwood (*Populus*) trees revealed a more-or-less regular pattern of establishment, until wolves were extirpated. A significant gap in establishment stretches from the 1920s through the 1990s, a period marked by intense herbivory by elk (Beschta 2003, 2005; Larsen and Ripple 2003). Released from herbivory, heights, and densities of riparian aspen, cottonwood, and willow (*Salix*) seedlings increased (Ripple et al. 2001; Ripple and Beschta 2003, 2006, 2007; Beyer et al. 2007). Declines in elk density probably contributed to this recovery, but only in part. Foraging elk exhibited increased vigilance (Fortin et al. 2004) and selected summer habitats in areas that offered increased protection from wolves (Mao et al. 2005). Recovery of vegetation was most pronounced on sites where predation risk was highest (Ripple and Beschta 2003), suggesting that trophic cascades were mediated through prey behavior (Ripple and Beschta 2004; Fortin et al. 2005). When elk detect wolf activity, they become more wary and tend to avoid high-risk sites. The combination of increased vigilance, forgoing feeding opportunities where they would be vulnerable to predation, and frequent movement through the landscape dilutes browsing pressure on the landscape, enabling woody plants that were suppressed by browsing to begin recovery (Gude et al. 2006).

Recovering vegetation can lead to additional indirect effects on other species (Berger and Smith 2005). Berger et al. (2001) compared the consequences of the loss of predators in Grand Teton National Park, where hunting was not permitted, to those in control sites with human hunting pressure. In addition to noting greater densities of moose and lower heights and densities of willows, densities of birds were lower in riparian areas without human hunting pressure. Hebblewhite et al. (2005) found similar patterns in Banff National Park: in areas where densities of wolves were low, they observed higher densities of elk, lower net twig production by willows, and lower density and diversity of songbirds. They further noted a decrease in abundance of beaver (*Castor canadensis*) lodges in areas with high densities of elk, providing evidence of a positive, indirect effect of wolves on beavers.

Top-down effects of wolves have not been uniform throughout the region. Garrott et al. (2005) noted that densities of elk varied substantially between two areas 30 km apart in the Madison watershed, despite both having resident wolf packs. Ripple and Beschta (2007) found significant recovery of aspen in riparian areas and wet meadows, but not in upland steppe habitats. In another study, they reported recovering willows were evident in valley bottoms, but not in upland riparian areas (Ripple and Beschta 2006). With respect to both prey and trophic cascades, the effect of wolves has been strongly context dependent.

In addition to generating top-down effects, wolves can generate lateral effects by influencing scavenger food webs. Prior to reintroduction of wolves, availability of carrion was a function of winter severity and consequently was concentrated in late winter (Wilmers et al. 2003). Following the recovery of wolves, carrion became more abundant throughout most of the winter, potentially benefiting populations of other scavengers, including ravens (*Corvus corax*), golden eagles (*Aquila chrysaetos*), bald eagles (*Haliaeetus leucocephalus*), red foxes (*Vulpes vulpes*), black-billed magpies (*Pica hudsonia*), and numerous species of carrion-feeding insects. By decreasing inter-annual variation in the availability of carcasses, wolves could contribute to larger population sizes of scavengers. Wilmers et al. (2003) proposed that the growth in other scavenger populations could decrease the amount of food wolves derive from each kill, forcing them to kill more frequently and therefore strengthening top-down effects.

13.2.2 Wolves of the Great Lakes States

Wolf populations showed significant increases in Minnesota in the 1960s, Wisconsin and northern Michigan in the 1990s and early 2000s (Mladenoff et al. 1997; Erb, this volume; Wydeven et al., this volume; Beyer et al., this volume). There are few examples of systems where wolves limit prey densities in the region, but one such situation occurs at the northern limit of the geographic range of white-tailed deer in Minnesota (Nelson and Mech 2006). Eberhardt (1997) noted that it was difficult to assess the effects of wolves on moose on Isle Royale, but concluded that it was possible for wolves to limit prey populations (*see also* Mech and Peterson 2003; Vucetich and Peterson, this volume).

Trophic cascades attributed to large carnivores have been detected in the Great Lakes region, although the amount of research has been limited (Ray et al. 2005). One of the first studies to convincingly demonstrate a trophic cascade in terrestrial ecosystems came from the wolf–moose–balsam fir (*Abies balsamea*) system on Isle Royale (McLaren and Peterson 1994). Here, greater annual-increment growth rings of balsam fir corresponded with periods of high density of wolves and low density of moose, indicating a release from browsing pressure. Qualitative observations from northeastern Minnesota revealed greater recruitment of saplings in the swath of a tornado in areas with few deer and large wolf populations, compared to adjacent areas with more deer (Nelson and Mech 2006). D. P. Anderson et al. (unpublished data) found that forbs, shrubs, and saplings had higher biomass in cedar swamps located within territories of wolf packs, but this effect was not detected in three other forest types. Their study was conducted in Wisconsin and parts of western Michigan, where white-tailed deer depress forage productivity and biomass in cedar swamps (Rooney et al. 2002). Evidence for trophic cascades following wolf recovery in the Great Lakes region has not been as strong as the evidence uncovered in the Rocky Mountains region, but this seems more to do with absence of evidence rather than evidence of absence. We are not aware of any studies in the Great Lakes region that examine trophic cascades beyond responses of vegetation.

There is compelling evidence that wolves generate behavioral shifts in their prey in the Great Lakes region, as reported for western North America. In his study of a rapid decline in prey density, Mech (1977) observed that the majority of surviving white-tailed deer were found at the edges of wolf-pack territories. Mech hypothesized that where wolf-pack territories abut, conflicts between packs create a buffer zone that can serve as a refuge for prey. Later research confirmed that these buffer zones do support higher densities of deer (Rogers et al. 1980) because wolves experience higher mortality rates near the edge of their territory as a result of conflicts with neighboring packs (Mech 1994). In their study of a small, reintroduced herd of elk, Anderson et al. (2005) reported that the location of wolf packs influenced habitat selection of individual elk at broad spatial scales; elk established home ranges in areas away from wolves.

There has been no research specific to the Great Lakes region on how wolves might generate lateral effects to influence scavenger food webs, but it is likely to differ substantially from what other researchers observed in western North America. Collisions with vehicles are an important source of mortality for deer in Wisconsin and Michigan, and probably play a more important role in subsidizing scavenger food webs than wolves. Species tolerant of edges and traffic like ravens and crows (*Corvus brachyrhynchos*) are more likely to benefit than less-tolerant species. Populations of wolves in the Great Lakes region still subsidize scavengers, but their importance to this food web is probably less than in the west.

13.3   Factors Modulating Tropic Interactions

Over the last decade, researchers have shifted from asking whether trophic cascades operate in terrestrial systems to understanding the strength and importance of trophic cascades in different contexts and under different conditions (Pace et al. 1999; Polis et al. 2000). Spatial variation in top-down effects appears both in western North American and in the Great Lakes ecosystems. In Yellowstone National Park, for example, some of this variation is due to perceived predation risk (Ripple and

Beschta 2003). In fact, several factors can act to modulate the strength of trophic interactions.

Variation in climate appears to be one such factor in the Great Lakes region. Wolf–moose dynamics on Isle Royale in winter are mediated by snowfall. In years with high snowfall, wolves hunt in larger packs and per capita kill rates rise. Moose density shows a negative correlation with the average size of wolf packs in the previous winter, and annual growth of balsam fir increases with decreasing moose density (Post et al. 1999). The coupling between climate and strength of trophic interactions has important consequences with respect to climate change. If climate change leads to milder winters with less snowfall, the magnitude of this trophic cascade will probably weaken, all else being equal.

Climate also affects scavenger food webs in Yellowstone National Park. Wilmers and Getz's (2005) model predicts that a reduction in snow depth as a result of climate change will enhance survival of elk in winter. As a consequence, availability of carrion in late winter may decline in the future relative to today. The model predicts more carrion with wolves than without, but climate could become an overriding factor.

Parasites also can play an important role in community dynamics of predators (Hatcher et al. 2006), potentially modulating the strength and importance of trophic cascades. For example, Wilmers et al. (2006) examined an outbreak of canine parvovirus (CPV) on Isle Royale and its effects on wolves, moose, and balsam fir. Following the introduction of CPV, the wolf population declined numerically. This decline in the abundance of wolves diminished the importance of predation in moose population dynamics, resulting in an increase in the relative importance of bottom-up population regulation. CPV has probably also contributed to reduced growth rates of balsam fir through its direct effects on wolves and indirect effects on moose (McLaren and Peterson 1994). Perhaps most interesting, Wilmers et al. (2006) found that the relative importance of climatic variation as a determinant of moose population growth rates doubled after the introduction of CPV, reflecting complex interactions involving climate, parasites, and trophic cascades.

Habitat productivity also can modulate the strength of trophic interactions and indirect effects in ecosystems. In their study of herbivore-generated indirect effects in a tropical savanna, Pringle et al. (2007) found strong direct effects of ungulates on productivity of vegetation and significant indirect effects on lizards and arthropods. The strength of these indirect effects increased with decreasing productivity. How wolf-generated trophic cascades might vary across productivity gradients has not been explored in detail.

Recent work by Schmitz et al. (2004) on arthropod assemblages highlights the importance of food web topology as a mediator of trophic cascades. Depending on the particular configuration of the food web, predators can have both positive and negative effects on plants in systems where predators influence prey behavior.

Finally, human activities can modulate the strength of trophic interactions through a variety of mechanisms, only a few of which we will highlight. In the Great Lakes region, white-tailed deer serve as the primary prey for wolves. Human hunters and vehicular collisions are major sources of deer mortality, but are unlikely to reduce the wolf's prey base substantially. It is not clear if the human toll causes additive or compensatory mortality, although research from Yellowstone National Park found compensatory mortality in the elk population subjected to depredation by both humans and wolves (Vucetich et al. 2005). Other activities, though, can enhance this prey base. The fragmentation of land ownership creates a mosaic in which some land is accessible to deer hunters, while other land is not. Furthermore, firearms ordinances create safe havens for deer in municipal areas. Humans can increase availability of forage for deer at multiple scales: recreational feeding increases concentrations of deer locally, intensive forestry and crop production can boost carrying capacity regionally. Finally, human activities influence many other modulators already mentioned, including climate, introductions of disease, and habitat productivity.

13.4   Trophic Interactions of the Recovered Great Lakes Wolf Population: Predictions and Research Needs

Ray (2005) questioned whether wolf recovery in eastern North America would influence deer populations and generate the trophic cascades now demonstrated in Yellowstone National Park. We believe that they will under some but not all conditions. Where trophic cascades do occur, we expect them to be almost exclusively through the behavioral effects on prey, rather than through a numerical effect. By increasing vigilance and movement of prey (Switalski 2003; Fortin et al. 2004; Gude et al. 2006), wolves may generate increases in biomass and productivity of some plant species in some places. However, the strength and importance of this effect will be modulated by climate, human activity, habitat productivity, and several other factors. The challenge will be predicting where and when trophic cascades will be important.

   Where should we look for trophic cascades? The recolonization of the Great Lakes region by wolves has superimposed a type of chronosequence on the land. Packs became established in some areas 15 years ago, other areas 10 years ago, and still others only 5 years ago. Some areas remain uncolonized and will likely remain so. By conducting studies close to the core of wolf-pack territories in each of these areas and holding habitat constant, we can begin to look at differences in sapling, shrub, and herbaceous vegetation as a function of time since colonization by wolves. Alternatively, we can study vegetation in the buffer zones between packs and compare it to vegetation within an identical habitat type within territories. Both approaches might be combined with deer exclosure experiments to examine differences in plant performance in areas with and without wolves. These approaches provide a natural experimental framework to identify what, if any, effects wolves are having on vegetation.

   The snow depth gradient that decreases from Lake Superior south to Wisconsin and Michigan's Upper Peninsula provides another research opportunity. Since per capita kill rates increase with snow depth, wolves might generate trophic cascades by reducing winter deer densities close to Lake Superior where snows are deepest.

What should we look for? The wolf-generated trophic cascades described thus far have been what Polis (1999) terms species cascades, that is, affecting one or a few specific species of plants. In some contexts, wolves have facilitated increased growth rates of aspen, willow, cottonwood (Ripple and Beschta 2003, 2006; Hebblewhite et al. 2005; Beyer et al. 2007), and balsam fir (McLaren and Peterson 1994). Researchers should initially identify trophic effects on the size or reproductive status of individual plant species impacted by deer overbrowsing. Ideally, these species should be fairly widespread. Candidate species include northern white cedar (*Thuja occidentalis*), bluebead lily (*Clintonia borealis*), hairy Solomon's seal (*Polygonatum pubescens*), and sessile bellwort (*Uvularia sessilifolia*). Alternatively, researchers might look to a guild of plants likely to generate indirect effects. Rooney et al. (2004) observed declines in the relative frequency of plant species with animal-pollinated flowers where impacts of deer browsing were most pronounced. This could have indirect consequences for insect pollinators. It seems unlikely that wolves will generate trophic cascades strong enough to affect overall primary productivity, as there are some groups of plants such as graminoids that benefit under intense browsing pressure (Boucher et al. 2004).

Should we look for trophic interactions that are mediated via a numerical or behavioral response to wolves? Evidence from Yellowstone National Park suggests both numerical and behavioral trophic effects. Prior to reintroduction of wolves into Yellowstone in 1995, the northern range of the park harbored a very large elk population (19,000 elk; White and Garrott 2005). This population was unhindered in its movement and foraging, which took its toll on sensitive vegetation communities (Ripple et al. 2001; Ripple and Beschta 2003). Concurrent with the precipitous decrease in numbers of elk following the reintroduction of wolves (8,335 elk in 2004; White and Garrott 2005) were changes in the movement and foraging behavior of elk (Fortin et al. 2005; Mao et al. 2005). Consequently, it remains unclear in Yellowstone if the strong trophic effects are mediated primarily via numerical or behavioral responses. In contrast, the population size of white-tailed deer in the forests of northern Wisconsin at the onset of recolonization by wolves was approximately 250,000 deer (WiDNR 1999). The population of deer has increased concurrently with the increasing wolf population to an estimated 365,120 deer in 2005 (Rolley 2005). The deer population is primarily controlled by winter severity and human harvest (Creed et al. 1984; WiDNR 1999), and with the possible exception of deer residing in the Lake Superior snow belt, the numerical effect of wolves on deer is clearly minimal. Trophic effects of wolves extending to plant communities in Wisconsin are, therefore, likely mediated via a behavioral response.

Despite decades of research, scientists are only beginning to understand the complex role wolves play in ecosystems. Even after 40 years, ongoing studies of the wolves and moose on Isle Royale yield surprising insights (cf. Post et al. 1999; Wilmers et al. 2006). The recent recovery of wolves in the Great Lakes region presents us with a unique opportunity, a chance to understand how a top predator influences biodiversity on a regional scale. We have not yet even scratched the surface, and the most exciting discoveries still lie ahead.

References

Anderson, D. P., Turner, M. G., Forester, J. D., Zhu, J., Boyce, M. S., Beyer, H., and Stowell, L. 2005. Scale-dependent summer resource selection by reintroduced elk in Wisconsin, USA. Journal of Wildlife Management 69:298–310.

Berger, J., and Smith, D. W. 2005. Restoring functionality in Yellowstone with recovering carnivores: gains and uncertainties. In Large carnivores and the conservation of biodiversity, eds. J. C. Ray, K. H. Radford, R. S. Steneck, and J. Berger, pp. 100–109. Washington, DC: Island Press.

Berger, J., Stacey, P. B., Bellis, L., and Johnson, M. P. 2001. A mammalian predator-prey disequilibrium: how the extinction of grizzly bears and wolves affects the diversity of avian neotropical migrants. Ecological Applications 11:947–960.

Beschta, R. L. 2003. Cottonwoods, elk, and wolves in the Lamar Valley of Yellowstone National Park. Ecological Applications 13:1295–1309.

Beschta, R. L. 2005. Reduced cottonwood recruitment following extirpation of wolves in Yellowstone's northern range. Ecology 86:391–403.

Beyer, H. L., Merrill, E. H., Varley, N., and Boyce, M. S. 2007. Willow on Yellowstone's Northern Range: evidence for a trophic cascade in a large mammalian predator-prey system? Ecological Applications 17:1563–1571.

Boucher, S., Crête, M., Ouellet, J.-P., Daigle, C., and Lesage, L. 2004. Large-scale trophic interactions: white-tailed deer growth and forest understory. Ecoscience 11:286–295.

Creed, W. A., Haberland, F., Kohn, B. E., and McCaffery, K. R. 1984. Harvest management: the Wisconsin experience. In White-tailed Deer Ecology and Management, ed. L. K. Halls, pp. 243–260. Harrisburg, PA: Stackpole Books.

Eberhardt, L. L. 1997. Is wolf predation ratio-dependent? Canadian Journal of Zoology 75:1940–1944.

Fortin, D., Boyce, M. S., Merrill, E. H., and Fryxell, J.M. 2004. Foraging costs of vigilance in large mammalian herbivores. Oikos 107:172–180.

Fortin, D., Beyer, H. L., Boyce, M. S., Smith, D. W., Duchesne, T., and Mao, J. S. 2005. Wolf influence elk movements: behavior shapes a trophic cascade in Yellowstone National Park. Ecology 86:1320–1330.

Garrott, R. A., Gude, J. A., Bergman, E. J., Gower, C., White, P. J., and Hamlin, K. L. 2005. Generalizing wolf effects across the Greater Yellowstone area: a cautionary note. Wildlife Society Bulletin 33:1245–1255.

Gordon, I. J. 2003. Browsing and grazing ruminants: are they different beasts? Forest Ecology and Management 181:13–21.

Gude J. A., Garrott, R. A., Borkowski, J. J., and King, F. 2006. Prey risk allocation in a grazing ecosystem. Ecological Applications 16:285–298.

Hatcher, M. J., Dick, J. T. A., and Dunn, A. M. 2006. How parasites affect interactions between competitors and predators. Ecology Letters 9:1253–1271.

Hebblewhite, M., White, C. A., Nietvelt, C. G., McKenzie, J. A., Hurd, T. E., Fryxell, J. M., Bayley, S. E., and Paquet, P. C. 2005. Human activity mediates a trophic cascade caused by wolves. Ecology 86:2135–2144.

Jackson, J. B. C. 1997. Reefs since Columbus. Coral Reefs 16: S23–S32.

Jackson, J. B. C., Kirby, M. X., Berger, W. H., Bjorndal, K. A., Botsford, L. W., Bourque, B. J., Bradbury, R. H., Cooke, R., Erlandson, J., Estes, J. A., Hughes, T. P., Kidwell, S., Lange, C. B., Lenihan, H. S., Pandolfi, J. M., Peterson, C. H., Steneck, R. S., Tegner, M. J., and Warner, R. R. 2001. Historical overfishing and the recent collapse of coastal ecosystems. Science 293:629–637.

Kearns, C. A., Inouye, D. W., and N. M. Waser, N. M. 1998. Endangered mutualisms: the conservation of plant-pollinator interactions. Annual Review of Ecology and Systematics 29:83–112.

Kurta, A. 1995. Mammals of the Great Lakes Region, revised edition. Ann Arbor, MI: University of Michigan Press.

Larsen, E. J., and W. J. Ripple, W. J. 2003. Aspen age structure in the northern Yellowstone Ecosystem, USA. Forest Ecology and Management 179:469–482.

Mao, J. S., Boyce, M. S., Smith, D. W., Singer, F. J., Vales, D. J., Vore, J. M., and Merrill, E. H. 2005. Habitat selection by elk before and after wolf reintroduction in Yellowstone National Park. Journal of Wildlife Management 69:1691–1707.

McLaren, B. E., and Peterson, R. O. 1994. Wolves, moose and tree rings on Isle Royale. Science 266:1555–1558.

McShea, W. J. 2005. Forest ecosystems without carnivores: when ungulates rule the world. In Large Carnivores and the Conservation of Biodiversity, eds. J. C. Ray, K. H. Radford, R. S. Steneck, and J. Berger, pp. 138–153. Washington, DC: Island Press.

Mech, L. D. 1977. Wolf-pack buffer zones as prey reservoirs. Science 198:320–321.

Mech, L. D. 1994. Buffer zones of territories of gray wolves as regions of intraspecific strife. Journal of Mammalogy 75:199–202.

Mech, L. D., and Peterson, R. O. 2003. Wolf-prey relations. In Wolves: Behavior, Ecology, and Conservation, eds. L. D. Mech and L. Boitani, pp. 131–160. Chicago, IL: University of Chicago Press.

Mladenoff, D. J., Haight, R. G., Sickley, T. A., and Wydeven, A. P. 1997. Causes and implications of species restoration in altered systems. BioScience 47:21–31.

Nelson, M. E., and Mech, L. D. 2006. A 3-decade dearth of deer (Odocoileus virginianus) in a wolf (Canis lupus)-dominated ecosystem. American Midland Naturalist 155:373–382.

Pace, M. L., Cole, J. J., Carpenter, S. R., and Kitchell, J. F. 1999. Trophic cascades revealed in diverse ecosystems. Trends in Ecology and Evolution 14:483–488.

Paine, R. T. 1966. Food web complexity and species diversity. American Naturalist 100:65–75.

Polis, G. A. 1999. Why are parts of the world green? Multiple factors control productivity and the distribution of biomass. Oikos 86:3–15.

Polis, G. A., Sears, A. L. W., Huxel, G. R., Strong, D. R., and Maron, J. 2000. When is a trophic cascade a trophic cascade? Trends in Ecology and Evolution 15:473–475.

Post, E., Peterson, R. O., Stenseth, N. C., and McLaren, B. E. 1999. Ecosystem consequences of wolf behavioral response to climate. Nature 401:905–907.

Pringle, R. M., Young, T. P., Rubenstein, D. I., and McCauley, D. J. 2007. Herbivore-initiated interaction cascades and their modulation by productivity in an African savanna. Proceedings of the National Academy of Sciences USA 104:193–197.

Ray, J. C. 2005. Large carnivorous animals as tools for conserving biodiversity: assumptions and uncertainties. In Large Carnivores and the Conservation of Biodiversity, eds. J. C. Ray, K. H. Radford, R. S. Steneck, and J. Berger, pp. 34–56. Washington, DC: Island Press.

Ray, J. C., Redford, K. H., Berger, J., and Steneck, R. 2005. Conclusion: Is large carnivore conservation equivalent to biodiversity conservation and how can we achieve both? In Large Carnivores and the Conservation of Biodiversity, J. C. Ray, K. H. Radford, R. S. Steneck, and J. Berger, pp. 400–427. Washington, DC: Island Press.

Ripple, W. J., and R. L. Beschta, R. L. 2003. Wolf reintroduction, predation risk, and cottonwood recovery in Yellowstone National Park. Forest Ecology and Management 184:299–313.

Ripple, W. J., and Beschta, R. L. 2004. Wolves and the ecology of fear: Can predation risk structure ecosystems? BioScience 54:755–766.

Ripple, W. J., and Beschta, R. L. 2006. Linking wolves to willows via risk-sensitive foraging by ungulates in the northern Yellowstone ecosystem. Forest Ecology and Management 230:96–106.

Ripple, W. J., and Beschta, R. L. 2007. Restoring Yellowstone's aspen with wolves. Biological Conservation 138:514–519.

Ripple, W. J., Larsen, E. J., Renkin, R. A., and Smith, D. W. 2001. Trophic cascades among wolves, elk, and aspen on Yellowstone National Park's northern range. Biological Conservation 102:227–234.

Rogers, L. L., Mech, L. D., Dawson, D. K., Peek, J. M., and Korb, M. 1980. Deer distribution in relation to wolf pack territory edges. Journal of Wildlife Management 44:253–258.

Rolley, R. E. 2005. White-tailed Deer Population Status 2005. Monona, WI: Wisconsin Department of Natural Resources.

Rooney, T. P., and Waller, D. M. 2003. Direct and indirect effects of deer in forest ecosystems. Forest Ecology and Management 181:165–176.

Rooney, T. P., Solheim, S. L., and Waller, D. M. 2002. Factors influencing the regeneration of northern white cedar in lowland forests of the Upper Great Lakes region, USA. Forest Ecology and Management 163:119–130.

Rooney, T. P., Wiegmann, S. M., Rogers, D. A., and Waller, D. M. 2004. Biotic impoverishment and homogenization in unfragmented forest understory communities. Conservation Biology 18:787–798.

Schmitz, O. J., Krivan, V., and Ovadia, O. 2004. Trophic cascades: the primacy of trait-mediated indirect interactions. Ecology Letters 7:153–163.

Smith, D.W., Peterson, R. O., and Houston, D. B. 2003. Yellowstone after wolves. BioScience 53:330–340.

Shurin J. B., Borer, E. T., Seabloom, E. W., Anderson, K., Blanchette, C. A., Broitman, B., Cooper, S. D., and Halpern, B. S. 2002. A cross-ecosystem comparison of the strength of trophic cascades. Ecology Letters 5:785–791.

Soulé, M. E., Estes, J. A., Berger, J., and Del Rio, C. M. 2003. Ecological effectiveness: conservation goals for interactive species. Conservation Biology 17:1238–1250.

Soulé, M. E., Estes, J. A., Miller, B., and Honnold, D. L. 2005. Strongly-interacting species: conservation policy, management, and ethics. BioScience 55:168–176.

Switalski, T.A. 2003. Coyote foraging ecology and vigilance in response to gray wolf reintroduction in Yellowstone National Park. Canadian Journal of Zoology 81: 985–993.

Vucetich, J. A., Smith, D. W., and Stahler, D. R. 2005. Influence of harvest, climate and wolf predation on Yellowstone elk, 1961–2004. Oikos 111:259–270.

White, P. J., and Garrott, R. A. 2005. Yellowstone's ungulates after wolves – expectations, realizations, and predictions. Biological Conservation 125:141–152.

WiDNR. 1999. Wisconsin Wolf Management Plan, Publ-ER-099 99. Wisconsin Department of Natural Resources, Madison.

Wilmers, C. C., Crabtree, R. L., Smith, D. W., Murphy, K. M., and Getz, W. M. 2003. Trophic facilitation by introduced top predators: grey wolf subsidies to scavengers in Yellowstone National Park. Journal of Animal Ecology 72:909–916.

Wilmers, C. C., and Getz, W. M. 2005. Gray wolves as climate change buffers in Yellowstone. PLOS Biology 3 issue 4:571–576.

Wilmers, C. C., Post, E., R. O. Peterson, R. O., and Vucetich, J. A. 2006. Predator disease outbreak modulates top-down, bottom-up, and climatic effects on herbivore population dynamics. Ecology Letters 9:383–389.

Chapter 14
Wolves, Roads, and Highway Development

Bruce E. Kohn, Eric M. Anderson, and Richard P. Thiel

14.1 Introduction

Roads are pervasive in the landscape of the United States and have profound ecological effects (Forman et al. 2003). Recent studies estimate that nearly 20% of the United States is ecologically impacted by the public road system and associated traffic (Forman 2000). In general, wildlife is impacted by roads in four major ways: (1) loss of habitat; (2) traffic mortality; (3) inaccessibility to required resources; and (4) division of populations into smaller, isolated subdivisions (Jaeger et al. 2005). Although gray wolves (*Canis lupus*) in the upper Great Lakes region are not impacted by all of these factors, their unique and highly variable relationship to humans creates even more complex relationships with roads than many other wildlife species.

Habitat suitability for gray wolves primarily is dependant on two variables: (1) availability of prey, and (2) tolerance of humans who live or recreate near wolves (Mech 1995; United States Fish and Wildlife Service 1992; Fuller 1995; Fritts et al. 2003). However, highways, roads, trails, and other paths created by humans alter that suitability in many ways. Some interactions result in behavioral changes among wolves, while others result in higher mortality. The additional mortality may be directly related to collisions with vehicles on roads, or indirectly related through the intentional or unintentional killing of wolves by those accessing areas inhabited by wolves.

Regardless of the mechanism, research in the Great Lakes region has shown that high road densities limit habitat suitability for wolves and that major highway corridors can slow range expansion in recovering wolf populations (Thiel 1985; Jensen et al. 1986; Mech et al. 1988; Mladenoff et al. 1995, 2006; Mladenoff and Sickley1998). In this chapter, we summarize what biologists in the upper Great Lakes region have learned over the past 30 years about the impacts of roads and road densities on wolves, and discuss modifications to future highway projects that may reduce impacts on wolf populations and range expansion.

Before we can examine the relationship of wolves to roads, a clear understanding of what constitutes a road is needed. The term "road" is a broad and generic term used to describe a human-created structure to convey vehicles. However, not all roads are perceived as equivalent by wolves. Roads included in this discussion include three distinct categories: (1) structures dressed with hard surfaces (blacktop, reinforced

A.P. Wydeven et al. (eds.), *Recovery of Gray Wolves in the Great Lakes*
Region of the United States,
DOI: 10.1007/978-0-387-85952-1_14, © Springer Science+Business Media, LLC 2009

concrete) designed for higher speeds and volumes of traffic (federal, state, and county systems), (2) secondary access (primarily county and rural municipalities) with lightly maintained surfaces (gravel), and (3) public-use unimproved roads, recreation trails, and some logging access trails (maintained by federal, Native American, state, and county conservation agencies). As we discuss later, the type of road combined with traffic volume and regularity of use influence both crossings and use of roads by wolves in the upper Great Lakes region.

Despite a myriad of mechanisms influencing wolf interactions with roads, wolves living in the upper Great Lakes region have revealed a surprising and variable adaptability to the presence of roads. The significance of road impacts depends on dynamic factors such as changing human attitudes toward wolves, amount of suitable habitat occupied by wolves, and viability of wolf populations. Therefore, data and analyses presented here are applicable to the upper Great Lakes region. We caution against applying them to other regions or in different circumstances without further testing and validation.

14.2   Interactions with Individual Roads

Roads and trails in an otherwise undeveloped environment create a primary means of access into wild spaces by humans, some of whom may be intolerant of wolves. These roads increase opportunities for wolf–human encounters that may result in accidental and intentional killing of wolves (Wydeven et al. 2001; Fritts et al. 2003). Examples include vehicle collisions, animals misidentified as coyotes and mistakenly killed (B. Kohn, *unpublished data*), and intentional (and illegal) killings by individuals who are intolerant of wolves.

Vehicle-caused mortalities kill individual wolves (Wydeven et al. 2001; Fritts et al. 2003), but may impact an entire population. Roads can generate elevated accidental mortality through wolf–vehicle collisions and, in extreme cases, may account for 75–90% of total mortality (Fritts et al. 2003). Since 1992, an average of 4% of all radiocollared wolves in Wisconsin died annually from collisions with vehicles and these accidents accounted for 17% of all wolf mortalities (A. P. Wydeven, Wisconsin DNR, personal communication). Two out of 18 (11%) mortalities of wolves radiocollared in Minnesota during 1994–2004 were results of collisions with vehicles (J. Erb, Minnesota DNR, personal communication).

Dispersing wolves frequently encounter roads in terrain unfamiliar to them (Mech et al. 1995; Merrill and Mech 2000). Roads and associated human traffic may intimidate some dispersing wolves and hinder dispersal. In such cases, roads diminish the potential for wolves to reach and colonize disjunct or patchy habitats, affecting genetic interchange between otherwise connected subpopulations (Frair 1999; Kohn et al. 2000; Mech et al. 1995; Merrill and Mech 2000).

Wolves regularly use dirt roads and trails as efficient routes for travel and hunting yet avoid more heavily traveled paved roads, especially during periods of high traffic volume (Thompson 1952; Thurber et al. 1994; Gehring 1995; Kohn et al. 2000).

Gehring (1995) tracked 74 km of wolf trails in northwestern Wisconsin and found that wolves traveled on dirt and asphalt-paved town roads as well as on logging roads and snowmobile trails. Wolves appeared to select areas with shallower snow depths, greater snow compaction, greater visibility, and lower density of plant stems. Thus, wolves may travel roads and trails to reduce energetic costs of traveling during winter. This supposition is supported anecdotally by observations of a weakened wolf suffering from mange and a subcutaneous fistula that spent its last 3 weeks of life exclusively traveling a network of improved dirt roads (B. E. Kohn, personal observation).

14.3   Road Densities

Road density is a measure of the magnitude of the road network in a landscape. If road networks become too dense, they may affect wolves on a population level. Fuller (1989, 1995) felt that the threshold for negative population impacts was a total annual mortality of around 35% and additive mortality associated with roads may push total mortality across this threshold. Biologists in the upper Great Lakes region, first to recognize and publish this hypothesis, have used road density to measure the suitability of a landscape to support viable wolf populations (Thiel 1985; Jensen et al. 1986; Mech et al. 1988; Mladenoff et al. 1995, 2006; Mladenoff and Sickley 1998). Wolf habitat potentially becomes fragmented into smaller and less inhabitable parcels as road networks develop and improve (Mladenoff et al. 1995).

Human tolerance of wolves varies spatially and temporally. Therefore, predictions of threshold road densities that will cause a wolf population to falter or disappear, or predictions of geographic regions that should or should not have wolves, will vary according to circumstances. This point is too often not considered by lay persons and professionals alike.

Thiel (1985, 1993) compared historic road densities in areas of Wisconsin occupied and not occupied by wolves during 1920–1950 when wolves were being extirpated, and during the early phase of wolf recolonization in the 1980s. He found that areas with road densities exceeding 0.58 km road per km^2 were not suitable for wolves. This early study generally was supported by observations from the Michigan/ Ontario border area (Jensen et al. 1986) and Minnesota (Mech et al. 1988) and seemed to provide a threshold for predicting wolf habitat suitability in the upper Great Lakes region during the 1980s.

Mech (1989) reported on an area in Minnesota's wolf range where wolves existed above this threshold. While he did not use these specific terms, the study area with high road density was a population sink adjacent to a source in a relatively roadless area. He concluded that although the road density threshold applies most directly to large reservoirs of occupied wolf range, relatively small areas of high road densities adjacent to those reservoirs can also be occupied by wolves. Mladenoff et al. (1995, 1999) agreed with these conclusions, and Haight et al. (1998), using computer simulations, reiterated that population persistence was likely where dispersing wolves could supplement wolves in marginal habitats.

Mladenoff et al. (1995) used radiocollar locations from Wisconsin wolf packs to identify habitat parameters that predict suitability for wolves over a large geographic region including portions of Michigan, Minnesota, and Wisconsin. They found that prey availability (specifically white-tailed deer) was a poor predictor of wolf presence since adequate deer numbers were available everywhere. However, road densities and affiliated data (human population densities, agricultural activity) reliably identified regions where wolf populations might persist. The resulting model showed that Wisconsin had the most fragmented and least suitable habitat, and that Minnesota had the highest quality and greatest amount of habitat of the three states. Wydeven et al. (2001), using additional data generated after Mladenoff et al. (1995, 1999), also found that wolves used landscapes with higher road density less frequently and suggested that human-caused mortalities in these areas were higher than in areas of low road density.

Mech (2006) claimed that Mladenoff et al.'s (1995) model was a poor predictor of wolf habitat suitability because the source data used were derived from early colonizers and did not take into account the wolf's ability to adapt to the presence of humans. Mladenoff et al. (2006) conceded that their model was based on data from early colonizers with little or no competition in selecting unoccupied space. They also stated that wolves were adaptable and the "best" quality habitat (lowest road density) would be occupied first followed by more marginal habitats not necessarily predicted by their model.

The Mladenoff et al. (1995) model is a useful predictive tool in the upper Great Lakes region provided users understand its limitations (Mladenoff et al. 1995, 1999, 2006; Wydeven et al. 2001). However, use of the model outside of the upper Great Lakes region should be rigorously tested prior to application (Haight et al. 1998; Mladenoff and Sickley 1998).

14.4   Hierarchical Selection of Territories

Interactions with roads can influence wolf behavior at a multitude of levels both in space and time. Assuming that wolves will choose resources that best satisfy their life requirements and that higher quality resources will be selected first (Manly et al. 1993), wolves must make a series of hierarchical decisions when establishing territories and home sites – each of which may be influenced by presence of roads. For colonizing wolves, an initial decision is territory location relative to road densities in unoccupied landscapes. If mortality rates in areas with higher road density prevent populations from persisting in those areas, the effect is identical to behavioral decisions to settle only in areas of lower road density. Untangling the relative influence of behavioral choices and demographic outcomes is difficult. Again, evidence suggests that areas with higher road density have a higher rate of human-caused mortality (Wydeven et al. 2001). Nonetheless, this does not preclude behavioral avoidance of highly roaded areas.

A related issue is the extent to which roads are enough of a barrier that they define boundaries of a pack's territory. If roads are actively avoided or create significant sources of mortality, wolf territories should become established between major traffic corridors and not straddle highly trafficked highways.

Once territories are established, additional relationships with roads become important. Are all types of road within a home range crossed with equivalent frequency or are some avoided more than others? Does traffic intensity on a road influence use of nearby habitat by wolves or their likelihood of crossing it? Finally, since wolves may be especially sensitive to human disturbance during reproductive periods, how are special-use areas within territories (den and rendezvous sites) selected relative to presence of roads? The following sections evaluate current understanding of these hierarchical levels of selection in the Great Lakes region.

14.4.1   Territory Placement and Road Density

Mladenoff et al. (1995, 1999) found that road densities were the best predictors of wolf territory placement in northern Wisconsin and that areas with lowest road densities were occupied first. They showed that areas with road densities <0.4 km/km² had a ≥50% chance of being occupied by wolves, and that areas with road densities >0.6 km/km² had <10% probability of being occupied by wolves.

Keenlance (2002) found that as the wolf population continued to increase in northwestern Wisconsin, wolves began establishing territories with higher road densities. He suggested that areas with <1.5 km/km² be considered as potentially suitable habitat. As numbers of packs in his study area increased from 3 to 16 (1992–1999), mean road densities within pack territories increased from 0.35 to 1.09 km/km². In spite of this, wolves still were establishing territories in unoccupied areas with lower road densities. He felt that high tolerance by humans and legal protection allowed wolves to survive in areas previously thought to be unsuitable. This could reverse if human tolerance declines and wolves become less protected.

In addition to their general preference for areas of lower road density, wolves may use roads with high traffic volume as boundaries to their territories. Frair (1999) found that almost half of the wolf territories she delineated (n = 18) in northwestern Wisconsin contained no major highways and, for at least four packs, territory boundaries paralleled but did not cross major highways despite numerous instances of packs at least temporarily straddling major highways (Kohn et al. 2000; Keenlance 2002).

14.4.2   Frequency of Within-Territory Road Crossings and Use Versus Traffic Intensity

The willingness of wolves to cross most all road types, including busy highways, has been demonstrated repeatedly (Mech et al. 1995; Kohn et al. 2000; Merrill and

Mech 2000). However, if wolf behavior is influenced by roads, there should be some demonstrated reluctance to crossing certain types of roads within established territories. Research in Alberta, Canada, suggested that wolves selected areas near low-use roads and trails while avoiding high-use trails (Whittington et al. 2005). In Alaska, wolves avoided roads with high human use but used closed roads as travel corridors (Thurber et al. 1994).

In northwestern Wisconsin, Frair (1999) found wolves crossing minor highways within their territory at a much lower rate than expected. In that same area, Shelley and Anderson (1995) noted that responses to roads appeared to vary by individual wolf. Locations of three of five radiocollared wolves were further from county highways than random points, but two wolves spent more time than expected closer to two-lane highways. The attraction of the two-lane highway may have been the availability of carrion, since both of the wolves were from the same pack and were observed on five separate occasions feeding on deer killed by vehicles. Similar behavior occurred along a major state highway in central Wisconsin (Heilhecker et al. in press).

Keenlance (2002) found that wolves in northwestern Wisconsin avoided areas within 250 m of roads during daylight. Over 40% of his study area was within 250 m of a road or trail and was being avoided during daylight hours. While wolves are most active during the night in the summer, they travel most extensively during the day in the winter (Mech 1970). Mech (1970) felt that this avoidance could limit wolf activities in areas of high road densities, particularly during winter.

14.4.3   Den and Rendezvous Sites

The location of den and rendezvous sites may determine the reproductive success of each wolf pack (Harrington and Mech 1982). Denning and pup-rearing seasons may be times when wolves are more sensitive to roads and associated human disturbance.

Gehring (1995) investigated five den sites in northwestern Wisconsin and found them nearly twice the distance from roads and trails ($\bar{x}$ = 740 m) than randomly selected sites ($\bar{x}$ = 300 m). Shelley and Anderson (1995) documented a rendezvous site <400 m from a major state highway in that same area. In central Wisconsin, Thiel et al. (1998) reported a wolf den within 800 m of an intensively used all-terrain vehicle (ATV) trail, and another originally established within 4 km (erroneously reported as 2 km in original document; R. Thiel, personal communication) of an interstate highway. The second den was eventually moved to a more remote location 9 km from the interstate. Heilhecker et al. (In press) documented an active den within 230 m of a high-traffic state highway and rendezvous sites in the rights of way of moderately traveled state highways in central Wisconsin.

Unger (1999) and Keenlance (2002) both found that wolves in northern Wisconsin most often selected den sites in the centers of their territories and in areas of lower road density. Unger (1999) also found that roads and their associated disturbance did not appear to be important considerations when selecting rendezvous sites (Unger et al., this volume).

14.5   Wisconsin's US Highway 53 Research Project: A Case Study

Higher road standards generally lead to greater traffic volume, speed, and development resulting in greater impacts on habitat suitability and populations of wolves (Fuller 1995). A Wisconsin Department of Transportation (WDOT) proposal to convert a 71-km segment of US 53 in northwestern Wisconsin from two lanes into four lanes created concerns over negative impacts to wolf recovery in Wisconsin because the project passed through wolf habitat, and bisected the main dispersal route for wolves coming from Minnesota into Wisconsin (Fig. 14.1). WDOT prepared a Biological

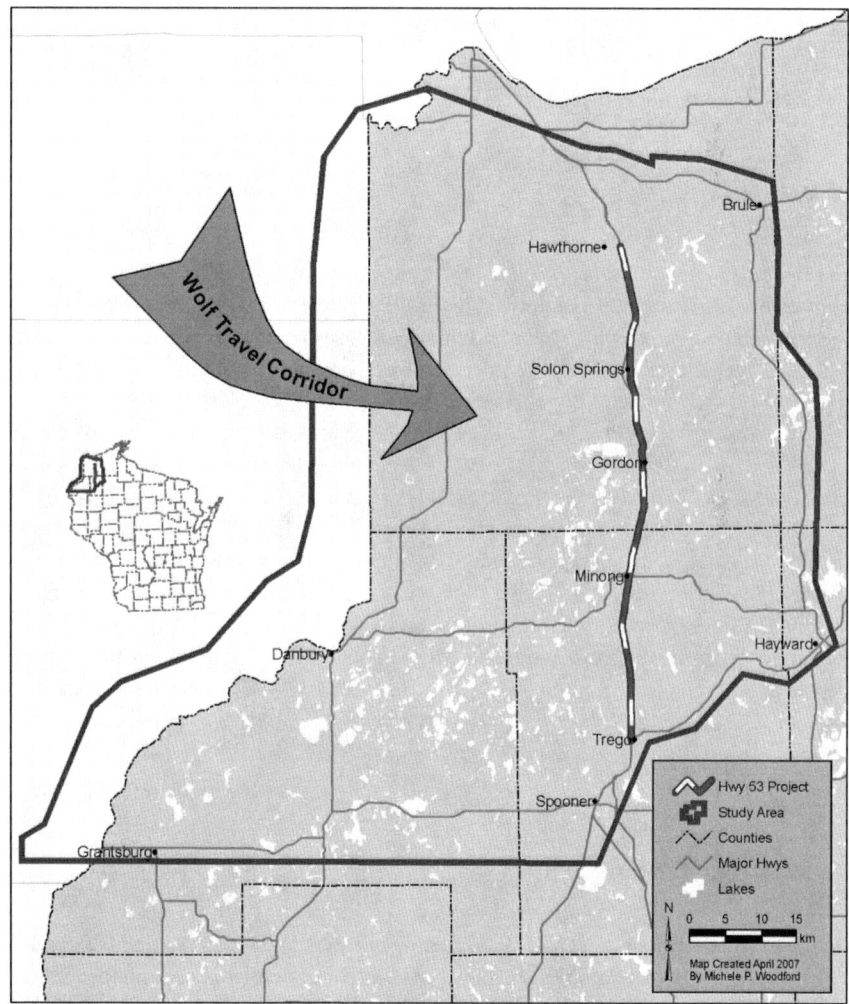

Fig. 14.1 Location of the US 53 Wolf Study Area. This area included the main dispersal route for wolves coming from Minnesota into Wisconsin

Assessment for the US 53 project that incorporated suggestions of regional wolf experts into the design for the expanded highway (WDOT 1990; Kohn et al. 2000).

The Biological Opinion for the US 53 project (USFWS 1991) concluded that the proposed highway was not likely to jeopardize the continued existence of wolves or adversely impact their habitat. However, it strongly recommended that the Federal Highway Administration and the WDOT fund a comprehensive study to determine how the addition of two highway lanes to the existing two-lane highway may affect wolf dispersal from Minnesota through Wisconsin, and how it may affect the movement, mortality, and recovery of Wisconsin's wolf population.

Research was conducted from May 1992 through June 1999 by WDOT, WDNR, University of Wisconsin – Stevens Point, and Michigan State University (Kohn et al. 2000; Keenlance 2002). Objectives were to: (1) determine the impacts of the US 53 expansion on dispersing and resident wolves, (2) determine the effectiveness of wolf crossing sites incorporated into highway design, and (3) develop criteria for identifying/mitigating any negative impacts of future highway projects on wolves.

14.5.1   Wolf Dispersal

The US 53 expansion project bisected the major travel corridor for wolves dispersing from Minnesota into Wisconsin (Mech et al. 1995). Mladenoff et al. (1995) felt that preserving the integrity of this travel corridor was necessary for successful maintenance of the wolf population in the Great Lakes region. Population viability analyses (Rolley et al. 1999) suggested that continued immigration of wolves from Minnesota greatly enhanced the persistence probabilities for a wolf population in Wisconsin.

Of 20 dispersing wolves radiocollared during the US 53 study, 13 (12 females, 1 male) encountered US 53 in their travels. All but one of them crossed the highway, often numerous times. Two wolves, including the one that didn't cross US 53 while dispersing, established new territories adjacent to the highway and crossed US 53 occasionally after settlement. Nine of these wolves attained alpha status either by acceptance into an existing pack, or through establishment of a new pack. One wolf was killed while dispersing, and researchers lost contact with the other three before their fates could be known. Evidently, the expanded US 53 was no barrier to dispersing wolves. All dispersers tracked for >1 year during the study eventually established new territories and became the dominant animals in those new packs.

14.5.2   Resident Wolves

The US 53 expansion project had no detectable negative impact on numbers of resident wolves (members of established packs) or quality of wolf habitat adjacent to the highway. The resident wolf population within the US 53 Study Area increased from 18 wolves in five packs in 1994 to 61 wolves in 16 packs in 1999 while US 53 was undergoing reconstruction. The population has continued to grow and, as of March 2006, included 80–86 wolves in about 20 packs (Wydeven et al. 2006). Moreover, 7 of 11 new pack territories established during the study were located immediately adjacent to US 53 and 2 included US 53 within their territories. US 53 formed the apparent physical boundary between six pack territories in 1999.

14.5.3   Wolf Crossings of US 53

Precise locations of 37 crossings of US 53 and 25 crossings of Wisconsin Highway (WI) 27 and WI 35 by radiocollared wolves were evaluated to determine habitats used by wolves for crossing highways (Frair 1999; Kohn et al. 2000). Most (81%) of the crossings of US 53 during 1992–1996 were made by dispersing animals. Crossings by dispersers peaked between late October and late December and again between late April and early June, whereas crossings by resident wolves were less time-specific.

Frair (1999) found that low patch density, an index to human-induced fragmentation, was the most significant and consistent landscape indicator of favorable wolf-crossing habitat in the US 53 Study Area. Wolves avoided developed lands, and did not cross highways in areas adjacent to homes, lakes, or large rivers. Wolves preferred to cross highways at points within large patches of homogeneous habitat. Lowland forest complexes were the most preferred habitats for crossing whereas large, unfragmented patches of upland forests and open types that provided adequate distance from human activity were used as crossing sites in proportion to their availability.

On a finer scale, wolves preferred crossing sites with greater visibility and ease of travel. Visual obscurity at eye level was lower at wolf crossing sites than adjacent habitats. The right-of-way along US 53 in the US 53 Study Area was wider than along WI 35 and appeared to provide the amount of visibility preferred by wolves at crossing sites. Conversely, wolves preferred to cross WI 35 where the right-of-way was wider than normal.

Frair (1999) backtracked nine trails made by wolves as they approached major highways. Sixteen percent of the total length of the trails followed (20 km) coincided with groomed snowmobile trails, plowed roads, or railroad tracks, 14% coincided with other linear features such as streams, ridgelines, or gas line rights-of-way, and 7% followed deer trails or individual ski/snowmobile tracks. Gehring (1995) found that snow was significantly more compact and shallower than expected at crossing sites, which related directly to ease of movement. Although highway-crossing sites analyzed in this study did not show preferential use of trails by wolves when crossing highways, it appeared that wolves opportunistically used trails that coincided with their intended direction of movement even if they led them across a highway.

Frair (1999) developed a model for identifying and rating potential crossing sites along major highways and tested it on US 53. That model identified "high," "moderate," and "low" potential crossing sites along US53 based on the patch density and percentages of open water, developed land, and wetlands within 200-ha sampling areas placed every 100 m along the highway. Fifty-nine percent of the known wolf crossings of US 53 occurred in areas labeled as "high potential crossing sites" and 34% occurred in areas labeled as "moderate potential crossing sites." "High" and "moderate potential crossing sites" comprised 20% and 48% of the US 53 corridor being studied. Only 7% of the wolf crossings of US 53 occurred in areas labeled as "low potential crossing sites."

14.5.4 Wolf Use of Ballooned Strips and Underpasses

Ballooned sections of divided highways are areas where the median is widened to include substantial amounts of natural habitat (Fig. 14.2). Seven ballooned sections totaling 12 km (17% of the highway project length) were incorporated into the US 53 highway design. Eighteen of 37 (49%) known wolf crossings of US 53 occurred in ballooned areas, and all three of the longer ballooned sections fell within or partially overlapped areas determined to be high probability wolf crossing sites. One dispersing wolf used ballooned sections to cross the highway at least six times, and a pack established a territory immediately adjacent to a ballooned section and occasionally used it to cross the highway.

In three cases, radiocollared dispersing wolves were monitored closely while approaching ballooned areas. The first wolf remained close to US 53 for 1–2 h and then trotted across the ballooned section during daylight. A second wolf remained near the highway for several hours during the daylight and finally crossed after dark when traffic was reduced. A third wolf crossed a ballooned area without hesitation during daylight. Wolves that were observed as they crossed highways easily avoided

Fig. 14.2 One of the seven "ballooned strips" constructed along US 53 to facilitate wolf crossings of the upgraded highway. Natural vegetation was maintained in a ≥100-m median between centerlines of the two lanes. Wolves crossing the highway in ballooned strips encountered traffic coming from only one direction at a time. (WDNR photo by Joseph F. Sprenger 2007)

vehicles coming from one direction but appeared confused when vehicles were coming from both directions. Ballooned sections minimized this problem because wolves encountered only one direction of traffic at a time. Resident wolves appeared less wary of vehicular traffic than dispersers when crossing US 53.

Researchers observed no wolf activity under bridges or overpasses along US 53. Deer and coyotes passed beneath US 53 at the Totagatic River bridge. Use of the underpass by coyotes suggested that bridges might be designed to provide safe crossings for wolves as well. However, wolves crossed US 53 twice within 0.4 km of that underpass without using it.

Highway underpasses and extended bridges, accompanied by fencing to funnel animals through the structures, have resulted in some reduction of highway-related impacts to wolf populations in Banff National Park, Alberta (Clevenger 1998). Clevenger and Waltho (2000) found that the amount of human activity around the underpass and the "openness" and length of the underpass were the best predictors of underpass use by wolves.

14.5.5 Wolf Mortalities on US 53

Three wolves were killed by vehicles on US 53 during June–October 1998. These included a radiocollared yearling female dispersing from the Frog Creek Pack, and a pup and a yearling male from the Stuntz Brook Pack whose territories overlapped US 53. The dispersing female crossed US 53 ≥ 7 times during her travels. All three vehicle-wolf collisions occurred in a 4.8 km segment south of Minong where US 53 crossed a large block of lowland habitat. This area was identified as a high potential crossing area by Frair (1999). The US 53 median is widened through much of this area and two of the mortalities occurred in the ballooned strip. Four lightly used forest roads and trails crossed the highway in this segment, and all three mortalities occurred near the intersections of the forest roads/trails and US 53. Seven other wolves were killed by vehicles on other highways and roads during the US 53 study.

Eight additional wolves were killed while crossing US 53 during 1999–2006 (A. P. Wydeven, Wisconsin DNR, personal communication). Ruediger et al. (1999) stated that highways with traffic volumes exceeding 4,000 vehicles per day increase habitat fragmentation and highway mortalities for large carnivores. Traffic volume on US 53 increased from an average of 4,500 vehicles per day in 1996 to 5,800 vehicles per day in 2005 (D. Lamont, Wisconsin Department of Transportation, personal communication). To date, these mortalities apparently have not impacted growth or persistence of the local wolf population.

14.5.6 Considerations for Future Highway Projects

We make the following recommendations for roadway construction in the upper Great Lakes region. More restrictive guidelines may be necessary in other regions where wolf populations are in very early stages of recovery, where suitable habitat is more fragmented due to human development or geological barriers, or for projects that will result in significantly higher traffic volume and speed than on US 53.

The status and distribution of wolves, road densities, and habitat connectivity should all be considered early in the planning phases for new highway projects. Intensive monitoring of wolves throughout the upper Great Lakes region has generated substantial information on wolf numbers, distribution, movement patterns, and habitat requirements. Spatially explicit predictive models can identify areas of potential problems. Frair (1999) found a maximum tolerance limit of 0.09 km/km² of major highways within wolf territories. Future highway projects should, wherever possible, follow existing road corridors to avoid increasing roadway density and altering habitat.

Existing wolf pack territories in the immediate area should be considered when selecting highway alignment alternatives. Size and shape of wolf pack territories can be sufficiently approximated during a winter of intensive track searches in areas where movement data from radiocollared wolves are not available (Wydeven et al. 1996). Unger (1999) stated that locating new highway alignments to avoid core areas of known pack territories, especially where ridges of upland habitat occur in wetlands and areas with lowest road densities, would protect most suitable den sites.

Frair's (1999) highway-crossing model can be used to identify moderate and high probability crossing sites for wolves during planning. Normally, these will occur within large, homogeneous landscapes, especially lowlands. Within them, wolves will usually cross in areas providing greater visibility and ease of travel. Winter track searches can also be used to verify crossing sites used by wolves.

Highways are often ballooned to preserve wetlands. This practice also protects areas most commonly used by wolves as highway crossing sites. Unfortunately, documented wolf crossings of US 53 were too few to estimate survival benefits of ballooned areas. Additional costs involved in ballooning a section of highway may be offset by benefits to both wetland protection and wolves.

The Biological Assessment for the US 53 Expansion Project (WDOT 1990) recommended maintaining cover as close to rights-of-way as possible. This was ultimately found unnecessary because wolves preferred crossing sites that afforded them greater visibility.

The Biological Opinion (USFWS 1991) stated that no additional road access sites were permitted to discourage associated human developments from becoming additional barriers to wolf movements. Data collected during the US 53 study clearly showed that wolves avoided habitats fragmented by human development. The opinion also prohibited erection of fences along rights-of-way to avoid impeding wolf movements. These measures probably aided dispersing wolves and should be considered in future highway projects in large blocks of undeveloped land where livestock grazing is minimal and snowmobile and ATV traffic on rights-of-way is prohibited.

Finally, the Biological Opinion for the US 53 project (USFWS 1991) also required that "Wildlife Crossing Area" (not "Wolf Crossing") signs be placed at bal-looned sites to further minimize wolf losses. This strategy may also be warranted for other high-probability crossing areas (Frair 1999), especially near intersections with forest roads and trails. However, permanent signs that warn of wildlife crossing areas appear to be ineffective in changing the speeds of vehicles traveling those marked sections and hence may be relatively useless in reducing wolf mortalities (Hedlund et al. 2004). Active signage, with intermittently flashing lights or specific changing messages, may draw more attention and result in more effective protection for crossing wildlife (Groot Bruinderink and Hazebroek 1996).

Deer killed by vehicles should be removed quickly to avoid attracting wolves to the highway. The planting of grasses less desirable to deer in the right-of-way may further reduce hazards of both species to collisions.

Ideally, the entire network of highways and railroads within the wolf range in Wisconsin, Minnesota, and Michigan should be considered when considering future development of existing highways or construction of new travel corridors. Their impacts on wolves should be addressed at a larger geographic scale rather than by individual highway segments. This would require a comprehensive planning process involving the USFWS, the Federal Highway Administration, the DOT and DNR agencies from each state, and the public.

14.6 Summary

Impacts of roads, road densities, and road standards on wolf populations vary with human tolerance for wolves and the viability of wolf populations. Negative impacts are most severe when human tolerance for wolves is low and wolf numbers are reduced. Conversely, impacts are least severe when public acceptance of wolves is high and populations are robust.

Mere presence of roads can cause individual wolf mortalities through collisions with vehicles and intentional or unintentional killings (shooting, trapping, etc.). Moreover, presence of roads within wolf territories may make some areas less suitable for wolf den sites and, perhaps, rendezvous sites. High road densities can impact wolf populations negatively by making areas unsuitable for wolf territories and by increasing wolf mortalities above sustainable levels. Therefore, road densities are good predictors of suitable habitat for wolves in the Great Lakes region, especially during early recolonization. Generally, wolves occupy areas with lowest road densities first when recolonizing. As relatively road-free areas become occupied and the wolf population becomes more secure, wolves pioneer areas with higher road densities and greater exposure to potential conflict with humans. Thus, human tolerance will ultimately be the arbiter in determining wolf abundance and distribution in the Great Lakes region.

The impacts of highway expansion on local wolf populations can be minimized if wolves are considered early in development and planning phases. New projects should follow existing travel corridors as closely as possible to minimize impact to existing wolf habitat and to maintain existing road densities. Access to new highways should be limited to existing levels to minimize additional human development. To protect den sites, new travel corridors should be located to avoid core areas of known wolf territories.

More aggressive modifications to normal highway design can be implemented where wolf populations require additional protection. These include ballooned sections and wolf under- and overpasses that facilitate safe crossings by dispersing and resident wolves. Unique or active warning signs located at known wolf highway crossing sites and removal of deer killed by vehicles along highways may also help minimize collisions with vehicles.

Acknowledgments University of Wisconsin – Stevens Point M.S. candidates Douglas P. Shelley, Thomas M. Gehring, David E. Unger, and Jacqui L. Frair, and Michigan State University Ph.D. candidate Paul W. Keenlance played major roles in the US Hwy 53 Research Project. They monitored the radiocollared wolves, conducted the preliminary analyses of those data, and reported the findings in their respective theses. Volunteers Joelle Gehring, Rebecca Montgomery, Alexa Spivy, Michelle Lassige, Lorrie Kohn, and Kelly Jones provided valuable assistance during the project.

References

Clevenger, A. P. 1998. Permeability of the Trans-Canada Highway to wildlife in Banff National Park: the importance of crossing structures and factors influencing their effectiveness. In Proceedings of the International Conference on Wildlife Ecology and Transportation. FL-ER-69-98, eds. G. L. Evink, P. Garrett, D. Zeigler, and J. Berry, pp. 109–111. Tallahassee, FL: Florida Department of Transportation.

Clevenger, A. P., and Waltho, N. 2000. Factors influencing the effectiveness of wildlife underpasses in Banff National Park, Alberta, Canada. Conservation Biology 14:47–56.

Forman, R. T. T. 2000. Estimate of the area affected ecologically by the road system in the United States. Conservation Biology 14:31–35.

Forman, R. T. T., Sperling, D., Bissonette, J. A., Clevenger, A. P., Cutshall, C. D., Dale, V. H., Fahrig, L., France, R., Goldman, C. R., Heanue, K., Jones, J. A. Swanson, F., J., Turrnetine, T., and Winter, T. C. 2003. Road ecology: science and solutions. Washington, DC: Island Press.

Frair, J. L. 1999. Crossing paths: gray wolves and highways in the Minnesota-Wisconsin border region. MS Thesis, University of Wisconsin – Stevens Point, WI.

Fritts, S. H., Stephenson, R. O., Hayes, R. D., and Boitani, L. 2003. Wolves and humans. In Wolves: Behavior, Ecology, and Conservation, eds. L. D. Mech and L. Boitani, pp. 289–316. Chicago, IL: University of Chicago Press.

Fuller, T. K. 1989. Population dynamics of wolves in north-central Minnesota. Wildlife Monograph 105.

Fuller, T. K. 1995. Guidelines for gray wolf management in the northern Great Lakes Region. Ely, MN: International Wolf Center Technical Publication 271.

Gehring, T. M. 1995. Winter wolf movements in northwestern Wisconsin and east-central Minnesota: a quantitative approach. MS Thesis, University of Wisconsin – Stevens Point, WI.

Groot Bruinderink, G. W. T. A., and Hazebroek, E. 1996. Ungulate traffic collisions in Europe. Conservation Biology 10:1059–1067.

Haight, R. G., Mladenoff, D. J., and Wydeven, A. P. 1998. Modeling disjunct gray wolf populations in semi-wild landscapes. Conservation Biology 12:879–888.

Harrington, F. H., and Mech, L. D. 1982. Patterns of homesite attendance in two Minnesota wolf packs. In Wolves of the World, eds. F. H. Harrington, and P. C. Paquet, pp. 81–105. Park Ridge, IL: Noyes Publishing.

Hedlund, J. H., Curtis, P. D., and Williams, A. F. 2004. Methods to reduce traffic crashes involving deer: what works and what does not. Traffic Injury Prevention 5:122–131.

Heilhecker, E., Thiel, R. P., and Hall Jr., W. In Press. Wolf, *Canis lupus*, behavior in areas of frequent human activity. Canadian Field-Naturalist.

Jaeger, J. A., Bowman, G. J., Brennan, J., Fahrig, L., Bert, D., Bouchard, J., Charbonneau, N., Frank, K., Gruber, B., and Tluk von Toschanowitz, K. 2005. Predicting when animal populations are at risk from roads: an interactive model of road avoidance behavior. Ecological Modeling 185:329–348.

Jensen, W. F., Fuller, T. K., and Robinson. W. L. 1986. Wolf (*Canis lupus*) distribution on the Ontario-Michigan border near Sault Ste. Marie. Canadian Field-Naturalist. 100:363–366.

Keenlance, P. W. 2002. Resource selection of recolonizing gray wolves in northwest Wisconsin. PhD Dissertation, Michigan State University, Lansing, MI.

Kohn, B. E., Frair, J. L., Unger, D. E., Gehring, T. M., Shelley, D. P., Anderson, E. M., and Keenlance, P. W. 2000. Impacts of the US Highway 53 expansion project on wolves in northwestern Wisconsin. Wisconsin Department of Natural Resources Final Report.

Manly, B. F., McDonald, L. L., and Thomas, D. L. 1993. Resource selection by animals, statistical design and analysis for field studies. London: Chapman and Hall.

Mech, L. D. 1970. The wolf: the ecology and behavior of an endangered species. Garden City, NY: Natural History Press.

Mech, L. D. 1989. Wolf population survival in an area of high road density. American Midland Naturalist 121:387–389.

Mech, L. D. 1995. The challenge and opportunity of recovering wolf populations. Conservation Biology 9:270–278.

Mech, L. D. 2006. Prediction failure of a wolf landscape model. Wildlife Society Bulletin 34:874–877.

Mech, L. D., Fritts, S. H., Radde, G. L., and Paul, W. J. 1988. Wolf distribution and road density in Minnesota. Wildlife Society Bulletin 16:85–87.

Mech, L. D., Fritts, S. H., and Wagner, D. 1995. Minnesota wolf dispersal into Wisconsin and Michigan. American Midland Naturalist 133:368–370.

Merrill, S. B., and Mech, L. D. 2000. Details of extensive movements by Minnesota wolves (*Canis lupus*). American Midland Naturalist 144:428–433.

Mladenoff, D. J., and Sickley, T. A. 1998. Assessing potential gray wolf habitat restoration in the northeastern United States: a special prediction of favorable habitat and potential population levels. Journal of Wildlife Management 62:1–10.

Mladenoff, D. J., Sickley, T. A., Haight, R. G., and Wydeven, A. P. 1995. A regional landscape analysis and prediction of favorable gray wolf habitat in the Northern Great Lakes Region. Conservation Biology 9:279–294.

Mladenoff, D. J., Sickley, T. A., and Wydeven, A. P. 1999. Predicting gray wolf landscape recolonization: logistic regression models vs. new field data. Ecological Applications 9:37–44.

Mladenoff, D. J., Clayton, M. K., Sickley, T. A., and Wydeven, A. P. 2006. L. D. Mech critique of our work lacks scientific validity. Wildlife Society Bulletin 34:878–881.

Rolley, R. E., Wydeven, A. P., Schultz, R. N., Thiel, R. T., and Kohn, B. E. 1999. Wolf viability analysis. In Wisconsin Wolf Management Plan – August 25, 1999, pp. 40–44. Madison, WI: Wisconsin Department of Natural Resources.

Ruediger, B., Claar, J. J., and Gore, J. G. 1999. Restoration of carnivore habitat connectivity in the northern Rocky Mountains. In Proceedings of the third international conference on wildlife ecology and transportation. FL-ER-73-99, eds. G. L. Evink, P. Garrett, and D. Zeigler, pp. 5–20. Tallahassee, FL: Florida Department of Transportation.

Shelley, D. P., and Anderson, E. M. 1995. Final report: impacts of US Hwy 53 expansion on timber wolves – baseline data. Report for Wisconsin Department of Natural Resources.

Thiel, R. P. 1985. The relationship between road densities and wolf habitat suitability in Wisconsin. American Midland Naturalist 113:404–407.

Thiel, R. P. 1993. The timber wolf in Wisconsin: the death and life of a majestic predator. Madison, WI: University of Wisconsin Press.

Thiel, R. P., Merrill, S. B., and Mech, L. D. 1998. Tolerance by denning wolves, *Canis lupus*, to human disturbance. Canadian Field-Naturalist 112:340–342.

Thompson, D. Q. 1952. Travel, range and food habits of timber wolves in Wisconsin. Journal of
    Mammalogy 33:429–442.
Thurber, J. A., Peterson, R. O., Drummer, T. D., and Thomasma, S. A. 1994. Gray wolf response
    to refuge boundaries and roads in Alaska. Wildlife Society Bulletin 22:61–68.
Unger, D. E. 1999. Timber wolf den and rendezvous site selection in northwestern Wisconsin and
    east-central Minnesota. MS Thesis, University of Wisconsin – Stevens Point, WI.
United States Fish and Wildlife Service. 1991. Biological opinion for construction of U.S.
    Highway 53 (Trego to Hawthorne), Washburn and Douglas Counties, Wisconsin. Log No.
    3-91-F-WI-GBFO.
United States Fish and Wildlife Service. 1992. Recovery plan for the eastern timber wolf. Twin
    Cities, MN.
Whittington, J., St. Clair, C. C., and Mercer, G. 2005. Spatial responses of wolves to roads and
    trails in mountain valleys. Ecological Applications 15:543–553.
Wisconsin Department of Transportation. 1990. Eastern timber wolf biological assessment for
    U.S. Highway 53. Federal Project F 018; I.D. 1198-01-01/02.
Wydeven, A. P., Schultz, R. N., and Megown, R. A. 1996. Guidelines for carnivore track surveys
    during winter in Wisconsin. Wisconsin Department of Natural Resources Endangered
    Resources Report 112.
Wydeven, A. P., Mladenoff, D. J., Sickley, T. A., Kohn, B. E., Thiel, R. P., and Hansen, J. L. 2001.
    Road density as a factor in habitat selection by wolves and other carnivores in the Great Lakes
    region. Endangered Species Update 18:110–114.
Wydeven, A. P., Wiedenhoft, J. E., Schultz, R. N., Thiel, R. P., Boles, S. R., and Heilhecker, E.
    2006. Progress report of wolf population monitoring in Wisconsin for the period October
    2005–March 2006. Wisconsin Department of Natural Resources Pub ER-2006.

Chapter 15
Taxonomy, Morphology, and Genetics of Wolves in the Great Lakes Region

Ronald M. Nowak

15.1 Wolves: Characters and Relationships

Wolves are animals of the Class Mammalia, Order Carnivora, and Family Canidae. Their genus, *Canis*, comprises at least seven living wild species, including the North American coyote (*C. latrans*) and the Old World jackals. Some taxonomists have "lumped" wolves in a single circumpolar species, *C. lupus*, and some have "split" them among a number of species. The domestic dog sometimes is regarded as the subspecies *C. lupus familiaris* and sometimes as a fully separate species, *C. familiaris*. Wolves resemble certain large breeds of the domestic dog, but have a narrower body, a tail that does not curl, relatively larger teeth, and a flatter forehead (Nowak 1979).

In both the Old and New worlds, small kinds of wolves are present all along the southern fringe of the range of *C. lupus*. Whether they represent components of *C. lupus* or some other entity is the most persistent problem in the systematics of modern wolves. One of those forms occupied the three southern main islands of Japan. Extinct for a century, it has been variously considered a full species, *C. hodophilax* (Imaizumi 1970a,b), or a distinctive subspecies, *C. lupus hodophilax* (Nakamura 1998, 2004). Another small wolf, *pallipes*, still occurs from central Israel to eastern India. Nowak (1995) found nearly complete statistical separation between *pallipes* and more northerly wolves of Eurasia. Sharma et al. (2004) concluded that the Indian population of *pallipes* had diverged from northern wolves over 400,000 years ago, but that another population, occurring from eastern Nepal, across northern India, to eastern Kashmir, and hitherto assigned to the subspecies *C. lupus chanco*, probably had been distinct for over 800,000 years. Aggarwal et al. (2007) suggested that both Indian populations warrant full specific rank.

To the south of *pallipes*, in the Arabian Peninsula and the Israeli Negev Desert, the wolf is even smaller but generally recognized as the subspecies *C. lupus arabs* (Harrison and Bates 1991; Hefner and Geffen 1999). It has been suggested that the form *lupaster* of the Sinai Peninsula and northern Egypt and Libya represents a continuation of the range of *C. lupus* (Ferguson 1981), though *lupaster* usually is considered a large subspecies of *C. aureus*, the golden jackal (Kurten 1974; Spassov 1989). All wolves of the Middle East and India have declined sharply and are in danger of extinction (Mech and Boitani 2004).

A.P. Wydeven et al. (eds.), *Recovery of Gray Wolves in the Great Lakes Region of the United States*,
DOI: 10.1007/978-0-387-85952-1_15, © Springer Science+Business Media, LLC 2009

Even farther south and more critically endangered is *C. simensis* of the Ethiopian highlands (Sillero-Zubiri and Marino 2004). Systematic studies have been somewhat contradictory, skull morphology suggesting it is the most distinctive species of *Canis* (Clutton-Brock ct al. 1976), but molecular analysis indicating it is more closely related to *C. lupus* than are the African jackals, though not so closely as is *C. latrans* (Wayne and Vilà 2003).

Goldman (1937, 1944) assigned the wolves of North America to two species, *C. lupus* (gray wolf) in most of the continent and *C. rufus* (red wolf) in the southeastern United States. Many authorities have accepted that arrangement (Nowak 1979, 1999; Kurten and Anderson 1980; Hall 1981), though others hold *rufus* to be at most a subspecies of *C. lupus* (Lawrence and Bosssert 1967, 1975; Wozencraft 2005), or even a modern hybrid of *C. lupus* and *C. latrans* (Wayne and Jenks 1991; Wayne et al. 1992; Roy et al. 1994b, 1996; Reich et al. 1999). A more recent proposal is that *rufus* and some wolf populations of the Great Lakes region form an independent species, *C. lycaon* (Wilson et al. 2000, 2003; Kyle et al. 2006).

Wolves certainly once occurred all around the Great Lakes, from Minnesota to New York. For about 30 years, authorities usually followed Goldman (1937, 1944), who referred all populations in that region to the single species and subspecies, *C. lupus lycaon*. Jackson (1961) characterized the original Wisconsin population as follows: total length of adult, 1,490–1,650 mm; tail length, 390–480 mm; hind foot, 255–290 mm; adult weight, 30–45 kg; skull length, 230–268 mm; skull width, 120–142 mm; ears moderate and less conspicuous than in the coyote; coat moderately dense, somewhat coarse; in typical full fall and winter pelage, upper parts generally grayish, more or less overlaid with black from nape to rump, under parts whitish to pale buff, head mixed with ochraceous or cinnamon, ears cinnamon to tawny, outer parts of legs cinnamon buff to cinnamon, forelegs with a more or less conspicuous black line; tail grayish above, suffused with black, buffy below, the tip blackish; no seasonal change except for the fading and sometimes more reddish color of old pelage in spring and early summer; other color variations ranging from very pale gray to near blackish.

Goldman (1944) recognized some general differences between the wolves of the western Great Lakes region and those of southeastern Ontario and southern Quebec. He noted the usual smaller size, narrower proportions, and darker coloration of animals in the latter region. Pimlott et al. (1969) reported that in southeastern Ontario's Algonquin Provincial Park the wolves are usually gray to dark gray in winter and grizzled red in summer, and that they weigh 6.8–9.1 kg less than western wolves; averages for Algonquin adults were 24.5 kg in 33 females, 27.7 kg in 40 males.

15.2 Changing Concepts of Taxonomy of Great Lakes Wolves

15.2.1 The Varying Concept of Lycaon

As explained by Goldman (1944), "*Canis lycaon*" was first used in a 1775 work by Von Schreber for an illustration he copied from a 1761 book by Buffon. The picture is of an animal regarded as a "black fox" by Von Schreber, though described as a "black wolf" by

Buffon, who indicated that it had been captured in Canada when very young and taken alive to Paris. Regarding that animal as the type specimen, Miller (1912a) designated *lycaon* the appropriate name for the wolf of eastern Canada and the northeastern United States. Goldman (1937) fixed the type locality of *lycaon* as the vicinity of Quebec City. Goldman (1944) observed that the skin of a dark-colored wolf, taken 80 km north of Quebec City in 1916, might resemble the type.

However, while *lycaon* sometimes is considered a relatively dark kind of wolf, fully black specimens are not well known (Mech and Frenzel 1971; Kolenosky and Standfield 1975). In contrast, melanistic examples of *C. rufus* were common; those seen in Florida by Bartram (1791) were the basis for his name *niger*, which formerly was applied to the red wolf. Audubon and Bachman (1851) reported black wolves from Texas to Indiana and the Carolinas, and Gregory (1935) photographed them in northeastern Louisiana. Partly on the basis of a skull collected in 1863 at Moosehead Lake, Maine, about 160 km southeast of Quebec City, Nowak (2002) thought that the range of *C. rufus* originally extended as far north as the St. Lawrence River. Therefore, it seems possible the type specimen of *lycaon* was taken from a population of the red wolf.

Whether scientific names of wolves have been applied at a specific or subspecific level seems for many years to have depended more on fashion than on careful study. Audubon and Bachman (1851) listed *lycaon*, *rufus*, *nubilus*, and all other named kinds of North American wolves (but not *latrans*) as varieties of *C. lupus*. Authorities of the late nineteenth and early twentieth centuries (e.g., Miller 1912a,b), generally treated *lycaon* and most other named kinds as full species. However, as noted by Kyle et al. (2006), it may be pertinent that Pocock (1935), who again united most of the world's wolves under the name *C. lupus*, did maintain *C. lycaon* as a separate species. Shortly thereafter, Goldman (1937, 1944) reduced *lycaon* to subspecific rank.

Goldman was a taxonomic splitter at the subspecies level, and named 11 of the 27 subspecies of Recent North American wolves listed by Hall (1981). He lumped into *lycaon* all wolves of the western and eastern Great Lakes regions, together with other populations extending as far south as Florida. That designated subspecies, then the most widespread in North America, comprised groups that now seem much more variable than do the seven subspecies of the western conterminous United States that Goldman (1944) accepted. He did acknowledge that in the western Great Lakes region, *lycaon* graded physically toward the neighboring subspecies, *C. lupus nubilus* of the Great Plains, while other specimens of *lycaon* showed close resemblance to *C. rufus* of the Southeast. Nonetheless, his arrangement was generally accepted and became fixed in United States law in 1967, when *C. lupus lycaon* was classified pursuant to the Endangered Species Protection Act and assigned a range from Minnesota to eastern Canada (United States Department of the Interior 1973).

Additional conservation interest may have been responsible for some initial challenge to Goldman's position. Mech and Frenzel (1971) pointed out that the Minnesota population might actually represent what was thought to be the otherwise extirpated subspecies *nubilus*. Their view centered primarily on observations of black or white pelage in Minnesota wolves, traits reportedly common for *nubilus* but supposedly not for *lycaon* in southeastern Ontario. Subsequently, based on color, size, and ecology, Van Ballenberghe (1977) concluded that *nubilus*, the wolf population of Minneosta, and the population farther north in western Ontario resembled each other and, to some extent, had been genetically isolated by the Great Lakes from *lycaon* of southeastern Ontario. Skeel and Carbyn (1977) also suggested morphological affinity between the population of northwestern Ontario (just north of Minnesota), the subspecies *nubilus* and *irreomtus* of the western conterminous United States, and possibly the subspecies *hudsonicus* of the region just west of Hudson Bay.

Meanwhile, Standfield (1970) and Kolenosky and Standfield (1975), citing examination of a large new collection of specimens, reported two "morphologically distinct types" of wolves in the Great Lakes region of Ontario. The "Boreal type," found from the Minnesota border in the west to about 47°N on the east side of Lake Superior, was said to be generally larger, to have a more massive skull, and to vary from pure white to jet black. The "Algonquin type," occurring east of Lake Superior from about 48°N to the vicinity of Algonquin Provincial Park (45°N), was relatively small and slender, had a narrow rostrum, and was invariably gray-fawn in color. Notably, no cline between the two types was recognized, and, while the two overlapped geographically (in the region between 47° and 48°N), there reportedly was "no conclusive evidence of their interbreeding." Such findings are practically suggestive of specific distinction, though a multiple discriminant analysis of skulls did show some statistical overlap. Further analyses by Schmitz and Kolenosky (1985) did indicate clinal variation of Algonquin and Boreal wolves, and that the two were more closely related to one another than either was to the Minnesota population or to *nubilus*; those conclusions seem incongruous with most other recent morphological study (*see* Nowak 1995), though might be assessed for compatibility with molecular approaches (*see* Kyle et al. 2006).

Using statistical analysis of multiple skull measurements, Nowak (1979, 1983, 1995, 2002, 2003) progressively corroborated earlier suggestions that the wolves of Minnesota and the boreal region of Ontario are closely related and should be assigned to the subspecies *nubilus*, together with most other named subspecies of the western conterminous United States and the subspecies *hudsonicus* farther north. He restricted the original range of *lycaon* to a relatively small part of southeastern Ontario and southern Quebec, and proposed that it had been affected by long-ago hybridization with *C. rufus*, but did conclude that it is a subspecies of *C. lupus* and does statistically intergrade with other populations of the latter.

15.2.2   Reexamination of Great Lakes and Great Plains Wolf Specimens

Nowak's (1979, 1983, 1995, 2002, 2003) previous work was part of a larger assessment of both intraspecific and interspecific relationships of living, historical, and fossil species and populations of *Canis* throughout North America and the world. Hence, a new analysis has been done, emphasizing roughly coeval series of Great Lakes wolves collected not long ago. This assessment involved subjecting ten cranial and dental measurements to canonical discriminant analysis using the Statistical Analysis System (SAS Institute 1987). The measurements, weighted by their ability to distinguish designated groups, assign each specimen a total abstract numerical value—the first canonical variable. The next best distinguishing combination of measurements, uncorrelated with the first, provides a second variable, and so on. Commonly, a single graphical position is plotted based on the first two canonical variables arranged as perpendicular axes. The ten measurements (numbered as in Table 15.1) are (1) greatest length of skull, (2) zygomatic width, (3) alveolar length from P1 to M2, (4) maximum width of rostrum across outer sides of P4, (5) palatal width between alveoli of P1, (6) width of frontal shield, (7) height from alveolus of M1 to most ventral point of orbit, (8) depth of jugal, (9) crown length of P4, and (10) greatest crown width of M2 [*see* Nowak (1995) for illustrations of the measurements and a more detailed explanation of statistical procedures].

As in several previous studies (Nowak 1995, 2002, 2003), only the skulls of fully and normally developed males were used in the analysis. Females tend to occur less frequently than do males in series of *Canis*. Earlier work (Nowak 1979) indicated that analysis of either sex produces about the same result.

Three groups of specimens were used: (1) 27 *C. lupus nubilus* collected prior to 1930 in Idaho, Kansas, Montana, Nebraska, Oklahoma, North Dakota, South Dakota, and Wyoming (26 now at United States National Museum, 1 at American Museum of Natural History); (2) 23 collected 1970–1975 in northern Minnesota (15 now at United States National Museum, 8 at University of Minnesota Museum of Natural History); and (3) 20 *lycaon* collected 1964–1965 during an effort to totally remove wolves from Algonquin Provincial Park, southeastern Ontario (examined at Natural Resources DNA and Forensic Profiling Centre, Trent University, Peterborough, Ontario). In addition, the following specimens were tested against the three groups as individuals: 16 collected before 1966 from the original wolf population of the Upper Peninsula of Michigan (9 now at United States National Museum, 3 at Michigan State Museum, 4 at University of Michigan

Table 15.1 Means of skull measurements (in millimeters and numbered as in text) for male *Canis lupus nubilus* taken before 1930, Minnesota specimens taken 1970–1975, and Algonquin *lycaon* taken 1964–1965

	nubilus	Minnesota	*lycaon*
	$n = 27$	$n = 23$	$n = 20$
1	256.85	256.30	245.10
2	139.59	140.13	132.05
3	86.24	86.25	82.53
4	82.35	81.82	76.16
5	31.89	31.97	27.05
6	64.57	64.69	60.69
7	39.95	39.67	37.27
8	19.73	20.00	17.20
9	25.70	25.09	24.51
10	13.39	14.20	14.34

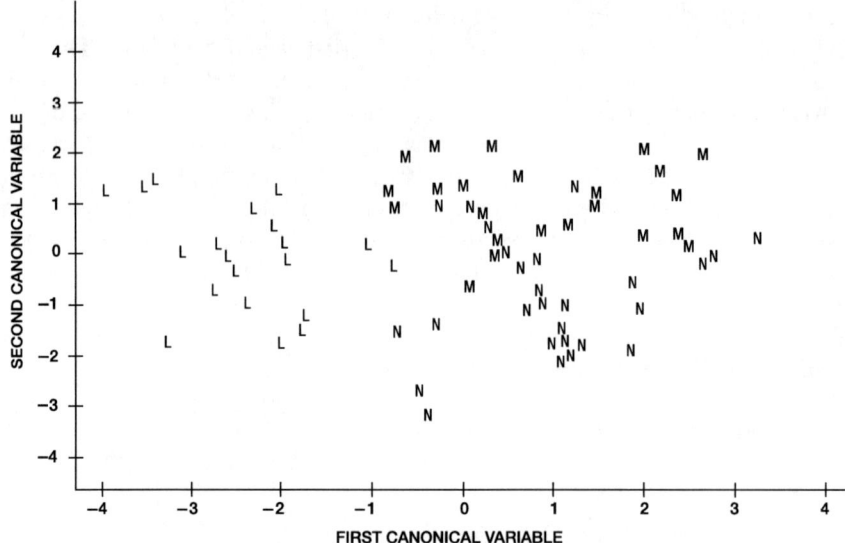

Fig. 15.1 Statistical distribution of three groups of North American male wolves (*Canis*), plotted on the first and second canonical variables. L's, *lycaon* collected 1964–1965 in Algonquin Provincial Park, southeastern Ontario; M's, specimens collected 1970–1975 in northern Minnesota; N's, *C. lupus nubilus* collected prior to 1930 in Idaho, Kansas, Montana, Nebraska, Oklahoma, North Dakota, South Dakota, and Wyoming

Museum of Zoology), and 6 collected 1958–1963, along an approximate east–west axis, between the Upper Peninsula of Michigan and Algonquin Provincial Park, at the Ontario towns of Sault Ste. Marie, Sudbury, and North Bay (examined at Natural Resources DNA and Forensic Profiling Centre, Trent University, Peterborough, Ontario).

15.2.3 Results of Recent Analysis

Western *C. lupus nubilus* and the Minnesota series show substantial statistical overlap, while each of those groups is completely distinct from *C. lupus lycaon* of Algonquin Provincial Park in southeastern Ontario (Fig. 15.1). The analysis thus supports assignment of the Minnesota wolf population to *nubilus*, not *lycaon*. However, individuals collected in the geographic region between Minnesota and Algonquin Park indicate morphological intergradation of *nubilus* and *lycaon* (Fig. 15.2). Hence, the analysis continues to support recognition of *lycaon* as a subspecies of *C. lupus*.

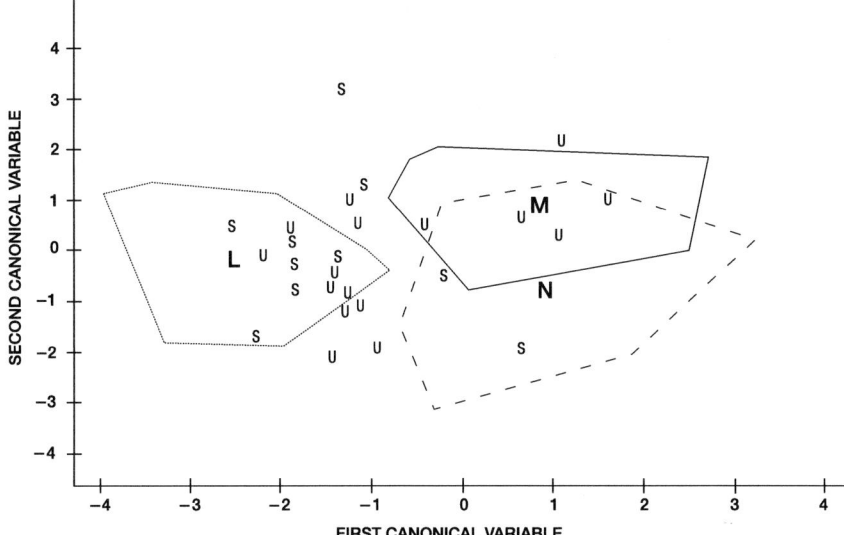

Fig. 15.2 Statistical distribution of three groups (the same depicted in Fig. 15.1) and certain individuals of North American male wolves (*Canis*), plotted on the first and second canonical variables. *Dotted lines*, limits of *lycaon* collected 1964–1965 in Algonquin Provincial Park, southeastern Ontario (letter L shows mean position); *solid lines*, limits of specimens collected 1970–1975 in northern Minnesota (letter M shows mean position); *dashed lines*, limits of *C. lupus nubilus* collected prior to 1930 in Idaho, Kansas, Montana, Nebraska, Oklahoma, North Dakota, South Dakota, and Wyoming (letter N shows mean position); U's, individuals collected before 1966 from the original wolf population of the Upper Peninsula of Michigan; S's, individuals collected 1958–1963, along an approximate east–west axis, between the Upper Peninsula of Michigan and Algonquin Provincial Park, at the Ontario towns of Sault Ste. Marie, Sudbury, and North Bay

15.3   The Molecular Recasting

Gray and red wolves, along with the domestic dog and coyote, have a diploid chromosome number of 78 (Wayne 1993). Molecular technology seeks to assess variation in the chromosomal (nuclear) DNA, or in the protein sequences that DNA specifies, to determine relationships of the individuals and populations involved. Such studies also sometimes use nucleotide variation in mitochondrial DNA, which is particularly applicable to phylogeny, as it has a very high mutation rate and, unlike nuclear DNA, is solely maternally inherited. Certain nuclear DNA loci, known as microsatellites, also have been found to have high mutation rates; their study has allowed broader evaluation of highly degraded DNA, as found in bones and old skins, and permitted identification of the two alleles inherited from the parents at each locus (Wayne and Vilà 2003).

Molecular genetic studies of *Canis* have proliferated, becoming far more common than systematic investigations using traditional morphological techniques. Wayne and

Vilà (2003) listed 35 such studies of *C. lupus* and related canids. Kyle et al. (2006) cited 11 more, particularly those applying to wolves in eastern North America.

An initial objective of this new methodology was a better means of identifying specimens of *C. rufus*. Wayne and Jenks (1991), however, reported the species to lack a unique identifying mitochondrial DNA genotype. Instead, it had only genotypes of *C. lupus* or *C. latrans* and thus was considered to have originated as a hybrid of those two species. Subsequent study of both mitochondrial and nuclear DNA supported that conclusion (Wayne 1992; Roy et al. 1994a,b, 1996; Wayne and Gittleman 1995; Wayne et al. 1995). Most recently, Reich et al. (1999: 143) reported that their comparison of microsatellite allele length distributions supported "the hypothesis of a recent hybridization between coyotes and grey wolves that may have been associated with the extensive agricultural cultivation of the southern United States by European settlers beginning around 250 years ago."

Meanwhile, Lehman et al. (1991) carried out a study of mitochondrial DNA of gray wolves and coyotes from localities throughout North America. They found all wolves within the presumed range of the "Algonquin type" of *lycaon* in southeastern Ontario and southern Quebec to have coyote genotypes. The wolves of Isle Royale, Michigan, and a majority of those in Minnesota and the adjacent part of western Ontario also were found to have coyote genotypes. This situation was seen to have led to formation of a "hybrid zone," paralleling the process with the red wolf, but not to be so far advanced. Essentially the same conclusions were reached by Roy et al. (1994a) using analysis of microsatellite loci.

The molecular genetic studies cited above, all associated with work initiated at the University of California, Los Angeles [except that of Kyle et al. (2006)], have not been universally accepted. Although wolf–coyote hybridization is known to have occurred in southeastern Canada (Kolenosky and Standfield 1975; Nowak 1979; Kyle et al. 2006), observations by field personnel in Minnesota and on Isle Royale indicate no change in the morphological, behavioral, or ecological characteristics and hence no evidence that introgression from *C. latrans* has spread to those areas (Nowak et al. 1995:413). Likewise, hybridization with coyotes long has been recognized as a factor in the *demise* of the red wolf (McCarley 1962; Nowak 1979), but hybrid *origin* of *rufus* has not been supported by morphometric analysis (Nowak 1979, 1992, 1995, 2002), by observation of living animals (Phillips and Henry 1992; Nowak et al. 1995), by several recent molecular studies (Bertorelle and Excoffier 1998; Hedrick et al. 2002; Mech and Federoff 2002), or by some other geneticists who have reviewed the issue (Dowling et al. 1992a,b; Cronin 1993).

Wilson et al. (2000), as amplified by Kyle et al. (2006), provided a completely new assessment and interpretation of molecular data relevant to all eastern wolves. Using both mitochondrial DNA and nuclear microsatellite loci, a close genetic relationship between *rufus* and *lycaon* was identified. That affinity was not caused by common introgression from *latrans*; hybrid origin for *C. rufus* was rejected. In addition, many of the genes found in both kinds of eastern wolves were not found in either western coyote or gray wolf populations, though seemed more closely associated with *latrans*. Genetic structure suggested *rufus* and *lycaon* were components of a single species, which, because of nomenclatural priority, would take the name *C. lycaon*. Previously reported molecular evidence of wolf–coyote hybridization in the western Great Lakes region now was considered to show presence of *C. lycaon*. That species would have diverged from *C. latrans* well after their common ancestor split from the line leading to *C. lupus*. On the basis of molecular analysis, the current range of *C. lycaon* was thought to include southern Quebec and the vicinity of Algonquin Provincial Park in southeastern Ontario. The species also occurred, together with *C. lupus*, around the north of Lake Superior to western Ontario and possibly southern Manitoba.

Wilson et al. (2003) examined genetic material from two wolf skins collected in the northeastern United States long before *C. latrans* had spread to the region. One was taken in northern New York around 1890, the other in Penobscot County, Maine in the 1880s. Neither specimen was identified as *C. lupus*. The Maine sample was found to have a genotype of the kind previously found in *lycaon* and *rufus*, while the New York sample was like that of western coyote populations. Those results were interpreted to be conducive to designation of original eastern wolves as an independent species, with affinity to *C. latrans*.

Such an assessment of eastern wolves has been supported with the assertion that *C. latrans* seems to readily hybridize with *lycaon* and *rufus*, at least during times of environmental disruption and when numbers of the latter two forms have been depleted, whereas hybridization between western *C. lupus* and *C. latrans* is unknown (Kyle et al. 2006). Actually, Nowak (1979) reported three specimens that statistically appeared to be possible hybrids of *C. latrans* and the small Mexican wolf *C. lupus baileyi*. A subsequent multivariate analysis by Bogan and Mehlhop (1983) suggested that the smallest and most coyote-like of the three did represent *baileyi*. That specimen was collected, probably in the 1800s, at Orizaba, Veracruz, far to the south of any substantive series of known *C. lupus*; the other two specimens also date back over a century and could be small examples of *baileyi*. Wayne and Vilà (2003) considered *baileyi* to be the most highly differentiated North American gray wolf taxon and was the only nominal subspecies of *C. lupus* they recognized based on their molecular studies.

In any case, even if the Mexican wolf did cross with the coyote on rare occasion, there is no morphological evidence of introgression from the latter in series of *baileyi* taken during intensive control operations in the twentieth century (Nowak 1979, 1995), or molecular evidence of such in the living population of *baileyi* (Wayne and Vilà 2003). There also is no sign of coyote introgression in the large series of *C. lupus* collected throughout the western United States in the early 1900s, when the wolf populations of that region were rapidly being fragmented and eliminated (Nowak 1979), or in the population of *C. lupus* that recently was introduced from western Canada to the northwestern United States (Pilgrim et al. 1998). Therefore, the molecular case for uniting *rufus* and Algonquin *lycaon* in a species separate from *C. lupus*, based on their readiness to hybridize with *C. latrans*, seems compelling.

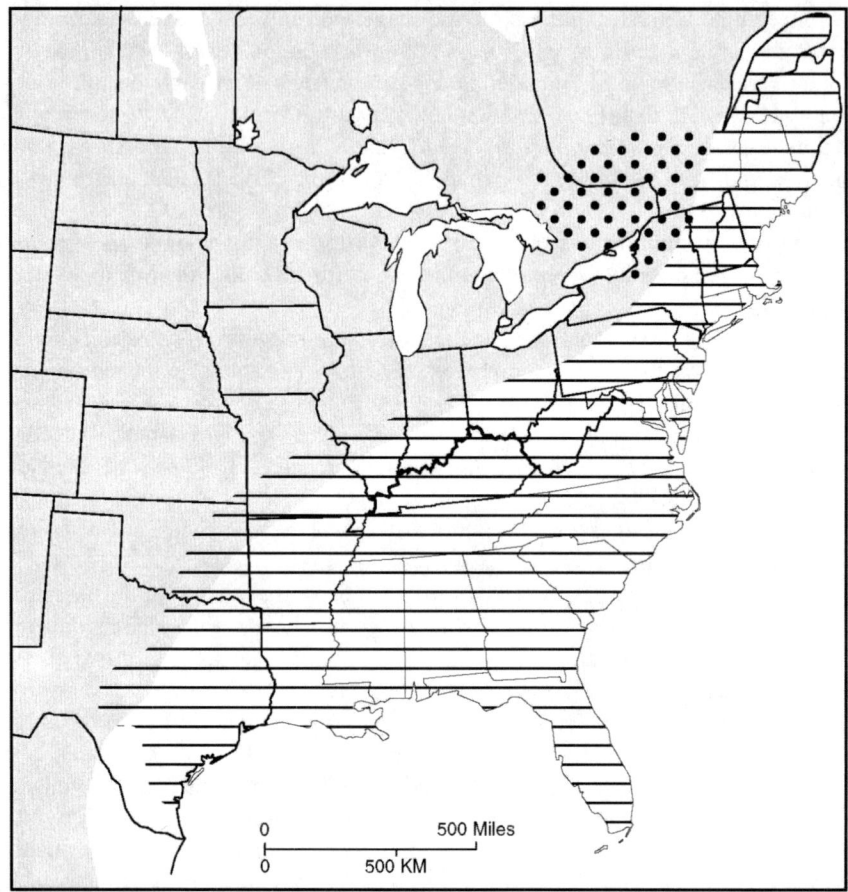

Fig. 15.3 Geographical distribution of wolves (*Canis*) in eastern North America, based primarily on morphological evidence (Nowak 2002, 2003). *Shading*, original range of *C. lupus*; *horizontal hatching*, original range of *C. rufus*; *stippling*, zone of possible hybridization between *C. lupus* and *C. rufus*, which may have contributed to development of the subspecies *C. lupus lycaon*

   Remarkably, if the views of Wilson et al. (2000, 2003) and Kyle et al. (2006) are accepted, the original geographic range of *lycaon* will be restored to much the same as that assigned by Goldman (1944), effectively undoing the reductions resulting from subsequent morphological assessment (Standfield 1970; Mech and Frenzel 1971; Kolenosky and Standfield 1975; Skeel and Carbyn 1977; Van Ballenberghe 1977; Nowak 1979, 1983, 1995, 2002, 2003). Not only would *lycaon* again be a name applicable throughout the Great Lakes region, it also would extend as far south as Miami, Florida, where a specimen was collected in 1854 that Goldman (1944) did indeed refer to *lycaon*, though Nowak (2002) included it within *C. rufus*. Figures 15.3 and 15.4 compare the overall distributions of eastern wolves suggested by the contending evidence.

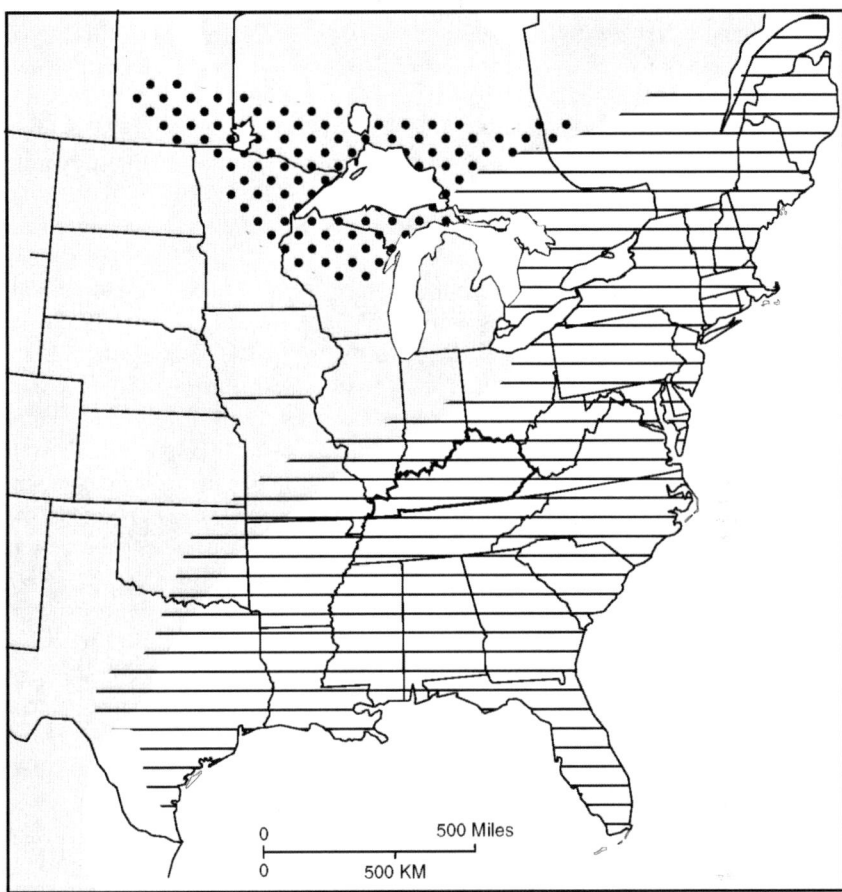

Fig. 15.4 Geographical distribution of wolves (*Canis*) in eastern North America, based primarily on molecular evidence (Wilson et al. 2000, 2003; Wayne and Vilà 2003; Kyle et al. 2006). *Shading*, original range of *C. lupus*; *horizontal hatching*, original range of *C. lycaon*; *stippling*, zone of possible hybridization between *C. lupus* and *C. lycaon*

15.4   Resolution?

As explained above, and in detail by Kyle et al. (2006), there are three rival systematic positions on eastern wolves. The most divergent are the two based on molecular studies, one designating the red wolf part of an ancient and distinct species, the other holding that the red wolf originated as a modern hybrid. However, as indicated by the advocates of the former view (Kyle et al. 2006), the advocates of the latter view (Wayne and Vilà 2003) now have acknowledged that there might have been a distinct red wolf-like species that migrated into Canada after the last glaciation and interbred with the gray wolf, which is exactly what was proposed by the advocate of the third, morphologically based, view (Nowak 2002). Wayne and Vilà (2003) used the name *C. lycaon* for that species, which would indeed have priority over *rufus*, if the animal named by Von Schreber in 1775 (see above) represented the red wolf, and/or if *rufus* and *lycaon* are accepted as conspecific.

While there thus has been a start to resolution, obstacles remain, one being the reported genotypes of *C. lupus* found in studies of both mitochondrial DNA and microsatellite loci of the red wolf (Wayne and Jenks 1991; Roy et al. 1994a,b, 1996). Wilson et al. (2000) suggested that the involved genetic samples were from actual specimens of *C. lupus nubilus, C. l. baileyi*, or *C. l. familiaris*. Such seems unlikely, as the samples were taken from specimens geographically and temporally well removed from then existing populations of *C. lupus*, and there is no evidence of introgression from the domestic dog into the wild canid populations of the involved region (Nowak 1979, 1995). A more plausible explanation for the presence of genotypes of *C. lupus* in specimens of red wolves taken in the south-central United States in the early twentieth century might be that the two species underwent limited hybridization when they were still in contact but coming under intense pressure from hunting and ecological disturbance. Such interaction would be closely comparable to what may have occurred more recently between *C. lupus* and *lycaon* in the western Great Lakes region.

There may be no trenchant disagreement between the assessments set forth by Nowak (2002) and Kyle et al. (2006). They concur that there was a distinct species of wolf in the east, though they have applied different names to it. Nowak also thought that its original range was primarily south of the St. Lawrence River, and that it underwent hybridization with *C. lupus* just to the north, in the Algonquin vicinity (Fig. 15.3). Kyle et al. (2006) believed the historical range to extend at least through southeastern Ontario and that hybridization with *C. lupus* occurred farther to the north and west (Fig. 15.4). Both sides might agree that genetic material of the eastern wolf, and perhaps the wolf itself, is spreading into the western Great Lakes region, possibly beyond, and that there could also be some opposite movement from *C. lupus*.

A more difficult problem involves evolutionary history. On the basis of divergence of genetic sequences, Wilson et al. (2000) concluded that the eastern wolf is a close relative of *C. latrans*, the two species separating only 150,000–300,000 years ago, while their common ancestor would have split from the line leading to *C. lupus* 1–2 million years ago. Nowak (1979) initially did postulate division of the ancestral stock of *rufus* and *lupus* at about that same time. Subsequently, however, Nowak (2002, 2003) noted that small wolves disappeared entirely from eastern North America for a period of about a million years, from the middle Irvingtonian to the late Rancholabrean. He suggested that the species *C. priscolatrans* (=*C. edwardii*) of the early Irvingtonian, previously thought to be the ancestor of *rufus*, actually gave rise to an archaic line of large New World wolves, culminating in *C. dirus* of the late Rancholabrean. He thought that *rufus* had arisen in the Old World from the same ancestral stock as *C. lupus*. Meanwhile, *C. latrans* had a separate history in North America since the early Irvingtonian.

An alternative was offered by Kurten (1974), who considered *C. priscolatrans* not a small wolf but part of a Holarctic group of coyotes that sometimes attained substantial size and preceded modern *C. latrans*. *C. priscolatrans* disappeared well before the division of *C. latrans* and *lycaon/rufus* hypothesized by Wilson et al. (2000), but a small Eurasian wolf, *C. mosbachensis*, may have persisted from the early Irvingtonian to the early Rancholabrean (Nowak 2003). Sotnikova (2001) suggested affinity among *mosbachensis*, *latrans*, *pallipes*, and *rufus*. However, while *mosbachensis* or *priscolatrans* might not be an unreasonable progenitor of a *lycaon/rufus* group, we still are left with the million-year gap in the fossil history of small wolves in eastern North America (Nowak 2002).

15.5 Conclusions and Conservation Implications

Notwithstanding debate on identity of *rufus* and *lycaon*, there is agreement that both hybridized with *C. latrans* as the latter species invaded the natural range of eastern wolves (Nowak 1979; Schmitz and Kolenosky 1985; Reich et al. 1999; Sears et al. 2003; Wayne and Vilà 2003; Kyle et al. 2006). That process was a critical factor in near extinction of the red wolf and remains a serious problem to the reintroduced population in North Carolina (Phillips et al. 2003; Fredrickson and Hedrick 2006). Hybridization also is a growing threat to surviving wolf populations from southern Quebec and southeastern Ontario to Minnesota (Lehman et al. 1991; Roy et al. 1994a). However, Kyle et al. (2006) suggested that presence of coyote-like genotypes in wolves of that region is partly reflective of some of those wolves representing a species (*C. lycaon*) that is closely related to *C. latrans*.

If Kyle et al. (2006) are correct, the extent of coyote introgression in Great Lakes wolves may be less imminent a threat than previously suggested, at least to more westerly populations. However, presence of a second species of wolf in the region would introduce entirely new issues of environmental usurpation and hybridization. Kyle et al. (2006) reported that the genotypic composition of wolves in that part of Ontario around Lake Superior represents both *C. lycaon* and *C. lupus*, thus indicating interbreeding between the two species. It may be that logging and other ecological changes have produced a habitat more favorable to a smaller, predominantly deer-eating wolf, *C. lycaon*, which is displacing and genetically swamping the larger *C. lupus*, which originally preyed mainly on moose. Interestingly, a remnant pocket of *C. lupus*, free of genetic material from *C. lycaon* or *C. latrans*, was found on the north shore of Lake Superior in Pukaskwa National Park, an area where boreal forest and moose still prevail.

It still seems uncertain as to whether and to what extent western (or boreal) *C. lupus* and eastern (or Algonquin) *lycaon* are intergrading as subspecies, interbreeding as distinct species, and/or behaving as sympatric entities. Kyle et al. (2006) suggested that *lycaon* is either phenotypically or genetically present throughout the involved region, perhaps essentially in its original form in the vicinity of Algonquin Park, but having extensively interbred with *C. lupus* to the north and west, and having crossed with *C. latrans* to the south to form a hybrid population sometimes designated the "Tweed wolf." Kyle et al. (2006) cautioned that presence of *lycaon* not only may threaten the integrity of *C. lupus* but may lead to overestimating the latter's population and thus to unwise mitigation of conservation measures. On the other hand, Kyle et al. (2006) believed that current hybridization should not always be viewed as negative, and that management policies should deemphasize preserving the phenotype of *lycaon* to allow continued adaptation to its anthropogenically modified environment.

That last suggestion is debatable. If *lycaon* is an ancient species, or even if it represents post-Pleistocene hybridization, its dissolution through human-induced ecological disruption and accelerated interbreeding would seem contrary to any program seeking to conserve examples of populations present before European colonization. *Lycaon* already has undergone extensive hybridization with *C. latrans* to the south of Algonquin Park, and probably elsewhere, and its continued integrity depends on proactive measures to maintain numbers and habitat. Moreover, if the still intact populations of *lycaon* undergo further introgression from *C. latrans*, and if *lycaon* is indeed spreading westward, physically or genetically, *C. lupus* would become exposed to intensified genetic introgression from *C. latrans*. Although the western gray wolf may not hybridize directly with the coyote, it apparently does interbreed or intergrade with *lycaon*. It may also have interbred with *rufus* in the south-central United States, but disappeared so early from that region that introgression from the coyote, through the red wolf, may never have developed.

The United States Department of the Interior did not discuss the hybridization threat, whether from *C. latrans* or from *C. lycaon*, in its recent rule removing the western Great Lakes population of *C. lupus* from the List of Endangered and Threatened Wildlife (Refsnider 2007). Presence on the List allowed *C. lupus* to rebuild numbers in Minnesota and recolonize Wisconsin and northern Michigan.

Under the new rules, states will for the most part continue to protect wolves on public forest lands and will mostly allow wolf populations to fluctuate with prey populations in these areas (A. P. Wydeven, personal communication). Special protections for den sites and management to maintain low road densities will continue on national forests and national parks, and many of the state and county forests will also provide such habitat protections. However, the states will oversee most management and maintenance of populations, and their ecological and genetic viability. In areas of agricultural land and mixed forest–farmland, wolf numbers will be more intensely controlled. Theberge and Theberge (2004) have indicated that hybridizations at least between *C. lycaon* and *C. latrans* are more likely in such fragmented habitats.

Introgressive hybridization in *Canis* was not known to science in the mid-twentieth century. The red wolf in the south-central United States then was subject to control and routinely reported to be common, while it vanished through interbreeding with the coyote (Nowak 1979). Hopefully, what has been learned since then, and follow-up studies by responsible agencies, will prevent repetition of that scenario for the gray wolf in the Great Lakes region.

Acknowledgments Tim Van Deelen very kindly volunteered to help me by running the canonical discriminant analysis. Cecilia A. Mauer prepared the figures.

References

Aggarwal, R. K., Kivisild, T., Ramadevil, J., and Singh, L. 2007. Mitochondrial DNA coding region sequences support the phylogenetic distinction of two Indian wolf species. Journal of Zoological Systematics and Evolutionary Research 45:163–172.

Audubon, J. J., and Bachman, J. 1851. The quadrupeds of North America, vol. 2. New York: V. G. Audubon.

Bartram, W. 1791. Travels. Philadelphia: James and Johnson.

Bertorelle, G., and Excoffier, L. 1998. Inferring admixture proportions from molecular data. Molecular Biology and Evolution 15:1298–1311.

Bogan, M. A., and Mehlhop, P. 1983. Systematic relationships of gray wolves (*Canis lupus*) in southwestern North America. University of New Mexico Museum of Southwestern Biology Occasional Papers 1:1–21.

Clutton-Brock, J., Corbett, G. B., and Hills, M. 1976. A review of the family Canidae, with a classification by numerical methods. Bulletin of the British Museum (Natural History) 29:119–199.

Cronin, M. A. 1993. Mitochondrial DNA in wildlife taxonomy and conservation biology: cautionary notes. Wildlife Society Bulletin 21:339–348.

Dowling, T. E., Demarais, B. D., Minckley, W. L., Douglas, M. E., and Marsh, P. C. 1992a. Use of genetic characters in conservation biology. Conservation Biology 6:7–8.

Dowling, T. E., Minckley, W. L., Douglas, M. E., Marsh, P. C., and Demarais, B. D. 1992b. Response to Wayne, Nowak, and Phillips and Henry: use of molecular characters in conservation biology. Conservation Biology 6:600–603.

Ferguson, W. W. 1981. The systematic position of *Canis aureus lupaster* (Carnivora: Canidae) and the occurrence of *Canis lupus* in North Africa, Egypt and Sinai. Mammalia 45:459–465.

Fredrickson, R. J., and Hedrick, P. W. 2006. Dynamics of hybridization and introgression in red wolves and coyotes. Conservation Biology 20:1272–1283.

Goldman, E. A. 1937. The wolves of North America. Journal of Mammalogy 18:37–45.

Goldman, E. A. 1944. Classification of wolves. In The Wolves of North America, S. P. Young and E. A. Goldman, pp. 389–636. Washington, D.C.: American Wildlife Institute.

Gregory, T. 1935. The black wolf of the Tensas. Chicago Academy of Sciences Program Activities 6(3):1–68.

Hall, E. R. 1981. The Mammals of North America. New York: John Wiley and Sons.

Harrison, D. L., and Bates, P. J. 1991. The Mammals of Arabia. Sevenoaks, Kent, England: Harrison Zoological Museum.

Hedrick, P. W., Lee, R. N., and Garrigan, D. 2002. Major histocompatibility complex variation in red wolves: evidence for common ancestry with coyotes and balancing selection. Molecular Ecology 11:1905–1913.

Hefner, R., and Geffen, E. 1999. Group size and home range of the Arabian wolf (*Canis lupus*) in southern Israel. Journal of Mammalogy 80: 611–619

Imaizumi, Y. 1970a. Systematic status of the extinct Japanese wolf, *Canis hodophilax*. Journal of the Mammal Society of Japan 5:27–32.

Imaizumi, Y. 1970b. Systematic status of the extinct Japanese wolf, *Canis hodophilax*. 2. Similarity relationship of *hodophilax* among the species of the genus *Canis*. Journal of the Mammal Society of Japan 5:62–66.

Jackson, H. H. T. 1961. Mammals of Wisconsin. Madison: University of Wisconsin Press.

Kolenosky, G. B., and Standfield, R. O. 1975. Morphological and ecological variation among gray wolves (*Canis lupus*) of Ontario, Canada. In The Wild Canids, Their Systematics, Behavioral Ecology and Evolution, ed. M. W. Fox, pp. 62–72. New York: Van Nostrand Reinhold.

Kurten, B. 1974. A history of coyote-like dogs (Canidae, Mammalia). Acta Zoologica Fennica 140:1–38.

Kurten, B., and Anderson, E. 1980. Pleistocene Mammals of North America. New York: Columbia University Press.

Kyle, C. J., Johnson, A. R., Patterson, B. R., Wilson, P. J., Shami, K., Grewal, S. K., and White, B. N. 2006. Genetic nature of eastern wolves: past, present and future. Conservation Genetics 7:273–287

Lawrence, B., and Bossert, W. H. 1967. Multiple character analysis of *Canis lupus, latrans*, and *familiaris*, with a discussion of the relationships of *Canis niger*. American Zoologist 7:223–232.

Lawrence, B., and Bossert, W. H. 1975. Relationships of North American *Canis* shown by a multiple character analysis of selected populations. In The Wild Canids, Their Systematics, Behavioral Ecology and Evolution, ed. M. W. Fox, pp. 73–86. New York: Van Nostrand Reinhold.

Lehman, N., Eisenhawer, A., Hansen, K., Mech, L. D., Peterson, R. O., Gogan, P. J. P., and Wayne, R. K. 1991. Introgression of coyote mitochondrial DNA into sympatric North American gray wolf populations. Evolution 45:104–119.

McCarley, H. 1962. The taxonomic status of wild *Canis* (Canidae) in the south central United States. Southwestern Naturalist 7:227–235.

Mech, L. D., and Boitani, L. 2004. Grey wolf. In Canids: Foxes, Wolves, Jackals and Dogs, Status Survey and Action Plan, eds. C. Sillero-Zubiri, M. Hoffmann, and D. W. Macdonald, pp. 124–129. Gland, Switzerland: International Union for Conservation of Nature and Natural Resources.

Mech, L. D., and Federoff, N. E. 2002. Alpha (1) – antitrypsin polymorphism and systematics of eastern North American wolves. Canadian Journal of Zoology 80:961–963.

Mech, L. D., and Frenzel, L. D. 1971. The possible occurrence of the Great Plains wolf in northeastern Minnesota. In Ecological Studies of the Timber Wolf in Northeastern Minnesota, eds. L. D. Mech and L. D. Frenzel, pp. 60–62. St. Paul: U.S. Forest Service Research Paper NC-52, North Central Forest Experimental Station.

Miller, G. S. 1912a. The names of two North American wolves. Proceedings of the Biological Society of Washington 25:95.

Miller, G. S. 1912b. The names of the large wolves of northern and western North America. Smithsonian Miscellaneous Collections 59(15):1–5.

Nakamura, K. 1998. A biogeographic look on the taxonomy of the Japanese wolves, *Canis lupus hodophilax* Temminck, 1839. Bulletin of the Kanagawa Prefectural Museum (Natural Science) 27: 49–60

Nakamura, K. 2004. Records of skull specimens of the Japanese wolf. Bulletin of the Kanagawa Prefectural Museum (Natural Science) 33:91–96.

Nowak, R. M. 1979. North American Quaternary *Canis*. Monograph of the Museum of Natural History, University of Kansas 6:1–154.

Nowak, R. M. 1983. A perspective on the taxonomy of wolves in North America. In Wolves in Canada and Alaska: Their Status, Biology, and Management, ed. L. N. Carbyn, pp. 10–19. Edmonton: Canadian Wildlife Service.

Nowak, R. M. 1992. The red wolf is not a hybrid. Conservation Biology 6:593–595.

Nowak, R. M. 1995. Another look at wolf taxonomy. In Ecology and Conservation of Wolves in a Changing World: Proceedings of the Second North American Symposium on Wolves, eds. L. N. Carbyn, S. H. Fritts, and D. R. Seip, pp. 375–397. Edmonton: Canadian Circumpolar Institute, University of Alberta.

Nowak, R. M. 1999. Walker's Mammals of the World. Baltimore: Johns Hopkins University Press.

Nowak, R. M. 2002. The original status of wolves in eastern North America. Southeastern Naturalist 1:95–130.

Nowak, R. M. 2003. Wolf evolution and taxonomy. In Wolves, Behavior, Ecology, and Conservation, eds. L. D. Mech and L. Boitani, pp. 239–258. Chicago: University of Chicago Press.

Nowak, R. M., Phillips, M. K., Henry, V. G., Hunter, W. C., and Smith, R. 1995. The origin and fate of the red wolf. In Ecology and Conservation of Wolves in a Changing World: Proceedings of the Second North American Symposium on Wolves, eds. L. N. Carbyn, S. H. Fritts, and D. R. Seip, pp. 409–416. Edmonton: Canadian Circumpolar Institute, University of Alberta.

Phillips, M. K., and Henry, V. G. 1992. Comments on red wolf taxonomy. Conservation Biology 6:596–599.

Phillips, M. K., Henry, V. G., and Kelly, B. T. 2003. Restoration of the red wolf. In Wolves, Behavior, Ecology, and Conservation, eds. L. D. Mech and L. Boitani, pp. 272–288. Chicago: University of Chicago Press.

Pilgrim K. L, Boyd, D. K, and Forbes, S. H. 1998. Testing for wolf–coyote hybridization in the Rocky Mountains using mitochondrial DNA. Journal of Wildlife Management 62:683–689.

Pimlott, D. H., Shannon, J. A., and Kolenosky, G. B. 1969. The ecology of the timber wolf in Algonquin Provincial Park. Maple: Ontario Department of Lands and Forests.

Pocock, R. I. 1935. The races of *Canis lupus*. Proceedings of the Zoological Society of London 1935(3):647–686.

Refsnider, R.L. 2007. Endangered and threatened wildlife and plants; final rule designating the western Great Lakes populations of gray wolves as a distinct population segment; removing the western Great Lakes distinct population segment of the gray wolf from the list of Endangered and threatened wildlife; final rule. Federal Register 72:6052–6103.

Reich, D. E., Wayne, R. K., and Goldstein, D. B. 1999. Genetic evidence for a recent origin by hybridization of red wolves. Molecular Ecology 8:139–144.

Roy, M. S., Geffen, E., Smith, D., Ostrander, E. A., and Wayne, R. K. 1994a. Patterns of differentiation and hybridization in North American wolflike canids, revealed by analysis of microsatellite loci. Molecular Biology and Evolution 11:533–570.

Roy, M. S., Geffen, E., Smith, D., and R. K. Wayne. 1996. Molecular genetics of pre-1940 red wolves. Conservation Biology 10:1413–1424.

Roy, M. S., Girman, D. J., and Wayne, R. K. 1994b. The use of museum specimens to reconstruct the genetic variability and relationships of extinct populations. Experientia 50:1–7.

SAS Institute.1987. SAS/STAT Guide for Personal Computers, Version 6. SAS Institute, Cary, North Carolina.

Schmitz, O. J., and Kolenosky, G. B. 1985. Wolves and coyotes in Ontario: morphological relationships and origins. Canadian Journal of Zoology 63:1130–1137.

Sears, H., Theberge, J. B., Theberge, M. T., Thornton, I., Campbell, G. D. 2003. Landscape influence on *Canis*. Morphological and ecological variation in a coyote-wolf *C. lupus* x *latrans* hybrid zone, southeastern Ontario. Canadian Field Naturalist 117: 589–600.

Sharma, D. K., Maldonado, J. E., Jhala, Y. V., and Fleischer, R. C. 2004. Ancient wolf lineages in India. Proceedings of the Royal Society, Biological Sciences, Series B (Supplement 3) 271:S1-S4.

Sillero-Zubiri, C., and Marino, J. 2004. Ethiopian wolf *Canis simensis* Rüppell, 1835. In Canids: Foxes, Wolves, Jackals and Dogs, Status Survey and Action Plan, eds. C. Sillero-Zubiri, M. Hoffmann, and D. W. Macdonald, pp. 124–129. Gland, Switzerland: International Union for Conservation of Nature and Natural Resources.

Skeel, M. A., and Carbyn, L. N. 1977. The morphological relationships of gray wolves (*Canis lupus*) in national parks of central Canada. Canadian Journal of Zoology 55:737–747.

Sotnikova, M. V. 2001. Remains of Canidae from the lower Pleistocene site of Untermassfeld. Monographien des Römisch-Germanischen Zentralmuseums Mainz 40:607–632.

Spassov, N. 1989. The position of jackals in the *Canis* genus and life-history of the golden jackal (*Canis aureus*) in Bulgaria and on the Balkans. Natural History of Bulgaria 1:44–56.

Standfield, R. 1970. Some considerations on the taxonomy of wolves in Ontario. In Proceedings of a Symposium on Wolf Management in Selected Areas of North America, eds. S. E. Jorgensen, C. E. Faulkner, and L. D. Mech, pp. 32–38. Twin Cities: United States Bureau of Sport Fisheries and Wildlife.

Theberge, J. B., and Theberge, M.T. 2004. The Wolves of Algonquin Park: A 12-Year Ecological Study. Waterloo: University of Waterloo.

United States Department of the Interior. 1973. Threatened Wildlife of the United States. Washington, D.C.: Bureau of Sport Fisheries and Wildlife, Resource Publication 114.

Van Ballenberghe, V. 1977. Physical characteristics of timber wolves in Minnesota. In Proceedings of the 1975 Predator Symposium, eds. R. L. Phillips and C. Jonkel, pp. 213–219. Bozeman: Montana Forest and Conservation Experiment Station.

Wayne, R. K. 1992. On the use of morphologic and molecular genetic characters to investigate species status. Conservation Biology 6:590–592.

Wayne, R. K. 1993. Molecular evolution of the dog family. Trends in Genetics 9:218–224.

Wayne, R. K., and Gittleman, J. L. 1995. The problematic red wolf. Scientific American 273(1):36–39.

Wayne, R. K., and Jenks, S. M. 1991. Mitochondrial DNA analysis implying extensive hybridization of the endangered red wolf, *Canis rufus*. Nature 351:565–568.

Wayne, R. K., and Vilà, C. 2003. Molecular genetic studies of wolves. In Wolves, Behavior, Ecology, and Conservation, eds. L. D. Mech and L. Boitani, pp. 218–238. Chicago, University of Chicago Press.

Wayne, R. K., Lehman, N., Allard, M. W., and Honeycutt, R. L. 1992. Mitochondrial DNA variability of the gray wolf: genetic consequences of population decline and habitat fragmentation. Conservation Biology 6:559–569.

Wayne, R. K., Lehman, N., and Fuller, T. K. 1995. Conservation genetics of the gray wolf. In Ecology and Conservation of Wolves in a Changing World: Proceedings of the Second North American Symposium on Wolves, eds. L. N. Carbyn, S. H. Fritts, and D. R. Seip, pp. 399–498. Edmonton: Canadian Circumpolar Institute, University of Alberta.

Wilson, P. J., Grewal, S., Lawford, I. D., Heal, J. N. M., Granacki, A. G., Pennock, D., Theberge, J. B., Theberge, M. T., Voigt, D. R., Waddell, W., Chambers, R. E., Paquet, P. C., Goulet, G., Cluff, D., and White, B. N. 2000. DNA profiles of the eastern Canadian wolf and the red wolf provide evidence for a common evolutionary history independent of the gray wolf. Canadian Journal of Zoology 78:2156–2166.

Wilson, P. J., Grewal, S., McFadden, T., Chambers, R. C., and White, B. N. 2003. Mitochondrial DNA extracted from eastern North American wolves killed in the 1800s is not of gray wolf origin. Canadian Journal of Zoology 81:936–940.

Wozencraft, W. C. 2005. Order Carnivora. In Mammal Species of the World, A Taxonomic and Geographic Reference, eds. D. E. Wilson and D. M. Reeder, pp. 532–628. Baltimore: Johns Hopkins University Press.

Chapter 16
Human Dimensions: Public Opinion Research Concerning Wolves in the Great Lakes States of Michigan, Minnesota, and Wisconsin

Kevin Schanning

16.1   Introduction

The recovery of the gray wolf (*Canis lupus*) in the western Great Lakes states is an exciting environmental success story. Changing social constructions are an essential part of the story of wolf recovery in the western Great Lakes region (Herda-Rapp and Goedeke 2005). In this chapter, I will examine the changing social construction (attitudes, beliefs, and perceptions) of wolves that set the stage for and facilitated the wolf's recovery across northern portions of Wisconsin, Michigan, and Minnesota. I focus on a social constructionist approach of how groups and individuals create and maintain various meanings concerning aspects to the world, in this case the wolf.

Social construction refers to the notion that beliefs, understandings, and values about the world are actively and creatively produced by human beings. Social constructionists hold that the world is made or invented, as opposed to merely given or taken for granted (Marshall 1998). Consequently, in the constructionist paradigm a "Wolf" is a socially constructed creature because certain cultural attitudes, social perceptions, and beliefs are assigned to it. The attitudes, beliefs, and ideologies that make up various social constructions of the wolf become as real as the teeth, bones, and fur that make up the biological mammal. In fact, these social constructions of the wolf are as important to the recovery and continued survival of wolves as are the biological and ecological factors that involved in their recovery. People's perceptions of wildlife shape their approval for policy and management. If policy goals or management techniques conflict with people's views of a species, they will ignore sanctions and incentives or actively attempt to undermine policy goals and the actions of managers. As Manfredo and Dayer (2004) put it, it is the thoughts and actions of humans that ultimately determine the course and resolution of human–wildlife conflict.

I begin this chapter with a brief discussion of the earliest Euro-American cultural constructions of wolves and the consequences of those constructions. Then I discuss the social scientific literature that documents human's perceptions and attitudes of wolves in the western Great Lakes region (Table 16.1). I conclude the chapter with an argument for the importance of using social attitude research in the management of wolves. I use the term "Great Lakes states" to refer to Michigan, Wisconsin, and Minnesota.

A.P. Wydeven et al. (eds.), *Recovery of Gray Wolves in the Great Lakes Region of the United States*,
DOI: 10.1007/978-0-387-85952-1_16, © Springer Science+Business Media, LLC 2009

Table 16.1 Attitudinal studies concerning wolves in the Great Lakes states of Michigan, Minnesota, and Wisconsin, 1974–2006

Year conducted	Publication	Methods and sample size
1972	Johnson 1974	Self-administered computer-assisted survey: convenience sample of 1,692 individuals
1974	Llewellyn 1978	Content analysis: 1,083 public comment letters; 700 from Minnesota residents
1981	Hook and Robinson 1982	Self-administered mail survey: stratified sample of 1,290 Michigan residents
1985	Kellert 1985a	Telephone survey: stratified random sample of 621 Minnesota residents
1985	Knight 1985	Self-administered survey: convenience sample of Wisconsin deer hunters
1988	Nelson and Franson 1988	Self-administered mail survey: purposive sample of 465 farmers and rural landowners in Wisconsin
1990	Kellert 1990	Self-administered mail survey: stratified sampling design of 639 Michigan residents
1997	Wilson 1999	Self-administered mail survey: stratified sampling of 1,101 Wisconsin endangered resources license plate holders
1999	Kellert 1999	Telephone survey: stratified random sample of 525 Minnesota residents
2001	Naughton-Treves et al. 2003	Self-administered mail survey: combined purposive and random sample of 658 Wisconsin residents
2002	Mertig 2004	Self-administered mail survey: random sample of 557 Michigan residents.
2003	Schanning 2003	Self-administered mail survey: stratified random sample of 644 Wisconsin residents
2004	Schanning 2004	Self-administered mail survey: stratified random sample of 1,017 Michigan residents
2005	Schanning 2005	Self-administered mail survey: stratified random sample of 909 Minnesota residents
2004 and 2005	Treves et al. 2007	Self-administered mail survey: stratified random sample of 1,364 Wisconsin residents
2005	Beyer et al. 2006	Self-administered mail survey: stratified random sample of 4,126 Michigan residents

16.2   The Earliest Social Constructs

The gray wolf was exterminated from much of the Great Lakes region as a result of the influx of European settlers that began in the early1800s. Bringing with them collections of mythologies and folklore, the European settlers' destructive attitudes toward predators often extended beyond the protection of their communities and livestock. The wolf was viewed symbolically and physically as a threat to progress and civilization, a threat these early settlers tried to eradicate (Lopez 1978).

Billing (1856), quoted in Young and Goldman (1944:122), in discussing the wolf's fears of man described it as "a cruel, savage, cowardly animal, with such a disposition that he will kill a whole flock of sheep merely for the sake of gratifying his thirst for blood...." The wolf is further described as being "...the most cowardly of animals, when caught in a trap, or wounded by a gun, or cornered so that they could not escape, [this person] invariably killed them with a club or tomahawk, and was never met with any resistance" (Young and Goldman 1944: 122). Scarff (1972) argues that in the American West, wolves were perceived as vicious murderers and thieves that deserved nothing but the most cruel death.

Possibly most important to the American social construction of wolves in the late nineteenth and early twentieth centuries are the numerous historical accounts of famous gray wolves summarized by Gipson and Ballard (1998). Gipson and Ballard (1998) described the accounts of 59 famous North American gray wolves and their exploits as notorious livestock killers and cunning escape artists. Despite the fact that a majority of these accounts were shown to be exaggerated, they served to shape people's perceptions of wolves. Such views show how the wolf was socially constructed in such a way as to legitimize and justify its extermination. Schlickeisen (2001: 61) argued, "... they [wolves] have occupied a special cultural niche in American society as the leading symbol of an evil wild nature, a demon to be conquered and extirpated as quickly as possible by any means available." For many, the wolf was socially constructed as a direct threat to one's livelihood and as an embodiment of evil and cruelty. Extermination of the wolf was a socially accepted practice in the United States, and the western Great Lakes states were no exception. Bounty systems were enacted in Michigan in 1838, Minnesota in 1849, and Wisconsin in 1839 (Van Ballenberghe 1974; Hammill 1993; Thiel 1993; McIntyre 1995). By 1960, no known breeding wolf populations were living in Wisconsin or Michigan, except for an isolated population living on Isle Royale (McIntyre 1995). The gray wolf of the western Great Lakes states had been mostly extirpated except for a small portion of northeastern Minnesota and Isle Royale in Lake Superior.

16.3   A Change is in the Air

Despite drastic reductions in numbers and strongly held negative attitudes about wolves, a small population remained in northern Minnesota because of the remoteness of the area (Van Ballenberghe 1974). As wolves were nearly extirpated from the lower 48 states of the United States, perceptions and behaviors began to change (Browne-Nunez and Taylor 2002). For instance, bounty systems were repealed: Wisconsin in 1957, Michigan in 1960, and Minnesota in 1965 (Van Ballenberghe 1974; Hammill 1993; Thiel 1993). With bounties removed, wolves in Minnesota had a chance to recover and recolonize parts of the Great

Lakes region. The natural ability of wolf populations to recover indicates that major limiting factors to wolf reestablishment are the social aspects, namely, human attitudes, beliefs, and the collective social constructions that shape acceptance of wolves (Mech 1995).

The Wisconsin legislature established a climate for wolf recovery when it gave the wolf full protection in 1957, with Michigan and Minnesota following shortly after by ending their wolf bounties. The first wolf sanctuary in Minnesota was created in 1970 in the Superior National Forest. In 1972, the Minnesota Department of Natural Resources (DNR), along with the U.S. Fish and Wildlife Service (USFWS), developed their first wolf management plan (Kellert 1985a). The Endangered Species Act (ESA) of 1973 listed the subspecies of gray wolf in the eastern United States as endangered in 1974, making it illegal to kill gray wolves in this region (USFWS 1992). The gray wolf as a species was listed as endangered throughout the continental United States in 1978, except Minnesota where it was listed as threatened. The act placed much of wolf recovery in the hands of the USFWS, which provided monetary and technical support to state and federal biologists. Investigations into everything from basic wolf physiology and pack structure to preferred habitat and range began in earnest (McIntyre 1995). The "demystifying of the wolf" became the goal of wildlife biologists, state officials, environmentalists, and wolf advocates alike (Lopez 1978). The information produced by these efforts was used to determine areas of suitable habitat for wolves, and to create a better-educated and more tolerant public. It was hoped that scientific discoveries about wolves would replace emotionally laden cultural myths with facts, and foster a greater respect toward wolves based on a scientific and ecological understanding of wolves as valuable members of an ecosystem.

16.4   Changing Attitudes: Early Protection and Recolonization (1974–1989)

Williams et al. (2002) indicated that the shift toward more positive attitudes toward wolves in the U.S. probably occurred between the 1930s and the 1970s. However, it was not until the most recent decades that it became apparent that an understanding of public attitudes about wolves and wolf management was as important to successful wolf recovery as was an understanding of wolf biology.

The first social scientific research related to human attitudes concerning wolves in the western Great Lakes region was conducted by Johnson (1974) during the 1972 Minnesota State Fair (Johnson 1974). Johnson's findings (1974) were based on a convenience sample of 1,692 individuals who visited a nature exhibit at the Minnesota State Fair. He argued that the traditional literary views of wolves that portrayed them as wicked, child-eating beasts were beginning to be challenged by a more scientific view that portrayed wolves as an important predator with a sophisticated and social pack structure. He also found age to be the strongest variable in determining attitudes about wolves, with children holding the strongest negative attitudes, most likely due to their exposure to the traditional negativist construction of wolves (e.g., in fairy tales) and a lack of exposure to scientific and ecological views. Johnson (1974) also found that 30% of respondents felt that wolves were dangerous to humans, 56% felt that wolves should be protected, and 90% felt that a wolf population was of value to Minnesota. Johnson (1974) showed that by the early 1970s at least two common attitudes toward wolves existed among Minnesota residents: a social construction that saw wolves as dangerous with no value and in need of extermination, and the construction that viewed wolves as valuable creatures in need of protection. These two attitudes toward wolves, the antagonistic anti-wolf construction and the more ecological pro-wolf construction, are found in varying degrees in all subsequent attitudinal studies related to wolves.

Llewellyn (1978) reported on an analysis of ~1,000 public comment letters received by the USFWS regarding the proposed reclassification of the gray wolf from endangered to threatened in Minnesota. Of the 700 letters from Minnesota residents, only 23% thought that wolves should continue to be classified as endangered, and 70% supported complete declassification (Llewellyn 1978). A split appeared between urban and rural residents, with approximately three-fourths of urban residents in favor of continuing the endangered status, while less than one-fourth of rural residents felt that the classification should continue. Those who supported the continued endangered classification did so out of an ecological, moralistic, or naturalistic attitude (Kellert 1985b) toward wolves, while those who did not support the continued classification did so more out of a utilitarian attitude (Brown-Nunez and Taylor 2002). The urban–rural split found by Llewellyn (1978) confirmed Johnson's findings (1974) that pro-wolf attitudes had developed, but that the traditional anti-wolf attitudes were still prevalent.

Efforts by the Michigan DNR in 1974 to reintroduce four wolves into the Upper Peninsula (UP) had failed (Weise et al. 1979). Three of the introduced animals were shot, and one was killed by a car within 1 year, indicating a lack of human acceptance for wolves (Weise et al. 1979). By 1976, other wolves that migrated naturally into the UP were found dead from trapping, shooting, or both, despite state and federal protection (Hook and Robinson 1982).

Hook and Robinson (1982: 383) wanted to assess the "extent of anti-wolf attitudes in Michigan and to determine their underlying causes." They mailed 3,382 questionnaires to a stratified random sample of residents of six Michigan counties, three in the UP and three in the Lower Peninsula (LP). The sampling frame was Michigan drivers' license files. The mailing resulted in 1,664 returned surveys, of which 1,290 were usable (Hook and Robinson 1982). Most of their analysis were based on attitudes about predators in general. They considered the wolf to be similar to other predators, but some of the antipredator attitudes found may not have been applicable to wolves. The two highest-ranking questions in Hook and Robinson's antipredator scale (1982: 386) (receiving the most "agree" responses) were "predators should be eliminated" and "a wolf is a varmint that should be eliminated."

The antipredator scale was used to assess different factors that affected whether an individual ranked high (strongly antipredator) or low (Hook and Robinson 1982). Only a small percentage of all respondents scored high on the scale (<8%).

Factors significantly related to predator attitudes included growing up in a rural area, level of fear toward predators, age and educational levels of the respondent, participation in hunting, concern for economic losses, and lack of knowledge about predators. Of these factors, it was found that lower educational level, rural background (i.e., being reared on a farm), and lack of knowledge about predators were highly correlated with antipredator responses. The dominant factor in determining whether a respondent scored high on the antipredator scale was the degree wolves were feared (35% of the variability within multiple regression analysis). The second most important factor for antipredator attitudes was having negative attitudes toward animals in general (9% of the variability). Hook and Robinson (1982) also found that hunters, or being from families with hunting histories, were less antipredator than nonhunters. Hook and Robinson (1982) were the first researchers to show that negative attitudes toward wolves were largely correlated with fear of wolves. Furthermore, those with rural backgrounds and those with less education were more likely to hold anti-wolf attitudes.

Other findings from Hook and Robinson (1982) included the majority of respondents supported restoration of wolves (55%), and a high percentage supported reintroduction of wolves (45%). While 15% of respondents said they would oppose wolf reintro-duction, the same percentage said they would actively support it. These results, with a low percentage of respondents having negative attitudes toward predators, revealed moderate public support for wolf protection and recovery. The findings from the study were the first based on a random sample that demonstrated a widely held alternative to the negative social construction of wolves. Hook and Robinson (1982: 394) con-cluded, "If this relatively small group of (anti-wolf) people can be favorably influenced, there may be hope for the restoration of the wolf in Upper Michigan and elsewhere."

Kellert (1985a) conducted the next major study of public sentiments toward wolves in Minnesota. Kellert administered a 45-minute telephone survey to a strati-fied random sample of Minnesota residents representing five groups: urban resi-dents of St. Paul and Minneapolis, residents from northern counties, deer hunters, trappers, and farmers. The survey revealed a public that was generally favorable to the presence of wolves (Kellert 1985a). Kellert (1985a) collected data on attitudes, knowledge, behaviors, and symbolic perceptions of wolves. This survey was based on his earlier work on attitudes toward wildlife (Kellert 1978, 1980). At the time of the survey, Minnesota had a wolf population of about 1,300–1,400 wolves living in the northern portions of the state (Erb and Don Carlos, this volume). Consequently, early surveys in Minnesota differed from those in Michigan and Wisconsin because management and livestock depredations were already affecting farmers.

Kellert (1985a) found support for the protection and conservation of the gray wolf in Minnesota; however, this support was conditional. Most respondents opposed limiting activities such as mineral extraction or the expansion of human settlements to preserve favorable wolf habitat. The majority of the public supported controlling and killing wolves that depredated livestock. The most preferred methods of control included eliminating individual wolves known to have caused damage, capturing and relocating problem wolves, compensating farmers for livestock, and training guard dogs to protect livestock (Kellert 1985a). The use of poisons to eliminate wolves was overwhelmingly rejected, likely due to education about the dangers of poisons (McIntyre 1995). Kellert (1985a) described a third social construction of wolves, one that was generally pro-wolf, but that also included a utilitarian or pragmatic component. Rather than simply viewing wolves as evil predators, or as aesthetically pleasing and ecologically necessary, this third social construction considered wolves in relation to competing human needs, wants, and desires. This utilitarian or pragmatic attitude of wolves seemed to be the result of humans having to live with wolves. The pro-wolf and anti-wolf constructions may be the product of a generation who did not live with wolves on the landscape.

Kellert (1985a: iii) found consistent support for timber wolves, except among farmers who "repeatedly viewed the timber wolf in highly negative, hostile, and unsympathetic ways." About 34% of farmers, deer hunters, and trappers indicated that they might shoot a wolf if encountered while deer hunting; consequently, it was feared that without adequate protection, excessive killing of wolves could occur (Kellert 1985a). About 12% of farmers and 17% of trappers indicated that they had killed a wolf themselves, and more than 40% of all respondents living in northern counties knew someone who had killed a wolf (Kellert 1985a). About one-third of the respondents indicated that they would be afraid if wolves lived near their home, or if they encountered a wolf in the woods (Kellert 1985a). Kellert (1985a) also found that fear of wolves was a major factor shaping attitudes toward wolves. Despite his concern, Kellert (1985a, b) felt that a consensus concerning how best to manage wolves in Minnesota was possible.

By 1985, studies of human attitudes concerning wolves in Minnesota and Michigan clearly demonstrated competing social constructions of the timber wolf. Some people, namely the elderly, children, less educated, rural residents, and farmers, held more strongly to the historic attitudes of wolves as dangerous. However, there also existed a more recent social construction of wolves as important members of an ecological community, as symbols of wilderness deserving of protection. This newer social construction of wolves was more likely to be held by educated, middle-aged, urban residents, whose livelihoods were not clearly put at risk by the presence of wolves. Finally, a third more utilitarian construction of wolves was also found, supporting the right of wolves to exist as long as they did not interfere with the activities of humans.

Two minor studies in the mid and late 1980s examined attitudes by hunters and farmers toward wolves in Wisconsin, as wolves began to recolonize the state. Knight (1985) found that 20% of hunters in two Wisconsin counties where wolves existed held very negative attitudes about wolves, but he also found that 69% of hunters said wolves should not be eliminated. Using a mail survey to 597 residents in six Wisconsin counties, Nelson and Franson (1988) explored the degree of support for wolf restoration along with people's general attitudes about wolves. Nelson and Franson (1988) determined that half of farmers in six Wisconsin counties within the wolf's range opposed wolf recovery, and that almost half of nonfarming landowners supported wolf restoration. Furthermore, they found that a significant percentage of both farmers (46%) and nonfarmers (64%) expressed an aesthetic appreciation for the wolf as a natural symbol of the beauty and wonder. However, Nelson and

Franson (1988) also found that neither group feared wolves and only a minority regarded them as a threat to livestock.

16.5   The Recovery Period (1990–1999)

The next major study of human attitudes toward wolves in the Great Lakes region was conducted in Michigan (Kellert 1990). A mail survey was distributed to a stratified sample of 639 UP and LP residents, hunters, trappers, and farmers. Kellert (1990) found substantial support for wolf restoration among the various groups sampled, with the exception of farmers. Despite this support, many groups still expressed fear of being attacked by a wolf: 67% of females, 72% of those with less than a high school education, 64% of those who lived in cities with >50,000 people, and 51% of those with household earnings between $15,000 and $24,999 (Kellert 1990). Overall, 41% believed wolves could be dangerous to humans, indicating that despite some change in attitudes, many people still feared wolves. Kellert (1990) found that many Michigan residents supported limiting wolf numbers in the state.

Similar to his study in Minnesota (Kellert 1985a), Michigan residents were not supportive of restrictions on economic development, including mining and forestry, or recreational activities such as hiking, fishing, camping, and use of off-road vehicles to aid the recovery of wolf populations (Kellert 1990). Respondents were generally supportive of providing information to farmers on protecting livestock from wolves, but less supportive of using government funds to pay for such protection (Kellert 1990). Respondents were supportive of paying farmers for livestock losses confirmed to have been killed by wolves but unsupportive of paying claims without confirmation (Kellert 1990). Thus, by 1990, many people had come to view wolves as creatures that had a right to exist and should be protected, but many also felt there should be limitations on degrees of protection. Kellert (1985a, 1990) began to paint a picture of complex and competing social constructions of wolves that were perhaps more fluid than one would suspect. As Kellert (1990: 100) concluded, "Our study found considerable public interest in the wolf and sympathy for its restoration, but also significant levels of ignorance, hostility and resistance to its reestablishment."

Wilson (1999) provided insight into Wisconsin residents' attitudes toward wolves from a mail survey distributed to a stratified random sample of about 1,100 vehicle license plate holders in Wisconsin in 1997. Wilson (1999) contrasted attitudes toward endangered wildlife between residents who had purchased special Endangered Resources (ER) license plates (these help fund management of endangered resources in the state) and regular license plate holders. At the time of his survey in 1997, the winter count of wolves was 148 in Wisconsin, and wolves were one of the endangered species considered. Wilson (1999) found that 97% of ER license plate holders and 80% of regular license plate holders agreed that it was important to protect rare predators like the wolf. About 90% of ER license plate holders and 50% of regular license plate holders supported the idea that the Wisconsin DNR should work to increase the number of wolves living in the state (Wilson 1999). Among hunters,

47% supported efforts to increase wolf numbers and 20% opposed increasing wolf numbers (Wilson 1999). As in studies in the adjacent states, Wilson (1999) found that a large segment of the society held positive attitudes toward wolves.

Williams et al. (2002) analyzed 38 attitude surveys about wolves done between 1972 and 2000 across Europe, Asia, and North America, and found that 51% of all respondents had positive attitudes toward wolves. They felt that attitudes toward wolves had not changed in the late twentieth century in the United States, but attitudes had probably changed earlier in the 1930s through 1970s. This is consistent with the idea that this was a period when wolves were mostly absent from the scene, allowing for pro-wolf attitudes to develop. As wolves again appeared on the landscape, more utilitarian social constructions were detected in the 1980s and 1990s. It appears that people's perceptions of the environment in general has changed over time, along with greater tolerance toward wolves, allowing populations to increase and recovery to occur (Hammill 1993; Mech 1995; Kellert et al. 1996; Nie 2001).

16.6   The Management Period (1999–2006)

Beginning in 1999, several attitude studies regarding wolves in the western Great Lakes states appeared (Kellert 1999; Naughton-Treves et al. 2003; Mertig 2004; Schanning 2003, 2004, 2005; Beyer et al. 2006; Treves et al. 2007). Furthermore, one summary (Williams et al. 2002) and one annotated bibliography (Browne-Nunez and Taylor 2002) of attitude studies related to wolves appeared during this period. The recent interest in human attitude studies relating to wolves in the western Great Lakes states was due to several factors. By 1999, gray wolves in the western Great Lakes region had met federal recovery goals (USFWS 1992), and wolves were being considered for delisting from the federal list of endangered and threatened species by the USFWS. All three states were developing state management plans for wolves (Michigan DNR 1997; Wisconsin DNR 1999; Minnesota DNR 2001), and managers recognized the importance of understanding human attitudes in planning wolf management (Jacobson and McDuff 1998). Working on the Michigan wolf plan, Hammill (1993) commented that the public must be included in the planning of wolf recovery and management to "feel ownership of the plan."

These most recent studies were conducted to ensure appropriate representation of interest or stakeholder groups (Naughton-Treves et al. 2003), representation of the general public (Mertig 2004; Schanning 2003, 2004, 2005), or both (Kellert 1999; Naughton et al. 2004; Beyer et al. 2006). The studies incorporated questions that directly addressed management or policy options relevant to managing wolves. All of these surveys were created in consultation with management agencies dealing with conflicts between wolves and humans. While wolf populations were recovering, management of the population became more complex due to issues such as compensation to pet and livestock owners, or issues following federal delisting such as landowner controls of problem wolves, wolf population management, and possible public harvest of wolves.

Kellert (1999) studied attitudes of Minnesota residents where wolves were facing possible federal delisting. He conducted a telephone survey of 525 randomly selected Minnesota residents that was stratified by northern residents, southern residents, and livestock farmers. Kellert (1999) confirmed findings of previous surveys that people were very accepting of wolves as ecologically important and aesthetically appealing. He also found that Minnesota residents believed wolves had a right to exist, but that wolf numbers and distribution needed to be limited to minimize conflicts with humans.

Comparing the results to his earlier work (Kellert 1985a), Kellert (1999) found that the pro-wolf attitudes had spread to more people in Minnesota and had become more deeply ingrained. It also seemed that the anti-wolf attitudes had declined. However, by 1999, it appeared that more Minnesota residents had become concerned about managing wolves to minimize conflicts between humans and wolves. This social construction of wolves that cannot easily be classified as pro-wolf or anti-wolf had become more widespread. Such a utilitarian or pragmatic social construction of wolves appeared to be related to the amount and type of wolf–human interactions.

The themes of social acceptance of wolves and concern over the management of wolves and wolf–human conflicts were major parts of the research of Naughton-Treves et al. (2003) in Wisconsin. Their 2001 study was a mailback questionnaire study of 658 individuals belonging to four groups: landowners who submitted wolf–livestock complaints to the Wisconsin Department of Natural Resources (WDNR), randomly sampled landowners, bear hunters who submitted wolf–hound complaints to the WDNR, and randomly sampled members of the Wisconsin Bear Hunters' Association. Among these stakeholder groups, there existed broad support for having wolves in Wisconsin, with only 17% of respondents wanting to eliminate the wolf population in Wisconsin (Naughton-Treves et al. 2003). As expected, support for wolves was less visible among those who had lost livestock, pets, or hounds to wolves than those who had not, but no differences were found between those who were compensated for their losses versus those who had not been compensated (Naughton-Treves et al. 2003). They found that some members of key interest groups continued to hold strong anti-wolf attitudes. Naughton-Treves et al. (2003) determined that most respondents from these groups were strongly in favor of compensation payments as a management strategy. Also, there was general support among these interest groups for the lethal control of problem wolves, especially on private lands (Naughton-Treves et al. 2003). They found that as wolf numbers increased in the region, so did calls for wolf management.

Mertig (2004) studied the attitudes, concerns, and opinions of the general public in Michigan. She conducted interviews of 557 randomly selected Michigan residents from both the UP and LP. Mertig (2004) found differences in the social constructions of wolves and preferences for management options among residents of Michigan. When compared to Kellert (1990), Mertig (2004: v) concluded, "Residents of the UP now appear less supportive of wolf recovery efforts,… while support for wolf recovery has remained steady or increased somewhat among residents of the lower peninsula." Generally, people living close to wolves tend to be less supportive of wolf recovery than those who do not live close to wolves (Williams et al. 2002).

Mertig (2004) detected an increase in utilitarian reasons for supporting wolf recovery, especially for residents of the UP. Similar to other studies, little support was found for imposing limits on human activities, especially recreation, to preserve wolves (Mertig 2004). Regarding management strategies, Mertig (2004) found relatively strong support for changing the endangered status of wolves in the UP, controlling the population of wolves, and dealing with nuisance animals.

As part of the larger "State of The Wolf Project" (a 5-year effort by the Sigurd Olson Institute at Northland College in Ashland, Wisconsin, to examine attitudes toward wolf management in the Great Lakes region), Schanning (2004) conducted a survey of Michigan residents that largely confirmed the findings reported by Mertig (2004). Schanning (2004) obtained a sample of 1,017 respondents from 5,000 surveys sent throughout Michigan resulting in a margin of error of ±3%. Schanning (2004) found that all three of the cultural constructions of wolves discussed previously existed among Michigan residents. Pro-wolf attitudes found included 68% of respondents that agreed that "the wolf is a symbol of the beauty and wonder of nature," 51% who agreed that "wolves are a part of our vanishing wilderness and should be protected," and 57% who agreed that "wolves are essential to maintaining the balance of nature" (Schanning 2004). However, anti-wolf attitudes were also found: 25% who disagreed that "wolves are part of our vanishing wilderness and should be protected," 19% who agreed that animals like wolves and rattlesnakes are naturally cruel, and 18% who agreed that wolves belong in places like Alaska, not in Michigan (Schanning 2004). Michigan residents supported a utilitarian construction of wolves. When asked about the killing of wolves to protect pets, 65% agreed, and nearly half (46%) agreed with a sport harvest if there are enough wolves. Respondents were almost evenly split regarding a public harvest, with ~40% supporting hunting and trapping seasons, 20% neutral, and 40% opposing such seasons (Schanning 2004). Only 4% of Michigan residents supported ending compensation for livestock owners.

Schanning (2004) found fear of wolves was still widely held by Michigan residents with 40% of residents expressing concerns for their own safety, 70% expressing concerns for the safety of children, 72% expressing concerns for the safety of pets, and 66% expressing concern for the safety of livestock. Most Michigan residents had limited knowledge about wolves, with most overestimating the number of wolves, the number of packs, and the average pack size in the state. Similar lack of knowledge and fear of wolves were found in Wisconsin (Schanning 2003) and Minnesota (Schanning 2005).

Using methods similar to those used in Michigan (Schanning 2004), studies were conducted in Wisconsin (Schanning 2003) and Minnesota (Schanning 2005). Among 644 respondents returning surveys (5,000 surveys sent), Wisconsin residents had slightly more pro-wolf attitudes than Michigan residents. In Wisconsin, 72% of respondents agreed that "the wolf is a symbol of the beauty and wonder of nature," 56% agreed that "wolves are a part of our vanishing wilderness and should be protected,"and 62% agreed that "wolves are essential to maintaining the balance of nature" (Schanning 2003). Negative attitudes toward wolves also existed in Wisconsin with 21% of respondents disagreeing that "wolves are part of our vanishing wilderness and should be protected," 19% agreeing that animals like wolves and rattlesnakes are naturally cruel, and 18% agreeing that wolves belong in places like Alaska, not in Wisconsin. Among Wisconsin residents, 41% agreed that killing wolves for sport should be allowed if there are enough wolves and 60% agreed that they would shoot a wolf if it threatened their pet.

Treves et al. (2007) conducted a mail survey using a stratified random sample of Wisconsin residents representing six zip codes in 2004 and 2005. Treves et al. (2007) focused on contributors and noncontributors to the Wisconsin DNR Endangered Resources (ER) Fund, as well as hunters, and livestock producers. They also found relatively pro-wolf attitudes among Wisconsin residents, especially among contributors to the ER Fund. Despite the overall support for wolves, 11% of respondents indicated they would shoot a wolf if they saw one while deer hunting. Treves et al. (2007) reported a total of 68% of respondents supporting a wolf harvest. Compared to noncontributors, ER contributors tended to favor nonlethal actions for problem wolves, were less in favor of a public harvest, and favored government agents as opposed to landowners conducting lethal controls. Despite providing their own money, ER contributors more strongly supported the use of the ER fund for providing reimbursement to livestock owners with wolf depredation (78%) than noncontributors (70%). In general, there was strong support from ER contributors (80%) and noncontributors (69%) for the continued reimbursements to livestock producers, but only about half of both groups supported payments for hunting dogs lost to wolves.

Schanning (2005) found pragmatic/utilitarian views toward wolves among 909 respondents in Minnesota (5,000 surveys sent), including strong support for compensation payments with only 4% of respondents wanting to stop compensating livestock owners entirely. A total of 71% of respondents supported the Minnesota DNR shooting problem wolves. Fear of wolves, as found in the other two states, was also present among Minnesota residents: 31% expressed fear for their own safety, 64% expressed fear for the safety of children, and 70% expressed fear for pets (Schanning 2005). Thus, the fear of wolves is likely an important factor in shaping attitudes among many residents in the western Great Lakes states.

The most recent study of human attitudes in the western Great Lakes region was conducted in 2005 in Michigan (Beyer et al. 2006). This survey was designed to assess the status of the Social Carrying Capacity (SCC), a notion that human society "represents a social environment capable of setting limits on the number and distribution of a wildlife species" (Beyer et al. 2006: 36). Stratified samples of five regions in Michigan were taken to ensure sufficient regional representation. A scale for assessing the SCC of wolves was used. The scale focused on the level of tolerance humans have for wolves given a variety of situations. The scale is based on the perceived range and number of wolves present, and on the type and amount of interactions between humans and wolves. People can be classified along a continuum of SCC that ranges from complete intolerance to complete tolerance. Beyer et al. (2006) found that 7% of citizens belonged to the intolerant group, 20% comprised the least tolerant group, 28% were in the midtolerant group, and 32% were in the most tolerant group. These findings supported the concept that a large number of people held the utilitarian, pragmatic construction of wolves and only a small percentage were anti-wolf.

The numerous studies of residents in the western Great Lakes states conducted on attitudes toward wolves between 1990 and 2006 reveal that public support for wolves and their recovery has remained relatively consistent from 1974 to the present. With wolf population numbers increasing in the western Great Lakes states, there continues to be strong support of wolves and their recovery; however, attitudes are tempered by concerns to reduce negative impacts of wolves on human activities. Although declining, fear of wolves still exists, causing a persistence of negative attitudes toward wolves among a few people.

16.7 Concluding Remarks on the Importance of Social Research

Wolf biologists have begun to understand the importance of managing human–wolf conflicts as a major part of wolf management. Wolf recolonization would not have been possible without the tolerance and cooperation of the general public (Kellert et al. 1996; Kellert 1999; Nie 2001; Beyer et al. 2006). The public has supported recovery through their actions, and to ensure that this support continues, it is important to bring public opinion to the "planning table" regarding wolf management. As Brown-Nunez and Taylor (2002: 1) stated, "Understanding the beliefs and attitudes of the public regarding natural resources management issues is key to making decisions that are more responsive to the public and, therefore, increase the effectiveness of resource management decisions."

Social research has also provided educational groups with new insight into the degree to which their programs and techniques have been successful over the years in changing attitudes about wolves and their recovery. Although the effectiveness of education programs in altering attitudes about wildlife and wolves, compared to other factors such as increased urbanization and occupational shifts, has been questioned (Manfredo and Dayer 2004), there is little doubt that education has played a role in altering attitudes. Knowing the social construction of groups and individuals toward wolves allows educators to focus their messages and to address misunderstandings and fear of wolves The extensive work done on understanding human attitudes concerning wolves provides educators additional tools to help tailor educational efforts. Naess and Mysterud (1987) and Nie (2001) concluded that education about wolves is the major factor in facilitating their recovery. Naess and Mysterud (1987) state that there are no established "rules for coexistence" held between wolves and humans, as wolves have only recently returned to areas where they have not existed for a long time. It is up to social science researchers to uncover and educate others about these emerging rules for coexistence. It is apparent that social science understanding of the human aspects of living with wolves will continue to be an essential part of managing wolves.

References

Beyer, D., Hogrefe, T., Peyton, R. B., Bull, P., Burroughs, J. P., and Lederle, P. 2006. Review of social and biological science relevant to wolf management in Michigan. Lansing: Michigan Department of Natural Resources.

Billings, E. 1856. The naturals history of the wolf (*Canis lupus*) and its varieties. The Canadian naturalist and geologist 1/3: 209–215.

Browne-Nunez, C., and Taylor, J. G. 2002. American's attitudes toward wolves and wolf reintroduction: an annotated bibliography. Information Technology Report USGS/BRD/ITR – 2002–0002. Denver: U.S. Government Printing Office.

Gipson, P. S., and Ballard, W. B. 1998. Accounts of famous North American wolves, *Canis lupus*. Canadian Field-Naturalist 112:724–739.

Hammill, J. 1993. Wolves in Michigan: a historical perspective. International Wolf 3:22–23.

Herda-Rapp, A., and Goedeke, T. L., eds. 2005. Mad About Wildlife. Boston: Brill.

Hook, R. A., and Robinson, W. L. 1982. Attitudes of Michigan citizens toward predators. In Wolves of the World, eds. F. H. Harrington and P. C. Paquet, pp. 382–394. Park Ridge: Noyes Publications.

Jacobson, S. K., and McDuff, M. D. 1998. Training idiot savants: the lack of human dimensions in conservation biology. Conservation Biology 12:263–267.

Johnson, R. T. 1974. On the spoor of the "Big Bad Wolf". Journal of Environmental Education 6:37–39.

Kellert, S. R. 1978. Characteristics and attitudes of hunters and anti-hunters. Transactions of the North American Wildlife and Natural Resources Conference 43:412–423.

Kellert, S. R. 1980. Americans' attitudes and knowledge of animals. Transactions of theNorth American Wildlife and Natural Resource Conference 45:11–124.

Kellert, S. R. 1985a. The Public and the Timber Wolf in Minnesota. New Haven: Yale University Press.

Kellert, S. R. 1985b. Public perceptions of predators, particularly the wolf and coyote. Biological Conservation 31:167–189.

Kellert, S. R. 1990. Public attitudes and beliefs about the wolf and its restoration in Michigan. New Haven: Yale University Press.

Kellert, S. R. 1999. The public and the wolf in Minnesota. Minneapolis: Report to the International Wolf Center.

Kellert, S. R., Black, M., Rush, C. R., and Bath, A. 1996. Human culture and large carnivore conservation in North America. Conservation Biology 10:977–990.

Knight, J. M. 1985. Survey of deer hunter attitudes toward wolves in two Wisconsin counties. Madison: MS Thesis, University of Wisconsin – Madison.

Llewellyn, L. 1978. Who speaks for the timber wolf? Transactions of the North American Wildlife and Natural Resource Conference 43:442–452.

Lopez, B. H. 1978. Of Wolves and Men. New York: Charles Scribner's Sons.

Manfredo, M. J., and Dayer, A. 2004. Concepts for exploring the social aspects of human-wildlife conflict in a global context. Human Dimensions of Wildlife 9:317–328.

Marshall, G. 1998. Oxford Dictionary of Sociology. New York: Oxford University Press.

Mech, L. D. 1995. The challenge and opportunity of recovering wolf populations. Conservation Biology 9:270–278.

Mertig, A. G. 2004. Attitudes about wolves in Michigan 2002. Lansing: Report to the Michigan Department of Natural Resources, Wildlife Division.

McIntyre, R. 1995. War Against the Wolf: America's Campaign to Exterminate the Wolf. Stillwater: Voyageur Press.

Michigan DNR. 1997. Michigan Gray Wolf Recovery and Management Plan. Lansing: Michigan Department of Natural Resources.

Minnesota DNR. 2001. Minnesota Wolf Management Plan. St. Paul: Minnesota Department of Natural Resources.

Naess, A., and Mysterud, I. 1987. Philosophy of wolf policies I: general principles and preliminary exploration of selected norms. Conservation Biology 1:22–34.

Naughton-Treves, L., Grossberg, R., and Treves, A. 2003. Paying for tolerance: rural citizens' attitudes toward wolf depredation and compensation. Conservation Biology 17:1500–1511.

Naughton, L., Treves, A., Grossberg, R., and Wilcove, D. 2004. Living with wolves. Wisconsin Department of Natural Resources, Madison.

Nelson, E., and Franson, D. 1988. Timber Wolf Recovery in Wisconsin: the attitudes of Northern Wisconsin farmers and landowners. Research Management Findings 13. Madison: Wisconsin Department of Natural Resources.

Nie, M. A. 2001. The sociopolitical dimensions of wolf management and restoration in the United States. Human Ecology Review 8:1–12.

Scarff, J. 1972. From beast to parka to friend. Defenders of Wildlife News 468–476.

Schanning, K. 2003. The state of the wolf project: Wisconsin survey results. Ashland: Sigurd Olson Environmental Institute, Northland College.

Schanning, K. 2004. The state of the wolf project: Michigan survey results. Ashland: Sigurd Olson Environmental Institute, Northland College.

Schanning, K. 2005. The state of the wolf project: Minnesota survey results. Ashland: Sigurd Olson Environmental Institute, Northland College.

Schlickeisen, R. 2001. Overcoming cultural barriers to wolf reintroduction. In Wolves and Human Communities: Biology, Politics and Ethics, eds. V. A. Sharpe, B. Norton, and S. Donnelley, pp. 61–73. Washington: Island Press.

Thiel, R. P. 1993. The timber wolf in Wisconsin: the death and life of a majestic predator. Madison: University of Wisconsin Press.

Treves, A., Naughton, L., Schanning, K., and Wydeven, A. P. 2007. Public opinion of wolf management in Wisconsin, 2001–2005. In Wisconsin wolf management plan, addendum 2006 and 2007, Wisconsin DNR, pp. 36–47. Madison: Wisconsin Department of Natural Resources.

USFWS 1992. Recovery Plan for the Eastern Timber Wolf. Twin Cities: U.S. Fish and Wildlife Service.

Van Ballenberghe, V. 1974. Wolf management in Minnesota: an endangered species case history. Transactions of the North American Wildlife and Natural Resources Conference 39:313–322.

Weise, T. F., Robinson, W. L., Hook, R. A., and Mech, L. D. 1979. An experimental translocation of the eastern timber wolf. In The Behavior and Ecology of Wolves, ed. E. Kilinghammer, pp. 346–419. New York: Garland Press.

Williams, C. K., Ericsson, G., and Heberlein, T. A. 2002. A quantitative summary of attitudes toward wolves and their reintroduction. Wildlife Society Bulletin 30:1–10.

Wilson, M. A. 1999. Public attitudes towards wolves in Wisconsin. In Wisconsin Wolf Management Plan, Wisconsin DNR, pp.66–70. PUB-ER-099 99, Madison: Wisconsin Department of Natural Resources.

Wisconsin DNR. 1999. Wisconsin Wolf Management Plan. PUBL-ER-099 99, Madison: Wisconsin Department of Natural Resources.

Young, S. P., and Goldman, E. A. 1944. The Wolves of North America. Washington: American Wildlife Institute.

Chapter 17
Ma'iingan and the Ojibwe

Peter David

Preface This chapter will attempt to explore the significance of wolf recovery in the western Great Lakes region to one group of people – those known to others as the Ojibwe or Chippewa, and to themselves as the Anishinabe. It is not written by an Ojibwe, but by an individual who has had the pleasure and privilege of working with and for the Ojibwe for over two decades. It does not purport to extend the concepts discussed to other Native American nations – even those others residing in the western Great Lakes region – though in some cases there will be similarities.

It also does not intend to suggest that it fully captures the complexities of the relationship that exists between the Ojibwe and the wolf – or even that a singular relationship exists. The connection that individual Ojibweg share with ma'iingan tends to be deep, significant, and personal; any suggestion in the essay below that implies otherwise reflects only the shortcomings of the author.

17.1 Introduction

The resurgence of the wolf in the western Great Lakes region holds great significance to many people, but the cultural meaning it holds for the Ojibwe is especially profound, for ma'iingan, or the wolf, is the one species in all of nature with whom the Ojibwe – as a people – feel the greatest common union.

The relationship with ma'iingan goes back nearly to the origin of the people themselves. Wolf enters the Ojibwe Creation Story early and dramatically. In that story, as related by Lac Courte Oreilles (LCO) spiritual leader Eddie Benton-Banai in *The Mishomis Book* (1988), Original Man is the last creature the Creator sends to earth. He is given the task of walking the world to give names to all its plants and animals. As he completed this task he observed that each animal held its own kind of wisdom. He also noticed that all the *other* animals came in pairs, while he was alone.

That was an observation worth mentioning to the Creator.

The Creator responded by providing not a lover, but a brother; not a woman, but a wolf: Ma'iingan. The Creator indicated that Original Man and Ma'iingan were to travel the world together, and visit all of its places. As the two undertook this great journey, they became very close. They grew to realize their unique brotherhood with each other, and with all of creation.

When their travels were over, however, the Great Spirit told them that they now had to go their separate paths. Despite this physical separation, He indicated that Man and Ma'iingan would forever be linked, telling them "What shall happen to one of you shall also happen to the other. Each of you will be feared, respected, and misunderstood by the people who will later join you on this earth."

This linkage of wolf and man is a central tenet of the traditional Ojibwe belief system. And for others who hope to understand the significance of the recovery of wolves in the western Great Lakes region to the Ojibwe, no other characteristic is as important – or perhaps even necessary – than being able to fully envision the natural consequences of holding this world view. Those who can conceptually embody this perspective will find it easy to understand why the Ojibwe's vision of wolf management often differs so significantly from those in the non-Indian community.

17.2   The Union

Although Ma'iingan's role in the Creation Story foretold the similar pathways that the Ojibwe and the wolf were to follow on the grand scale, it does not portray the remarkable similarity of existence that also occured on a daily basis.

The relatively harsh environment of the western Great Lakes could alternatively provide great abundance or meager provision. Ojibwe survival depended on understanding the biotic community that enveloped them, and that understanding was often gained through the thoughtful observation of their spiritual brothers – the animals and plants – in that community. Of all the species in nature that the keen collective eye of the Ojibwe fell upon, none resonated so closely with life as ma'iingan.

The list of similarities is long and has frequently been noted (Lopez 1978). Some of the most notable: both are significant predators who shared common prey and in some instances, hunting techniques; both shared similar social organization, living in extended family groups in which all adults act as parents toward the young; larger Ojibwe tribes lived within a territorial distribution on the landscape in juxtaposition with other tribes, and similar to wolves, these territories often had buffer areas between them.

These similarities in nature led to a very different perspective toward wolves than was commonly found in livestock-raising societies, where wolves – not surprisingly – were vilified as a threat to livelihood. The Ojibwe, in contrast, recognized that wolves embodied many of the qualities that they themselves needed to survive on a demanding landscape. While a person of European decent is likely to be insulted by being called a wolf, an Ojibwe may take this as extreme compliment, for who has greater knowledge of the natural world, who hunts with greater stamina and skill, who works in greater cooperation, and who goes to greater extreme to provide for their young than ma'iingan? The Ojibwe who was truly wolf-like was one who was likely to survive and flourish.

It is striking that the Ojibwe did not appear to view ma'iingan as a competitor, although both depended on some of the same resources for survival. The sighting of wolf tracks that causes many contemporary deer hunters to conclude that prey will be reduced or absent from an area triggered just the opposite reaction from the Ojibwe hunter (who was hunting a landscape where wolves had not yet been targeted by eradication efforts). Wolf sign was good sign, for an area that could support wolves could likely support them as well. Where wolf sign was lacking, the Ojibwe were likely to face difficulty meeting their own needs.

This relationship between the abundance of wolves and game also was recognized by early European explorers to the region. On August 9, 1831, Henry Schoolcraft (1975) was canoeing about 18 miles south of what is currently Rice Lake, Wisconsin. He noted in his journal that "During the night wolves set up their howls near our camp, a sure sign that we were in deer country."

This lack of animosity toward wolves does not mean that negative interactions never took place between ma'iingan and the Ojibwe. Although written records are not common, wolves were occasionally harvested by tribal members. Danziger (1979:13) includes a reference to wolves impaling themselves "while snatching hungrily at baited hooks suspended about five feet off the ground" (though some Ojibwe contend harvesting may have been spurred by European contact). Wolves also negatively impacted the Ojibwe at times – becoming bold around camps or taking animals captured in Ojibwe traps (Tanner 1994). Based on their absence from available records, however, more significant negative impacts – such as attacks directly on Ojibwe people – appear to have been rare or nonexistent.

There may be an explanation of the perhaps surprising lack of wolf attacks on humans in the western hemisphere in a story retained in the Ojibwe oral tradition, and preserved in ink by Basil Johnson (1990). In this story, the first humans made the animal beings do all their work for them, and the animals – who at this time could converse freely among themselves and with humans – eventually convened as a group to address this unfair situation. The options contemplated by the angry group went so far as to include killing the humans. A dog that slunk off to tell the humans of this discussion was followed, captured, and returned to the group by a wolf. The group was outraged by the dog's act of treason, but spared its life. Makwa (the bear), speaking on behalf of the group, told the dog "For your betrayal, you shall no longer be regarded as a brother among us. Instead of man, we shall attack you. Worse than this, from now on you shall eat only what man has left, sleep in the cold and rain, and receive his kicks as a reward for your fidelity."

Perhaps this event displaced any aggression between ma'iingan and the Ojibwe, and helped maintain a positive relationship between the two. This story may have also been used to teach Ojibwe children that the relationship with ma'iingan – and other animals – is better founded in brotherhood than dominion.

17.3   The Contemporary Relationship

Although rooted in ancient understandings, the relationship the Ojibwe have with ma'iingan is not constrained to an ancient expression. The application of the old teachings on a contemporary landscape plays out with some unusual clarity in the recovery of wolves in the western Great Lakes region. Wisconsin can be looked at as a particular example.

In a coincidence – or not – of time and geography, the recovery of wolves in Wisconsin occurred in a neat time-step with the affirmation of Ojibwe off-reservation treaty rights in the state.

On March 8, 1974, Wisconsin was believed to be wolf-less. This fact likely was not on the forefront of the minds of Fred and Mike Tribble, two members of the LCO Band who set out that morning to spear fish on Chief Lake. Chief Lake borders the LCO reservation; part of the lake is within the boundary, part is outside. This human construct meant little to the fish swimming below the ice on Chief Lake, but it meant a great deal to the state wardens who arrested the Tribbles for being on the wrong side of the line. Still, the arrest came as no surprise to the Tribbles, because they had notified the Department of Natural Resources (DNR) of their intent before going out that morning.

The two brothers were initially found guilty. However, a year later, the LCO Band of Lake Superior Chippewa filed suit in federal court on behalf of all its members, arguing that the tribe had never relinquished the right to hunt, fish, and gather from the lands they had ceded to the United States government (LCO Band of Chippewa Indians v. Voigt). The treaties of 1837 and 1842, for example, which ceded lands that eventually became parts of Wisconsin, Minnesota, and Michigan (Fig. 17.1), both contain articles that provide for the reservation of these rights, which the tribes held at the time the treaties were signed. The state, they argued, did not have the right to enforce its law upon the Tribbles.

Interestingly, in the year that passed between the initial arrest and the appeal, another seemingly small event occurred that involved the crossing of a political boundary: several wolves ventured into northwestern Wisconsin from the growing population in neighboring Minnesota. It was the first documented presence of wolves in the state in over a decade and a half (Wisconsin Department of Natural Resources 1999).

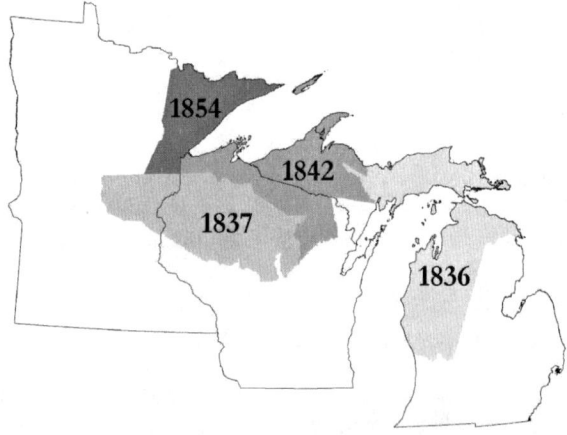

Fig. 17.1 Lands ceded by the Ojibwe in the treaties of 1836, 1837, 1842, and 1854. Ceded territory boundaries are representations and may not be the legally binding boundaries

Both the court case and the wolves would advance slowly. Four years passed before Judge James Doyle ruled against the Band and its treaty rights claim. Over this time, Wisconsin's wolf population increased to only about 25 animals, most of whom remained bunched near the original pack territory in Douglas County. The LCO Band appealed the Doyle decision to the U.S. Court of Appeals, Seventh Circuit, instigating an additional decade of litigation – a decade during which Wisconsin's wolf population, too, would struggle.

In late January of 1983, the federal Appeals Court ruled in favor of the Band, in what became known as the Voigt decision. By the time this decision was handed down, the state's wolf population had actually declined to about 19 animals.

Later that year, the U.S. Supreme court refused to hear an appeal of the Voigt decision, affirming the Appeals Court ruling. Litigation continued through a series of trials to define the extent of the Ojibwe treaty rights, to determine the permissible scope of state regulation, and to establish the extent of the tribes' regulatory authority. During this period, the LCO Band, and the other Ojibwe bands that had been signatories to the treaties and had joined the litigation, began exercising their treaty rights for the first time in recent history, on off-reservation lands and waters in the ceded territories. These exercises were conducted under annual interim agreements hammered out in negotiation with the state. These would prove to be difficult times for both the wolf and the Ojibwe.

Many non-Indians in the state regarded the reaffirmation of the treaty rights the same way they regarded the recovery of the wolf: not as the restoration of a lost part of the region's heritage, but as the resurrection of a dangerous threat to the state's natural resources. The quiet woods of northern Wisconsin began to make national news as tribal members once again began exercising old practices, including the taking of fish by spear at night. (The Lac du Flambeau, or "Lake of the Torches" Band, takes its name from this practice.) Although the tribal harvest was tightly regulated and coordinated with state harvest levels to ensure resource protection, tribal fish-spearers were frequently met with crowds of angry protestors, and bumper stickers appeared stating "Save a Deer, Shoot an Indian" (Satz 1991). Fish-spearers were pelted with rocks, gunshots were fired over their heads, and an individual with a pipe bomb was arrested near a boat landing. Meanwhile, Wisconsin's wolves were getting similar treatment, except that the bullets were aimed to kill. With annual mortality rates in excess of 35%, the wolf population struggled to maintain itself. The leading cause of death among wolves: gunshot wounds (Wisconsin Department of Natural Resources 1999).

In 1989, opposition to the exercise of treaty rights was peaking, and wolf populations continued to flounder. In spring of that year, Governor Tommy Thompson went to federal court to seek an emergency halt to the exercise of treaty rights, indicating that he would rather shut down spearing than bring in the National Guard to act as peacekeepers. Judge Barbara Crabb denied his request, stating "If this court holds that violent and lawless protests can determine the rights of the citizens of this state, what message will that send? Will that not encourage others to seek to resolve disputes by physical intimidation?" (Whaley with Bresette 1994).

In October of that year, with tension over the exercise of treaty rights running high throughout the state, members of the Lac du Flambeau Band shocked many state officials and residents by rejecting, in a 439 to 366 vote, a multimillion dollar offer from the state to greatly curtail the exercise of the rights (Satz 1991). This vote silenced many critics who believed that the tribes were exercising the treaty rights primarily to drive up the amount of what they expected would be an eventual monetary settlement from the state.

By May of 1991, the trials to determine the scope of the treaty rights had finally run their gamut, and the Voigt litigation concluded rather quietly when neither party choose to appeal Judge Crabb's summary judgment. Protests at the boat landings began to wane. Coincidentally, for the first time in recent history, the Wisconsin wolf population reached two score animals. For both the tribes and the ma'iingan, the future was more encouraging than at any time in decades, but the fight for either also was hardly over.

Although the issue of Ojibwe treaty rights had run its course in the courts of Wisconsin, similar litigation had begun unfolding in Minnesota in 1990. This included litigation addressing the same 1837 treaty that the Wisconsin case had discussed, as this treaty ceded lands that also became part of present-day Minnesota (Fig. 17.1). The six Ojibwe bands in Wisconsin that were cosignatories to the 1837 Treaty intervened in the case in March of 1995. Around this time Wisconsin's wolf population began to rise in earnest. The courts in Minnesota were busy in 1997, with a ruling in late January in favor of the tribes, a ruling in April by the Appellate Court to agree to hear an appeal, and a decision by the Appellate Court in August upholding the original ruling by the District Court. Meanwhile, at a wolf den somewhere in Wisconsin – quite possibly within the 1837 treaty area – a pup was born that pushed Wisconsin's wolf population into triple digits.

In 1998, Minnesota made a final appeal to the U.S. Supreme Court, which the court agreed to hear. However, on March 24, 1999, all doubt about the existence of the treaty rights for Wisconsin's Ojibwe bands was laid to rest when the Supreme Court ruled in favor of the tribes. Wisconsin's wolf population was now growing at a healthy rate. By year's end, it was estimated at nearly 200 animals, and the Wisconsin DNR downgraded the status of wolves under state law from endangered to threatened.

By 8 years later (2007), Wisconsin's wolf population likely exceeded 540 animals (Wydeven et al. this volume), a figure that suggests ma'iingan may be nearing its biological carrying capacity in the state. And 33 years and 4 days after the arrest of Fred and Mike Tribble, the U.S. Fish and Wildlife Service officially delisted wolves in its Western Great Lakes Distinct Population Segment. Most of the wolves living in the delisted region reside in areas originally ceded by the Ojibwe to the U.S. government.

While many people are aware that the affirmation of the treaty rights restored the opportunity for the tribes to hunt, fish, and gather in the ceded territories under their own regulation, many are unaware that these court cases also restored the tribes' opportunity to help manage those off-reservation resources. These resources include species, such as wolves, which were listed as endangered under federal, state, and tribal regulations at the time of litigation. As a result, over the last 20 years, tribal involvement in off-reservation management of ma'iingan greatly exceeded what would likely have occurred in the absence of affirmation of the treaty rights. This involvement in management of ma'iingan was deeply important to the tribes, in part because Wisconsin's Ojibwe reservations are too small to hold more than a few wolves. One form of that involvement consisted of participation in the development of the state wolf management plans for Minnesota, Wisconsin, and Michigan.

The increased involvement by the Ojibwe in off-reservation management of wolves has frequently juxtaposed conflicting cultural perspectives, and created challenges for managers attempting to bridge these differences. It is notable that while many see the exercise of treaty rights as primarily harvest-driven, there may be no area with a greater divergence in management perspectives between the states and the tribes than in the arena of this (at least to date) unharvested species. At the same time, within the realm of off-reservation management, ma'iingan has proved to be one of the most challenging species for the tribes themselves to address.

17.4   Contemporary Management Issues

Following the affirmation of the off-reservation treaty rights, the tribes created the Great Lakes Indian Fish and Wildlife Commission (GLIFWC) to assist them in the exercise of those rights. Within GLIFWC, a board known as the Voigt Intertribal Task Force makes decisions regarding inland fish and wildlife issues in the 1837 and 1842 ceded territories. The task force, which takes its name from the Voigt decision, consists of representatives from 9 of GLIFWC's 11 member tribes. This is the body that ultimately has to formulate and enunciate the positions of various Ojibwe bands on contemporary ma'iingan management.

For a variety of reasons, this task has not been easy. Foremost among these is the cultural significance of ma'iingan. The creation story gives ma'iingan a stature that is unique even among other clan animals. In fact, the spiritual significance of the wolf is so profound that many tribal members feel a certain degree of discomfort discussing it, considering ma'iingan a topic best addressed only by those most intimate with tribal philosophy. Even members of the task force, who routinely address management issues on a wide array of fish and wildlife species, sometimes indicated that certain matters concerning wolves were best resolved not by themselves but by select tribal spiritual leaders.

The bands' efforts to formulate positions on wolf management also were complicated – as they were for state and federal managers – by the changing biological status of wolves. Although wolves were never abundant in comparison to most other species the task force addressed, wolf populations were changing annually. This was especially true in the period following litigation, when the bands, coincidentally, were able to shift their energies from reestablishing the treaty right, to exercising it – all while expanding their role in the management of off-reservation resources. Although wolf populations were trending in a direction much desired by the tribes, the growth also meant that management issues were in a relatively constant state of flux.

Finally, the Ojibwe's world view toward ma'iingan is so markedly different from that held by most state, federal, and private interests that it was sometimes difficult to find even a common language with which to discuss management issues. This point can be exemplified by examining a few of the issues central to most non-Indian discussions of wolf management: federal legal status, population goals, and management in response to depredation.

17.4.1   Federal Legal Status

Despite the significant gains that wolves made in the western Great Lakes region, GLIFWC's member bands opposed the federal delisting of wolves in the Western Great Lakes Distinct Population Segment. This position may have been based less on a sense that regional wolves remained threatened with extinction than on a belief that the cultural significance of wolves was so great that it was difficult for the bands to consider the species "recovered" just because this minimal threshold had been reached. Indeed, even after full recovery occurs, the types of protection afforded by the Endangered Species Act (ESA) are likely to be seen from the Ojibwe perspective as remaining appropriate for an animal with the cultural significance of ma'iingan. In short, the ESA is a non-Indian construct, and the Act's definition of "recovered" simply does not reflect the Ojibwe's meaning as applied to ma'iingan.

The position of the Objibwe regarding wolves is similar to the one held by many non-Indian Americans in regard to eagles. Eagles, particularly bald eagles, maintain a unique status as a symbol of the country. This status is recognized legally in the Bald Eagle and Golden Eagle Protection Act, which provides special protection for these birds not because they are unique biologically, but culturally. Ma'iingan holds an even higher place of cultural significance to the Ojibwe, and if similar, post-delisting protection had existed for wolves, there may have been greater acceptance of the loss of protections that the ESA provided. This is particularly true because the tribes held real concerns over certain aspects of the state management plans that would be setting much of the region's direction in wolf management following delisting. One significant example is Wisconsin's wolf population goal.

17.4.2   Population Goals

Population goals are a frequent cornerstone of state management plans, yet there is probably no topic for which the language of discussion between the state and the tribes has less common ground. The differences are clear to those who can truly understand the Ojibwe ontology of *What happens to the wolf, happens to you.* How different might state management plans be if they had been written in the belief that they would be shaping not only the future of wolves but the human community as well?

One does not need to look far. The first sentence of the Minnesota Wolf Management Plan (Minnesota Department of Natural Resources 2001) reads: "The goal of this plan is to ensure the long-term survival of wolves in Minnesota while addressing the wolf–human conflicts that inevitably result when wolves and people live in the same vicinity." Clearly, this is not drafted from the perspective of the Ojibwe, who do not consider simple survival an adequate goal for ma'iingan (or themselves), and who view occasional "conflict" with wolves as being as natural – and about as significant – as occasional disagreements with your brother.

Those who can embrace this perspective will also understand why discussions of basic wildlife management parameters such as "minimum viable populations" and "population caps" sound offensive to one holding the Ojibwe world view, when applied to ma'iingan. Like the wolf, many Ojibwe populations have struggled to maintain themselves. They do not use these terms when discussing the future of their own communities, and feel they are inappropriate in the discussion of ma'iingan's future as well.

Wisconsin's management plan lists a population goal of 350 animals, and indicates that public harvest could be considered when that threshold is surpassed. This goal differs significantly in number and in nature from the goals in Minnesota – with a minimum population goal of 1,600 and no consideration of a general public take for the first 5 years following federal delisting – or proposed in Michigan's draft management plan, which does not set a numeric goal or make any recommendation regarding general public take (Michigan Department of Natural Resources 2007).

The current goal for Wisconsin also differs significantly from versions proposed in earlier drafts of the plan. An initial proposal of 300–500 animals evolved into a goal of a minimum of 350 wolves (without a stated maximum), before finally being established at simply 350 animals. According to the plan, the goal of 350 animals was settled on "as a reasonable first attempt at assessment of social tolerance" (Wisconsin Department of Natural Resources 1999:16).

Clearly, the community whose social tolerance was being assessed was not Ojibwe. Although the earlier proposed goals may have had greater acceptance by the tribes than the one ultimately adopted, the preferred tribal alternative was often stated simply as allowing wolves to reach their "natural population level." This approach is akin to setting the goal on the basis of the biological carrying capacity for ma'iingan, as opposed to some variable (and estimated) human social carrying capacity.

The expression "natural level" is a remarkably succinct description of the general Ojibwe perspective toward ma'iingan management, and the desire to reach this goal overshadowed other considerations by the bands. Topics such as public take have yet to be appreciably explored by the Voigt Task Force, in large part because of the strong feeling that it was grossly premature – perhaps even morally wrong – to discuss these topics before the population had fully recovered by reaching its natural level.

17.4.3 Lethal Depredation Control

The issue of lethal control of wolves that depredated livestock was an uncomfortable one for the tribes to address. Although lethal control tends to have rather broad acceptance among the non-Indian society and is a fairly standard component of non-Indian management regimes, it has far less support among the Ojibwe community.

It is an interesting coincidence that the three western Great Lakes states that support viable wolf populations have all prohibited capital punishment for a century or more. It appears that many in the non-Indian community are uncomfortable killing their brothers, even when the crimes they committed have been deadly themselves. It should not be surprising that many traditional tribal members feel the same unwillingness to apply the death penalty to brother wolf – especially when the actions for which the wolves are being persecuted are "wrong" only from a particular human perspective.

It is also important to note that one of most common justifications for lethal control programs – that they can increase public support for higher wolf populations – is essentially moot when applied to the Ojibwe public, who do not feel that social carrying capacity should determine population levels of ma'iingan in the first place. Indeed, one tribal member told me, "Depredation is basically a non-Indian issue, and it should be addressed by non-Indians."

GLIFWC's member bands, in exercising their off-reservation authority, ultimately decided not to oppose the judicious application of lethal control. However, the decision was not made without great discussion, and the task force was not unanimous in its decision. The bands also strongly desired that high levels of verification of wolf depredation be required, that control efforts be targeted as much as possible toward individual animals that have been verified as depredators of livestock, and that nonlethal methods of control remain the preferred alternative whenever possible.

The bands also feel strongly that depredation control remains just that: a response to individual wolves in individual situations, to provide relief to people experiencing losses to their livelihood. It must never creep toward a de facto form of population control.

17.5   Conclusion

The rebound of ma'iingan populations in the western Great Lakes region holds great meaning to the Ojibwe, who understand that their future is intertwined with that of the wolf. While this rebound brings hope to many in the Ojibwe community, the great intimacy of this relationship with ma'iingan also means that wolf "recovery" will only be realized from an Ojibwe perspective when the ma'iingan population reaches its natural level on the landscape, and becomes a fully integrated and accepted component of the community.

Acknowledgments I am grateful to Charles Rasmussen, Jason Stark, James Zorn, and James St. Arnold for review of earlier drafts of this document. Special contributions made by Patty Loew, Ed Heske, and Lisa David were especially appreciated. My sincere thanks go out to you all.

References

Benton-Banai, E. 1988. The Mishomis Book. Hayward, Wisconsin: Indian Country Communications, Inc.
Danziger, E. J., Jr. 1979. The Chippewas of Lake Superior. Norman, Oklahoma: University of Oklahoma Press.

Johnson, B. 1990. Ojibway Heritage. Lincoln, Nebraska: University of Nebraska Press.

Lopez, B. H. 1978. Of Wolves and Men. New York, New York: Charles Scribner's Sons, Inc.

Michigan Department of Natural Resources. 2007. Draft Michigan Wolf Management Plan. Michigan Department of Natural Resources, Wildlife Division, Lansing, Michigan.

Minnesota Department of Natural Resources. 2001. Minnesota State Wolf Management Plan. Minnesota Department of Natural Resources, Division of Wildlife, Minneapolis, Minnesota.

Satz, R. N. 1991. Chippewa Treaty Rights. Transactions of the Wisconsin Academy of Science, Arts and Letters. 79(1):1–251.

Schoolcraft, H. R. 1975. Thirty Years with the Indian Tribes. New York, New York: Arno Press.

Tanner, J. 1994. The Falcon. Harmondsworth, Middlesex, U. K.: Penguin Books Ltd.

Whaley, R., with W. Bresette. 1994. Walleye Warriors. Philadelphia, Pennsylvania: New Society Publishers.

Wisconsin Department of Natural Resources. 1999. Wisconsin Wolf Management Plan. Wisconsin Department of Natural Resources, Madison, Wisconsin.

Chapter 18
Wolf–Human Conflicts and Management in Minnesota, Wisconsin, and Michigan

David B. Ruid, William J. Paul, Brian J. Roell, Adrian P. Wydeven, Robert C. Willging, Randy L. Jurewicz, and Donald H. Lonsway

18.1 Introduction

Recovery of gray wolves (*Canis lupus*) in the Great Lakes region has been accompanied by an increase in wolf–human conflicts. The interface between owners of domestic animals and wolf recovery presents unique challenges for wildlife management. Investigating wolf complaints, explaining wolf ecology, conservation goals, and litigation that has impacted wolf management to people who have had domestic animals killed by wolves are challenges faced by those involved with managing wolf–human conflicts. In this chapter, we describe wolf–human conflicts and management, focusing on the period 1974–2006, when wolves were protected under the Federal Endangered Species Act (ESA).

The patterns of European settlement and wolf persecution were similar in Minnesota, Wisconsin, and Michigan. Minnesota maintained a bounty system for wolves from 1849 to 1965, aerial hunting of wolves persisted until 1956, from 1965 to 1973 wolves could be harvested for fur, and depredation control existed through a state program until May 1974, removing ~250 wolves per year (Minnesota Department of Natural Resources [MNDNR] 2001; United States Fish Wildlife Service [USFWS] 2007). Wisconsin maintained a bounty system for predators, including wolves, from 1839 to 1957. A wolf bounty was the ninth law passed by the first Michigan legislature in 1838. By 1910, wolves were extirpated from Michigan's Lower Peninsula. The bounty continued until 1922. From 1922 to 1935, a state trapper system was in effect. The bounty was reinstated in 1935 and repealed in 1960, after wolves were nearly extirpated from Michigan. In 1915, the United States Congress appropriated funds for a federal wolf control program administered by the United States Department of Agriculture, Bureau of Biological Survey (Young and Goldman 1944).

Managers quickly recognized that public acceptance and effective depredation management were necessary for wolf recovery (USFWS 1978a, 1992; Peek et al. 1991). During recovery, depredation management was an important component of federal and state wolf management plans (USFWS 1992; Michigan Department of Natural Resources [MIDNR] 1997; Wisconsin Department of Natural Resources [WDNR] 1999, 2006; MNDNR 2001).

A.P. Wydeven et al. (eds.), *Recovery of Gray Wolves in the Great Lakes Region of the United States*,
DOI: 10.1007/978-0-387-85952-1_18, © Springer Science+Business Media, LLC 2009

The gray wolf was listed as an endangered species in 1967 by the USFWS and again in 1974 under the 1973 ESA. The federal wolf depredation management program (administered by the United States Department of Interior from 1974 to 1986, and the United States Department of Agriculture, Wildlife Services [WS] from 1986 to 2006) was initiated in 1974 and has existed in similar fashion until wolves were delisted from the ESA for this region (Fritts et al. 1992).

18.2   Wolf Complaints

Quick and professional responses to wolf conflicts in this region have been important for wolf recovery. Congress appropriated funds for wolf depredation management in Minnesota in 1974 and additional funds for Minnesota, Wisconsin, and Michigan in 2003. In Minnesota, local conservation officers, the USFWS, or WS managed wolf conflicts. The WDNR, MIDNR, and WS (through cooperative agreements), have comanaged wolf complaints since 1990 and 2000, respectively.

Wolf complaints in the Great Lakes region are categorized as: (1) confirmed, (2) probable, (3) confirmed non-wolf, and (4) unconfirmed. Confirmed complaints require evidence such as canine tooth punctures and hemorrhaging on the depredated carcass and presence of wolf tracks and scat. Probable complaints have inconclusive evidence on the depredated carcass despite evidence of a struggle, blood on the ground, and wolf tracks/scat at the depredation site. Confirmed non-wolf complaints are determinations that another species depredated the domestic animal. Unconfirmed complaints are complaints not meeting these criteria. In addition to depredations, complaints involving wolves harassing livestock, perceived threats to human safety, property damage, and nuisances have been recorded.

Responses to confirmed or probable wolf depredations varied as wolf classification changed under federal and state ESAs and state wolf management plans (Refsnider, this volume). Generally, lethal control of wolves was only allowed when wolves were classified as *threatened* except in Wisconsin and Michigan during 2005 and 2006 under special permit from the USFWS. One wolf was euthanized in Wisconsin in 1999 after repeated depredations inside a captive white-tailed deer (*Odocoileus virginianus*) farm. States afforded wolves additional legal protections under state ESAs and through wolf management zones with different criteria for responding to wolf complaints. Zones of primary wolf habitat contained more restrictions for implementing lethal control to insure long-term viability of wolf populations (WDNR 1999; MNDNR 2001).

From 1974 to 2006, managers in Wisconsin and Minnesota documented 4,724 (MN = 3,896, WI = 896) wolf complaints of which 2,406 (MN = 2,046, WI = 360) were classified as either confirmed or probable (Fig. 18.1). We combined wolf population estimates for Minnesota (USFWS 1978b; Berg and Kuehn 1982; Fuller et al. 1992; Berg and Benson 1999; Erb and Benson 2004) with annual minimum wolf counts from Wisconsin to derive a population estimate for both states from 1975 to 2006. Wolf population estimates in Wisconsin and Minnesota were correlated

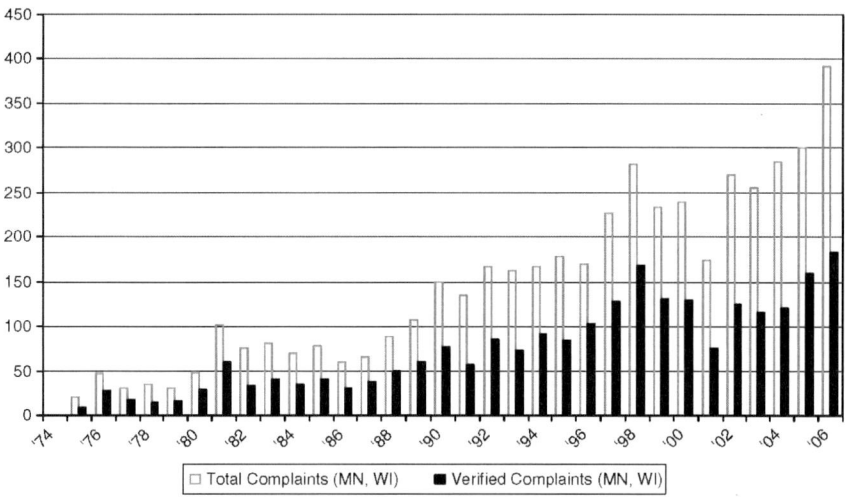

Fig. 18.1 Total wolf complaints and verified wolf complaints in Minnesota and Wisconsin, 1974–2006

with total wolf complaints ($r^2 = 0.91$, $P < 0.001$), verified wolf complaints ($r^2 = 0.82$, $P < 0.001$), and farms with verified livestock depredations ($r^2 = 0.78$, $P < 0.001$).

From 2004 to 2006, the MIDNR received 209 wolf complaints. Perceived threats to human safety (30%) and livestock depredation (20%) were most commonly reported. Home or property owners (50%) filed most reports, followed by livestock producers (33%). Forty-four percent were confirmed as wolves, 12% as coyotes (*C. latrans*) or dogs (*C. familaris*), and 44% were unconfirmed.

18.3 Wolf-Livestock Depredations in the Great Lakes Region

In the Great Lakes region, wolf depredation on livestock is well documented (Fritts 1982; Treves et al. 2002; Beyer et al. 2006). Cattle, mostly calves, are the most common livestock depredated, excluding fowl (Table 18.1). Wolf depredations usually involve 1–2 calves per incident. Depredations on sheep or free-ranging poultry may involve surplus killing (killing more prey than can be immediately consumed). Depredations on flocks of domestic turkeys on open ranges in Minnesota have resulted in 50–300 turkeys lost per night.

Wolves may live near livestock without causing depredations (Fritts and Mech 1981), and proportionally few wolf packs cause depredations (Wydeven et al. 2004). In other instances, wolves cause significant chronic losses at certain farms. Generally, wolves kill livestock opportunistically when they find livestock in close association with wild prey. Livestock production in this region includes pastures and agricultural crops that attract white-tailed deer, which may attract wolves. However, wolves in the Central Forest of Wisconsin, a region of mixed forest and agriculture,

Table 18.1 Cattle, sheep, horses, and fowl killed by wolves in Minnesota, Wisconsin, and Michigan, 1976–2006

Year	Minnesota				Wisconsin				Michigan			
	Cattle	Sheep	Horse	Fowl	Cattle	Sheep	Horse	Fowl	Cattle	Sheep	Horse	Fowl
1976					1							
1979	17	1		1								
1980	16	56	1	56	1							
1981	30	110		577	3							
1982	24	12		50								
1983	35	29	1	127								
1984	10	92	1	295								
1985	23	75			2							
1986	26	13		285								
1987	24	9		1,754								
1988	31	68		267	1	1						
1989	40	47		1,636	3							
1990	37	112		697								
1991	35	31		984		1		115				
1992	55	38	2	132	1	8						
1993	57	23		682				25				
1994	82	14	1	143								
1995	63	15	2	93	11							
1996	74	21	1	1,614								
1997	101	35		1,140	10							
1998	118	33	4	140	20				3			
1999	96	3	1	899	6			44	1			
2000	95	19	1	512	6			4	2	1		8
2001	64	5	1	85	11			68	3			
2002	97	58	2	6	36	7	2	2	4			21
2003	63	14	2	313	20	15			11	2		13
2004	66	15	3	101	29	5			7	3		
2005	92	39		207	32	2	3		2	7		1
2006	85	17	1	554	35	6		50	9	4		35
Total	1,556	1,004	24	13,350	225	48	5	308	42	17	0	78

persisted for a decade before the first confirmed depredation on livestock. This occurred in 2005 despite high white-tailed deer and livestock densities.

Abundant prey populations and decreased physical condition of prey may buffer wolf–livestock conflicts (Fritts et al. 1992). Mild winters resulting in increased physical condition of fawns were correlated with increased depredation in Minnesota (Mech et al. 1988; Fritts et al. 1992). Farms with chronic depredations usually occur within primary wolf habitat and suffer chronic depredations despite persistent wolf control. Most depredations are from single packs, but chronic depredations may result from multiple packs utilizing a single farm.

An analysis of wolf control efforts in Minnesota from 1979 to 1998 indicated that few farms had additional depredations during the same year, while 23% of farms had depredations the following year (Harper et al. 2008) either due to remaining pack members or recolonization by other wolves (Harper et al. 2005). Other factors that predispose farms to depredations include farm size, number of livestock pastured, and distance livestock are grazed from human dwellings (Mech et al. 2000). Improper disposal of livestock carcasses may condition wolves to prey on livestock (Fritts 1982); however, Bradley and Pletscher (2005) found no relationship between livestock carcasses and depredations in Montana and Idaho, and Mech et al. (2000) reported this relationship was inconclusive.

Wolf depredation on livestock mostly occurs between April and October, when livestock are pastured (Fritts et al. 1992; Musiani et al. 2005; Wydeven et al. 2006). Wolf depredations occur in all habitat types including edges of densely populated urban areas. Researchers have used road density to predict suitable wolf habitat (Mladenoff et al. 1995; Erb and Benson 2004), and the probability of livestock depredation (Treves et al. 2004). Population growth and range expansion of wolves has resulted in wolves occupying agricultural areas and increasing wolf–livestock conflicts (Treves et al. 2002; Harper et al. 2005). During expansion of wolf range in this region, wolves have proven adaptable at occupying or colonizing human-disturbed areas.

The mean number of Minnesota farms with livestock depredation during 1975–2006 was 47.7 (SD = 27.2). However, during 1975–1984, 1985–1996, and 1997–2006, mean numbers increased from 18.7 (SD = 9.8) to 48.4 (SD = 17.2) to 75.8 (SD = 16.9), respectively. During early wolf recovery in Minnesota, there were an estimated 7,200 livestock farms. Thus, during the first decade of ESA-regulated wolf recovery, 0.3% of farms had verified livestock depredations annually (Fritts et al. 1992). During 1997–2006, the number of livestock farms in Minnesota's wolf range increased to ~8,500, of which 0.9% annually had verified wolf depredations (W. J. Paul, unpublished data). In Wisconsin, verified depredations on livestock averaged 2.8 farms annually during the 1990s, but increased to 15 farms annually between 2000 and 2006. By 2006, the number of farms with depredations had increased to 25, which is similar to the number of farms with depredations in Minnesota when there were ~1,400 wolves. In northern Wisconsin, there are ~2,000 beef cow/calf producers (National Agricultural Statistics Service [NASS] 2002), of which about 1% have depredations annually. In Michigan's Upper Peninsula during 1998–2006, 56 verified wolf depredations occurred on 39 farms.

Livestock depredations are a direct economic loss for producers but other non-depredation conflicts occur that are difficult to quantify and may be underreported. Livestock producers report missing livestock as an economic burden associated with wolves (Fritts et al. 1992). Missing livestock are animals possibly depredated by wolves that are undetected (Bjorge and Gunson 1981; Oakleaf et al. 2003). Oakleaf et al. (2003) reported that on large cattle grazing allotments in Idaho, livestock producers detected one of eight calves depredated by wolves and one of 11.5 calves dying from other causes. Losses associated with missing calves prompted the WDNR to initiate a compensation program for missing livestock (below). Other impacts (non-depredation) wolves have on livestock production include harassment, damage to fences, stress to animals, difficult animal handling, time spent searching for animals, and possibly induced abortions, weight loss, and disease transmission (Howery and DeLiberto 2004; Shelton 2004; Lehmkuhler et al. 2007).

In the Great Lakes region, black bears (*Ursus americanus*), bald eagles (*Haliaeetus leucocephalus*), coyotes, domestic dogs, and, occasionally, bobcats (*Lynx rufus*) kill

livestock. While investigating wolf complaints in Wisconsin (2002–2006), WS
verified 277 depredations of livestock animals (non-fowl) by wolves, coyotes,
bears, and domestic dogs. Of these, 69% were depredated by wolves, 28% by coyotes,
and 3% by dogs or bears. In Wisconsin, livestock producers have an incentive to report
wolf and bear depredation because of compensation whereas coyote depredation is
not compensated.

18.4 Wolf-Domestic Dog Depredations in the Great Lakes Region

Between 1974 and 2006, at least 340 dogs were killed and 134 were injured by
wolves in the Great Lakes region (Table 18.2). Most depredations on dogs occurred
in Minnesota. However, Minnesota's wolf population is six times larger than
Wisconsin's and Michigan's. During 2001–2006, Wisconsin had more depredations
on dogs than Minnesota, probably because of Wisconsin's compensation for hunting

Table 18.2 Dogs killed and injured by wolves in Minnesota, Wisconsin, and Michigan, 1979–2006

	Minnesota		Wisconsin		Michigan	
Year	Injured	Killed	Injured	Killed	Injured	Killed
1979		1				
1980		1				
1981		3				
1982		2				
1983		4				
1984	3	6				
1985	3	2				
1986		1		1		
1987	2	2				
1988	2	3				
1989	1	10		1		
1990	1	11	2			
1991		9				
1992	6	5		2		
1993	2	6				
1994	5	8		2		
1995	4	8				
1996	4	10	3	5		1
1997	5	12	1	5		
1998	15	25	5	10		
1999	5	16	2	2	1	2
2000	9	17		5		
2001	4	6	1	17		3
2002	1	6	4	10	2	4
2003	4	2	4	6	3	8
2004	4	4	2	15		4
2005	3	7	6	17	1	2
2006	4	2	10	25		4
Total	87	189	40	123	7	28

dogs. The first wolf depredation on a dog in Wisconsin occurred in 1986, about 10 years after wolf colonization and in Michigan in 1996, 6 years after a breeding wolf pack was detected.

Most reported wolf attacks on dogs in Minnesota occurred near people's homes (Fritts and Paul 1989), but in Wisconsin and Michigan most attacks were on dogs used in hunting or being trained for hunting. Higher rates of attacks on hunting dogs occurred because Wisconsin and Michigan allow bear hunting with dogs, which is prohibited in Minnesota. More attacks occur in Wisconsin than in Michigan, despite similar wolf numbers. This probably occurs because Wisconsin has an earlier start to bear hound training, more bear hunters, and more reporting (Wisconsin reimburses for wolf depredations on dogs). Bear hunters commonly use baits to harvest bears. In Wisconsin, bait sites can be established as early as April 14 each year, compared to August 10, in Michigan. Wolves using bear bait sites for food have been documented by trail cameras, tracks, and the stomach contents of a captured wolf (D. Ruid, unpublished data). Wolves may defend bait sites from other predators, including bear hounds.

Because of better reporting, we used data from Wisconsin as a case study of wolf depredations on dogs. Dogs (all hounds) attacked during hunting or training for hunting, were compared to pet dogs attacked near homes. From 1986 to 2006, 163 dogs were attacked, including 123 (75.5%) killed by wolves (Table 18.3). One hundred thirty-one (80.4%) attacks occurred during hunting or training (Table 18.3). In Wisconsin, training of bear hounds occurs during July and August, and bear hunting occurs from early September through early October. Dogs attacked during

Table 18.3 Depredations on dogs and number of wolf packs and percentage of wolf packs that depredated dogs in Wisconsin, 1986–2006

| Year | Hunting hounds[a] | | | Pet dogs[b] | | | All dogs | | |
	Killed	Injured	No. and percentage of packs	Killed	Injured	No. and percentage of packs	Killed	Injured	No. and percentage of packs
1986	1		1 (25)				1		1 (25)
1989	1		1 (14)				1		1 (14)
1990		2	0 (loner)					2	0 (loner)
1992	2		2 (15)		`		2		2 (15)
1994	2		0 (loner)				2		0 (loner)
1996	5	2	3 (10)		1	0 (hybrid)	5	3	3 (10)
1997	5		3 (9)		1	1 (3)	5	1	4 (11)
1998	9	3	7 (15)	1	2	3 (6)	10	5	10 (21)
1999	2	1	2 (3)		1	0 (loner)	2	2	2 (3)
2000	3		2 (3)	2		1 (2)	5		3 (3)
2001	17		6 (9)		1	1 (1)	17	1	7 (10)
2002	9	1	4 (5)	1	3	3 (3)	10	4	7 (8)
2003	6	1	3 (3)		3	3 (3)	6	4	6 (6)
2004	13	1	5 (5)	2	1	1 (1)	15	2	6 (6)
2005	13	3	7 (6)	4	3	5 (4)	17	6	12 (11)
2006	23	6	11 (9)	2	4	3 (3)	25	10	14 (12)
Total	111	20		12	20		123	40	

[a]Hunting hounds includes dogs killed while hunting or training to hunt
[b]Pet dogs included dogs killed on or near the property of the owner and not in hunting situations

bear hunting and training included 103 (63.2%) attacks and 89 (72.3%) killed. Wolves use rendezvous sites during July through early October (Fritts and Mech 1981), and wolves aggressively defend these sites (Ballard et al. 2003). Wolf attacks on dogs during other hunting seasons included 12 during bobcat hunting (December), nine during coyote hunting (fall and winter), and seven during snowshoe hare (*Lepus americanus*) hunting (winter). At least 23 dog breeds were attacked including 52 Walker coonhounds, 34 plott hounds, 18 bluetick coonhounds, nine redbone coonhounds, nine Labrador retrievers, eight beagles, three black and tan hounds, and three huskies. No dogs in bird hunting situations were depredated by wolves. Wolves killing dogs during attacks in hunting and training situations is proportionally higher (85%) than during attacks on pet dogs (38%). Wolf attacks during hunting and training generally occurred on lands open to public hunting while hunters were ≥200 m away.

A small percentage of wolf packs depredate dogs (Wydeven et al. 2004; Table 18.3). Eleven attacks were by lone wolves, possibly by wolf–dog hybrids, and these occurred mainly at people's homes. Larger packs were more likely to attack hunting dogs and attack in subsequent years (Wydeven et al. 2004). Because dogs used for bear hunting run in groups of ≤6 dogs, wolf packs of ≤4 adults (Wydeven et al., this volume) likely avoid dog groups. Larger wolf packs might attack dogs to secure territories, defend pups, or defend bait/kill sites. Despite this, over half the dogs killed were either partially or completely consumed.

18.5   Wolf Depredation Management

The ESA prohibited killing of depredating wolves from 1975 to 1978. As an alternative, 104 wolves were translocated 50–317 km to areas of northern Minnesota. Survival of wolves was unaffected by translocation (Fritts et al. 1985), but Fritts et al. (1984) concluded that translocation was unsuccessful because wolves left their release sites, traveled through agricultural areas, and were often recaptured at depredation sites. In Wisconsin and Michigan, 57 wolves were translocated during 1991–2002 an average of 138 km ($n = 33$) and 121 km ($n = 24$), respectively. The leading source of mortality for translocated wolves was humans.

Federal classification of wolves in Minnesota changed from endangered to threatened in 1978 and provided greater flexibility for managing wolf–livestock conflicts. This change allowed authorized personnel to kill livestock-depredating wolves in Minnesota following significant depredations, if killing was humane (USFWS 1978b). A "significant depredation" was later defined by the USFWS as "the killing or serious maiming of one or more domestic animals by wolves where the imminent threat of additional domestic animals being killed or severely maimed by wolves is apparent" (USFWS 1978b). During 1978, environmental groups accused the USFWS of not following its own regulations and filed suit (Fund for Animals v. Andrus, MN Federal District Court 1978) to limit depredation control activities. The court ruling ordered that trapping and killing of wolves must occur only after a significant depredation and trapping must, as nearly as possible, be directed toward capture of the problem wolves. To reduce captures of non-depredating wolves, trapping was restricted to within 0.4 km of affected farms and prohibited killing wolf pups on or before September 1 because pups were not considered depredating animals (USFWS 1985). In 1983, the USFWS amended regulations in Minnesota to allow public taking of wolves in certain areas and modified the guidelines for wolf depredation management, which did not provide protection for pups and increased the distance wolves could be taken at depredation sites from 0.4 km to 0.8 km (USFWS 1983). Environmental groups (Sierra Club and Defenders of Wildlife v. Clark, MN Federal District Court, 1984, 1985) challenged the rulemaking and the court ordered the USFWS to amend their regulations, which prohibited a public harvest of Minnesota wolves, allowed trapping of depredating wolves <0.8 km of affected farms, and allowed the taking of wolf pups captured after August 1 (USFWS 1985).

Following the court rulings in 1984 and 1985, and the 2003 reclassification of wolves from endangered to threatened in Michigan and Wisconsin and under special permit during 2005 and 2006 (Refsnider, this volume), lethal control for wolves occurred under five specific conditions: (1) presence of a carcass or wounded animal, (2) evidence that wolves were responsible for damage, (3) reasonable expectation of additional losses if wolves were not removed, (4) trapping activities restricted to <0.8 km from affected farms, and (5) wolf pups captured at farms on or before August 1 are released on-site. Thus, the strategy for wolf control in the Great Lakes region prior to federal delisting in 2007 was selective removal of problem wolves where depredations were verified. Since federal delisting of wolves in the Western Great Lakes Distinct Population Segment on March 12, 2007, wolf depredation management is governed by the states' wolf management plans.

Approved control methods included foothold traps, cable restraints, and calling/shooting. Traps (10–25) were set along travel routes, near depredated carcasses (>9.1 m), and near artificial and natural scent stations for 10–15 days. Trapping was extended to 30–45 days at sites with chronic depredations.

During 1979–2006, the federal wolf depredation control program in Minnesota killed 6–216 wolves, annually (Table 18.4). Annual averages and percentage of statewide populations for wolves killed were 26 (2%) from 1979 to 1984, 49 (3%) from 1985 to 1989, 115 (6%) from 1990 to 1994, 152 (7%) from 1995 to 1999, and 127 (4%) from 2000 to 2006.

During 2003–2006, Wisconsin's wolf population increased annually by an average of 10% while annual depredation control efforts removed 6% of the population. Ninety-one wolves were killed in Wisconsin during 2003–2006; 87 were removed from farms with livestock depredation and four had depredated dogs (Table 18.4). Average pack size of wolves depredating livestock was 3.5, similar to average pack size in the state during the same period (Wydeven et al. 2004).

Michigan's wolf population increased by 16% despite having 1% of the wolf population removed during 2003–2006. From 2003 to 2006, 22 wolves were killed in Michigan, 18 for livestock protection, and four because of threats to human or pet safety (Table 18.4). During 33 years of protection under the ESA, 3,192 wolves were captured of which 2,773 wolves were euthanized for depredation control

Table 18.4 Wolves captured and killed for depredation management in Minnesota, Wisconsin, and Michigan, 1975–2006

Year	Minnesota		Wisconsin		Michigan	
	Captured	Killed	Captured	Killed	Captured	Killed
1975	17					
1976	51					
1977	59					
1978	40	34				
1979	15	6				
1980	26	21				
1981	42	29				
1982	24	20				
1983	49	42				
1984	47	36				
1985	36	31				
1986	31	31				
1987	45	43				
1988	64	59				
1989	95	81				
1990	91	91				
1991	63	54	1			
1992	122	118	0			
1993	145	139	0			
1994	175	172	0			
1995	78	78	0			
1996	167	154	0			
1997	227	216	2			
1998	166	161	4		7	
1999	163	151	2	1	2	
2000	153	148	2			
2001	114	109	8		5	
2002	163	146	19		9	
2003	129	125	17	17	6	4
2004	115	105	27	24	17	6
2005	148	134	36	32	8	5
2006	131	122	22	18	7	7
Total	2,991	2656	140	92	61	22

(Table 18.4). Despite lethal removal of wolves, Great Lakes wolves have increased to roughly 4,000 animals and have colonized most of the suitable habitat (Erb and DonCarlos, this volume; Beyer et al., this volume; Wydeven et al., this volume). Continued wolf colonization of agricultural areas will likely necessitate increases in wolf removal.

An assessment of wolf depredation management in Minnesota during 1979–1986 concluded that when trapping was successful (≥1 wolf removed), 34% of farms had continued depredations during the same year. Conversely, when no wolves were captured, 23% of farms had continued depredations (Fritts et al. 1992). The difference in subsequent depredations between successfully and unsuccessfully trapped farms was likely related to farm characteristics and juxtaposition within wolf habitat (Fritts et al. 1992). Harper et al. (2008) analyzed the efficacy of wolf depredation management in Minnesota during 1979–1998. Their analysis concluded that management was more effective during the year depredations occur but not necessarily during subsequent years. Trapping was more effective in newly colonized areas than in areas with high wolf densities. Trapping wolf pups or pups and adults (especially adult males) was the most effective. Attempting to trap appeared more effective than not trapping at reducing livestock depredation. Farms with cattle had fewer subsequent depredations than farms with sheep or turkeys.

18.6 Nonlethal Methods

Early use of nonlethal or preventative methods in North America consisted of fencing and burning to remove vegetation near livestock (Young and Goldman 1944). The development of practical nonlethal techniques that reduce conflicts at appropriate spatial scales (Treves et al. 2004; Shivik 2006) may increase public acceptance of wolves, use of nonlethal techniques, and wolf depredation management (Gehring and Potter 2005). Prior to restoration of wolves in this region and the Northern Rocky Mountains, similar research focused on developing nonlethal methods to prevent coyotes from depredating livestock (Shivik 2004).

Nonlethal techniques are varied (Cluff and Murray 1995; Shivik et al. 2003; Shivik 2004, 2006) and include animal husbandry (e.g., changing pastures, night confinement, and changing birthing dates), disruptive stimuli (fladry, lights, and sirens), aversive stimuli (shock collars and taste aversion), sterilization of wolves, use of livestock-guarding animals, and translocation of problem wolves. Cluff and Murray (1995) suggested that nonlethal methods had higher public acceptance than killing wolves to resolve depredation complaints (USFWS 1978c; Arthur 1981; Andelt 1987; Cluff and Murray 1995).

Fladry (flagging) was developed in Europe as a technique to capture wolves (Musiani and Visalberghi 2001). More recently fladry has been used to restrict wolves' access livestock pastures (Shivik 2006). Fladry is now produced commercially and consists of 50 cm × 10 cm nylon flags sewn 50 cm apart onto 3-mm diameter ropes and is suspended ~50 cm above the ground. From 2004 to 2006, WS installed fladry on nine farms in Wisconsin after wolves depredated (four farms) or harassed livestock (five farms). After 60–180 days on each property, anecdotal evidence suggested that wolves only crossed this visual barrier once and never depredated livestock within an enclosed pasture (D. Ruid, unpublished data).

Fritts (1982) experimented with solar-powered flashing lights in Minnesota where wolves had depredated livestock and reported mixed results. In Wisconsin (2004–2006), flashing lights were installed around livestock pastures on 13 farms after wolves depredated or harassed livestock. Lights were spaced ~50 m apart for an average of 74 days. Results in Wisconsin were similar to those reported by Fritts (1982); subsequent depredations occurred on two farms.

Shock collars and surgical sterilization of wolves have been assessed for their feasibility in preventing wolf depredations (Mech et al. 1996; Spence et al. 1999; Hawley 2005; Schultz et al. 2005). Mech et al. (1996) sterilized five adult male wolves in Minnesota (by vasectomy) and Spence et al. (1999) sterilized both adult male and female wolves to evaluate the effects on pair bonding and territory retention. Neither study related sterilization to livestock depredations although Bromley and Gese (2001) documented that sterilized coyotes depredated fewer sheep than intact coyotes. However, breeding wolves die or can be displaced by other wolves and territories shift (Mech 1970); this would require additional wolves be sterilized and if a wolf harvest was implemented it would require protection of sterilized individuals.

Schultz et al. (2005) fitted wolves with shock collars (used for training dogs) in Wisconsin where wolves had depredated cattle during 1999–2001. Negative stimulus associated with shock collars kept collared wolves away from the farm, although other pack members without collars caused depredation during 2 of 3 years. Hawley (2005) documented fewer visits to sites baited with deer carcasses by wolves wearing shock collars compared to wolves without shock collars.

Proper disposal of livestock carcasses or use of liming or composting may prevent some depredations (Fritts et al. 1992; Mech et al. 2000; Bradley and Pletscher et al. 2005). Use of nonlethal methods is a prerequisite to wolf recovery; however, as wolves in the Great Lakes region expand their range and conflicts increase, lethal removal of wolves will remain necessary.

18.7   Michigan's Compensation Program

Compensation for wolf depredation benefits wolf conservation in the United States and other countries (Fritts et al. 2003). Annual compensation payments have been much lower in Michigan than in Minnesota or Wisconsin (Table 18.5). In Michigan, compensation is available only for livestock as defined by the Michigan Department of Agriculture (MIDA). Compensation in Michigan for wolf depredation requires verification by MIDNR or WS and is limited to $4,000 per animal. In addition to verifying cause of death, agency personnel assist livestock producers by providing technical information on animal husbandry that may prevent future depredations (Beyer et al. 2006).

Currently, Michigan has two sources of funding to compensate livestock losses. State funding, administered by the MIDA first became available in 1998. Despite an annual appropriation, legislation allows MIDA to seek reimbursement from the MIDNR for indemnification payments and until 2005 MIDA had not requested reimbursement. A private conservation group, Defenders of Wildlife, donated $10,000 to pay the difference between values at time of loss and fall market value. This fund is administered for the MIDNR by the International Wolf Center (Ely, MN).

18.8   Minnesota's Compensation Program

Minnesota Department of Agriculture (MNDA) compensates only for livestock depredated or severely injured by wolves (Table 18.5). Owners are entitled to fair market value of depredated livestock as determined by the MNDA. University extension

Table 18.5 Wolf compensation payments (US dollars) in Minnesota, Wisconsin, and Michigan, 1977–2006

Year	Minnesota (US$)	Wisconsin (US$)	Michigan (US$)
1977	8,667.50		
1978	22,482.08		
1979	20,773.22		
1980	20,459.00		
1981	38,605.60		
1982	18,971.04		
1983	24,868.66		
1984	19,457.74		
1985	23,558.50	200.00	
1986	14,444.19		
1987	24,233.64	2,500.00	
1988	28,109.90		
1989	43,663.92	400.00	
1990	42,739.04	2,500.00	
1991	32,205.67	1,038.55	
1992	23,339.10	684.00	
1993	31,182.38	1,600.00	
1994	31,223.84	6,125.00	
1995	34,096.77	1,509.75	
1996	43,579.68	11,918.82	
1997	50,262.50	11,850.00	
1998	71,766.55	9,340.16	612.50
1999	68,479.50	84,279.47	400.00
2000	88,097.19	18,630.00	850.00
2001	59,456.76	47,454.14	2,200.00
2002	78,217.70	56,997.10	3,648.50
2003	53,852.85	30,106.89	4,720.00
2004	55,931.00	109,941.60	5,435.00
2005	63,503.00	72,355.40	1,890.00
2006	70,000.00	114,799.52	2,590.00
Total	1,206,228.52	584,230.40	22,346.00

agents, conservation officers, or WS must inspect the site and the MNDA determines whether livestock were depredated by wolves based on the investigation, and whether deficiencies in the owner's husbandry contributed to wolf depredation.

18.9 Wisconsin's Compensation Program

The WDNR has compensated for depredation on domestic animals by wolves since 1985. Wisconsin's program pays for all domestic animals killed or injured including pets and hunting dogs. In addition, WDNR compensates for some calves that are missing at the end of the grazing season beyond those that would normally die according to a normal calf mortality rate determined by the NASS. The WDNR relies heavily on WS to investigate and verify wolf depredations. Wisconsin pays a maximum of $2,500 for hunting dogs and pets. Three agriculture experts annually determine maximum values for livestock. Funding for Wisconsin's compensation program comes from voluntary contributions to the Wisconsin Endangered Resources Fund on the Wisconsin Income Tax form and through the sale of Endangered Resources license plates.

The WDNR reimbursement program for wolf depredation paid 299 claims (1985–2006, Table 18.5). From 1985 to 2006, reimbursement for hunting hounds accounted for 37% of all payments, calves 33%, farm-raised deer 14%, cows 5%, horses 3%, pet dogs 2%, veterinary bills 3%, sheep 2%, and poultry 1%.

18.10 Summary

Following removal of wolves from ESA protection on March 12, 2007, wolf management in Minnesota, Wisconsin, and Michigan has changed slightly. All three states have substantial commitments to wolf conservation. The MIDNR, MNDNR, and WDNR developed new wolf depredation control guidelines that allow greater flexibility for managing wolf–human conflicts. Minnesota, Wisconsin, and Michigan have ~3,000, 540, and 500 wolves, respectively, thus removal of problem wolves (currently about 5% annually) will not jeopardize population viability. Conflict management needs to be flexible because depredation scenarios are multifaceted. Agency guidelines and polices to manage wolf conflicts, while necessary, cannot fully anticipate the adaptability of this species. Lethal removals are appropriate when wolves are actively harassing or hunting livestock and consideration for stakeholders who are negatively impacted by wolves must accompany wolf recovery. Removal of human-habituated wolves will become more important as wolves continue to colonize unsuitable areas. Incorporating recent advances in technology has improved nonlethal devices (Shivik 2006).

Wolf recovery and conflict management will remain controversial. Maintaining wild places for wolves to exist should be a priority for this region and other areas with suitable wolf habitat. More flexible guidelines for wolf depredation management, while possibly increasing the number of wolves killed annually, will not increase the number of packs targeted because guidelines do not allow for targeting non-depredating packs. Nor will it jeopardize the viability of the region's wolf population.

References

Andelt, W. F. 1987. Coyote predation. In Wild furbearer management and conservation in North America, eds. M. Novak, J. A. Baker, M. E. Obbard, and B. Malloch, pp. 128–140. North Bay, Ontario, Canada: Ontario Trappers Association.
Arthur, L. M. 1981. Coyote control: the public response. Journal of Range Management 34:14–15.
Ballard, W. B., Carbyn, L. N., and Smith, D. W. 2003. Wolf interactions with non-prey. In Wolves behavior, ecology, and conservation, eds. L. Boitani and L. D. Mech, pp. 289–316. Chicago, IL: The University of Chicago Press.

Berg, W. E., and Benson, S. 1999. Updated wolf population estimate for Minnesota, 1997–1998. Minnesota Department of Natural Resources Report. Grand Rapids, MN.

Berg, W. E., and Kuehn, D. W. 1982. Ecology of wolves in north-central Minnesota. In Wolves of the world: perspectives of behavior, ecology and conservation, eds. F. H. Harrington and P. C. Paquet, pp. 4–11. Park Ridge, IL: Noyes Publications.

Beyer, D., Hogrefe, T., Peyton, R. B., Bull, P., Burroughs, J. P., and Lederle, P. 2006. Review of social and biological science relevant to wolf management in Michigan. Lansing, MI: Michigan Department of Natural Resources.

Bjorge, R. R., and Gunson, J. R. 1981. Wolf predation of cattle on the Simonette River pastures in northwestern Alberta. In Wolves in Canada and Alaska: their status, biology, and management, ed. L. N. Carbyn, pp. 106–111. Ottawa, Canada: Supply and Services Canada.

Bradley, E. H., and Pletscher, D. H. 2005. Assessing factors related to wolf depredation of cattle in fenced pastures in Montana and Idaho. Wildlife Society Bulletin 33:1256–1265.

Bromley, C., and Gese, E. M. 2001. Effect of sterilization on territory fidelity and maintenance, pair bonds, and survival rates of free-ranging coyotes. Canadian Journal of Zoology 79:386–392.

Cluff, H. D., and Murray, D. L. 1995. Review of wolf control methods in North America. In Ecology and conservation of wolves in a changing world, eds. L. N. Carbyn, S. H. Fritts, and D. R. Seip, pp. 491–504. Edmonton, AB, Canada: Circumpolar Institute.

Erb, J., and Benson, S. 2004. Distribution and abundance of wolves in Minnesota, 2003–2004. Grand Rapids, MN: Minnesota Department of Natural Resources.

Fritts, S. H., and Mech, L. D. 1981. Dynamics, movements, and feeding ecology of a newly-protected wolf population in northwestern Minnesota. Wildlife Monographs 80:1–79.

Fritts, S. H. 1982. Wolf depredation on livestock in Minnesota. Resource Publication 145. Washington, DC: United States Fish and Wildlife Service.

Fritts, S. H., Paul, W. J., and Mech, L. D. 1984. Movements of translocated wolves in Minnesota. Journal of Wildlife Management 48:709–721.

Fritts, S. H., Paul, W. J., and Mech, L. D. 1985. Can relocated wolves survive? Wildlife Society Bulletin 13:459–463.

Fritts, S. H., and Paul, W. J. 1989. Interactions of wolves and dogs in Minnesota. Wildlife Society Bulletin 17:121–123.

Fritts, S. H., Paul, W. J., Mech, L. D., and Scott, D. P. 1992. Trends and management of wolf-livestock conflicts in Minnesota. Resource Publication 181. Washington, DC: United States Fish and Wildlife Service.

Fritts, S. H., Stephenson, R. O., Hayes, R. D., and Boitani, L. 2003. Wolves and humans. In Wolves behavior, ecology, and conservation, eds. L. Boitani and L. D. Mech, pp. 289–316. Chicago, IL: University of Chicago Press.

Fuller, T. K., Berg, W. E., Radde, G. L., Lenarz, M. S., and Joselyn, G. B. 1992. A history and current estimate of wolf distribution and numbers in Minnesota. Wildlife Society Bulletin 20:42–55.

Gehring, T. M., and Potter, B. A. 2005. Wolf habitat analysis in Michigan: an example of the need for proactive land management for carnivore species. Wildlife Society Bulletin 33:1237–1244.

Harper, E. K., Paul, W. J., and Mech, L. D. 2005. Causes of wolf depredation increase in Minnesota from 1979–1998. Wildlife Society Bulletin 33: 888–896.

Harper, E. K., Paul, W. J., and Mech, L. D. 2008. Effectiveness of lethal, directed wolf-depredation control in Minnesota. Journal of Wildlife Management 72:778–784.

Hawley, J. A. 2005. Experimental assessment of shock collars as a non-lethal control method for free-ranging wolves in Wisconsin. MS Thesis. Central Michigan University, Mount Pleasant, MI.

Howery, L. D., and DeLiberto, T. J. 2004. Indirect effects of carnivores on livestock foraging behavior and production. Sheep and Goat Research Journal 19:64–71.

Lehmkuhler, J., Palmquist, G., Ruid, D., Willging, R., and Wydeven, A. P. 2007. Effects of wolves and other predators on farms in Wisconsin: beyond verified losses. Publication ER-658–2007. Madison, WI: Wisconsin Department of Natural Resources.

Mladenoff, D. J., Sickley, T. A., Haight, R. G., and Wydeven, A. P. 1995. A regional landscape analysis and prediction of favorable gray wolf habitat in the Northern Great Lakes region. Conservation Biology 9:279–294.

Mech, L. D. 1970. The wolf. Minneapolis, MN: University of Minnesota.

Mech, L. D., Fritts, S. H., and Paul, W. J. 1988. Relationship between winter severity and wolf depredations on domestic animals in Minnesota. Wildlife Society Bulletin 16:269–272.

Mech, L. D., Fritts, S. H., and Nelson, M. E. 1996. Wolf management in the 21st century, from public input to sterilization. Journal of Wildlife Research 1:195–198.

Mech, L. D., Harper, E. K., Meier, T. J., and Paul, W. J. 2000. Assessing factors that may predispose Minnesota farms to wolf depredations on cattle. Wildlife Society Bulletin 28:623–629.

Michigan Department of Natural Resources. 1997. Michigan gray wolf recovery and management plan. Lansing, MI: Michigan Department of Natural Resources.

Minnesota Department of Natural Resources. 2001. Minnesota wolf management plan. St. Paul, MN: Minnesota Department of Natural Resources.

Musiani, M, and Visalberghi, E. 2001. Effectiveness of fladry on wolves in captivity. Wildlife Society Bulletin 29:91–98.

Musiani, M., Muhly, T., Cormack-Gates, C., Callaghan, C., Smith, M. E., and Tosoni, E. 2005. Seasonality and reoccurrence of depredation and wolf control in western North America. Wildlife Society Bulletin 33:876–887.

National Agricultural Statistics Service. 2002. Census of agriculture-county data. Washington DC: United States Department of Agriculture.

Oakleaf, J. K., Mack, C., and Murray, D. L. 2003. Effects of wolves on livestock calf survival and movements in Central Idaho. Journal of Wildlife Management 67:299–306.

Peek, J. M., Brown, D. E., Kellert, S. R., Mech, L. D., Shaw, J. H., and Van Ballenberghe, V. 1991. Restoration of wolves in North America. Technical Review 91. Bethesda, MD: The Wildlife Society.

Shivik, J. A., Treves, A., and Callahan, P. 2003. Non-lethal techniques for managing predation: primary and secondary repellents. Conservation Biology 17:1531–1537.

Shivik, J. A. 2004. Non-lethal alternatives for predation management. Sheep and Goat Research Journal 19:64–71.

Shivik, J. A. 2006. Tools for the edge: what's new for conserving carnivores. Bioscience 56:253–259.

Schultz, R. N., Jonas, K. W., Skuldt, L. H., and Wydeven, A. P. 2005. Experimental use of dog-training shock collars to deter depredations by gray wolves. Wildlife Society Bulletin 33:142–148.

Shelton, M. 2004. Predation and livestock production perspective and overview. Sheep and Goat Research Journal 14:2–5.

Spence, C. E., Kenyon, J. E., Smith, D. R., Hayes, R. D., and Baer, A. M. 1999. Surgical sterilization of free-ranging wolves. Canadian Veterinarian Journal 40:118–121.

Treves, A., Jurewitz, R. R., Naughton, L., Rose, R. A., Willging, R. C., and Wydeven, A. P. 2002. Wolf depredation on domestic animals: control and compensation in Wisconsin, 1976–2000. Wildlife Society Bulletin 30:231–241.

Treves, A., Naughton-Treves, L., Harper, E. K., Mladenoff, D. J., Rose, R. A., Sickley, T. A., and Wydeven, A. P. 2004. Predicting human-carnivore conflict: a spatial model based on 25 years of wolf depredation on livestock. Conservation Biology 18:114–125.

United States Fish and Wildlife Service. 1978a. Recovery plan for the eastern timber wolf. Washington, DC: United States Fish and Wildlife Service.

United States Fish and Wildlife Service. 1978b. Wildlife and Fisheries: reclassification of the gray wolf in the United States and Mexico, with determination of critical habitat in Michigan and Minnesota. Federal Register 43:9607–9615.

United States Fish and Wildlife Service. 1978c. Predator damage in the west: a study of coyote management alternatives. Washington, DC: United States Fish and Wildlife Service.

United States Fish and Wildlife Service. 1983. Endangered and threatened wildlife and plants: regulation governing the gray wolf in Minnesota. Federal Register 50:36256–36266.

United States Fish and Wildlife Service. 1985. Endangered and threatened wildlife and plants: regulations governing the gray wolf in Minnesota. Federal Register 50:50792–50793.

United States Fish and Wildlife Service. 1992. Recovery plan for the eastern timber wolf. Minneapolis-St. Paul, MN: United States Fish and Wildlife Service.

United States Fish and Wildlife Service. 2007. Endangered and threatened wildlife and plants: final rule designating the Western Great Lakes populations of gray wolves as a distinct population segment: removing the Western Great Lakes distinct population segment of the gray wolf from the list of endangered and threatened wildlife. Federal Register 72:6051–6103.

Wisconsin Department of Natural Resources. 1999. Wisconsin Wolf Management Plan. Wisconsin Department of Natural Resources PUBL-ER-099 99. Madison, WI: Wisconsin Department of Natural Resources.

Wisconsin Department of Natural Resources. 2006. Wisconsin Wolf Management Plan, Addendum. Madison, WI: Wisconsin Department of Natural Resources.

Wydeven, A. P., Treves, A., Brost, B., and Wiedenhoeft, J. E. 2004. Characteristics of wolf packs in Wisconsin: identification of traits influencing depredation. In Predators and people: from conflict to coexistence, eds. N. Fascione, A. Delach, and M. E. Smith, pp. 28–50. Washington DC: Island Press.

Wydeven, A. P., Willging, R. C., Ruid, D. B., and Jurewicz, R. L. 2006. Wolf depredations in Wisconsin through 2005, Appendix A-2. In Wisconsin Wolf Management Plan, Addendum 2006. Madison, WI: Wisconsin Department of Natural Resources.

Young, S. P., and Goldman, E. A. 1944. The wolves of North America. Parts 1 and 2. New York: Dover Publication Incorporated.

Chapter 19
Education and Outreach Efforts in Support of Wolf Conservation in the Great Lakes Region

Pamela S. Troxell, Karlyn Atkinson Berg, Holly Jaycox,
Andrea Lorek Strauss, Peggy Struhsacker, and Peggy Callahan

19.1 Introduction

A key component to the recovery of gray wolves (*Canis lupus*) in the Great Lakes region has been educational efforts about wolves done within the region. All four US Wolf Recovery Plans include recommendations to use public education to promote wolf conservation (Fritts et al. 2003). The importance of education also surfaced as a key component of the initial recovery plan for wolves in Wisconsin (Thiel and Valen 1995), and continued to be an important aspect of all wolf management plans in the Great Lakes region. The objective presentation of wolves is considered necessary by most wolf biologists for sustaining recovery (Fritts et al. 2003). Agencies responsible for wolf recovery have been involved in promoting wolf conservation, but have also relied heavily on nongovernmental organizations (NGOs), and volunteers.

In this chapter, we discuss the changing attitudes toward wolves in the Great Lakes region, and examine how education has responded and helped change those attitudes. There are many approaches that can be used to educate people about wolves (Fritts et al. 2003), and we examined the approaches used by the six organizations we represent, but we want to stress that there are other approaches and other organizations that have also been involved with educating people about wolves in this region. We conclude with suggestions of how education may continue to be used to promote conservation and living with wolves in the areas where wolf populations have recovered in the western Great Lakes states.

Our goals for education include (1) provide information about wolf biology, natural history, and ecology; (2) connect the public with scientific research on wolves as well as other species (using wolves as a focus because of the high level of interest in this species); (3) help people understand how scientists gather information about wolves, so they can better judge the credibility of popular sources of information; (4) help people to make informed decisions based on science, not emotions – whether those emotions are positive or negative; and (5) give people practical suggestions of strategies for coexisting with wolves to minimize conflicts. Those educating about wolves must be careful to inspire appreciation for

A.P. Wydeven et al. (eds.), *Recovery of Gray Wolves in the Great Lakes*
Region of the United States,
DOI: 10.1007/978-0-387-85952-1_19, © Springer Science + Business Media, LLC 2009

the wolf and its important role in the ecosystem, without leading the public to have an unrealistic view of the species. Education about wolves needs to help people to wrestle with difficult and controversial questions – but should not try to provide them with the answers.

19.2   Education and the Changing Attitudes Toward Wolves

Early European settlers responded to the wolves inhabiting the New World with the same malevolence they embraced in their homeland. They carried with them centuries of fairy tales, myths, and legends of wolves as evil figures that devour children, devils that steal souls, and beasts that steal livestock. Along with the myths were fears of wolves attacking livestock and competing for limited resources in this wilderness, where survival was already tenuous. Because of concerns about wolves, the first bounty was enacted as law in 1630 by the Massachusetts Bay Colony, with the hunter receiving 1-cent per wolf (McIntyre 1995).

Wolves, along with snakes, rats, bats, coyotes, and other medium and large carnivores, were frequently viewed as intrinsically unworthy (Kellert et al. 1996). Lopez (1978, p. 139) wrote, "… the wolf is fundamentally different because the history of killing wolves shows far less restraint and far more perversity. A lot of people did not just kill wolves; they tortured them." Through intense control programs, humans rid the prairies, mountains, and forests of wolves until only handfuls survived in the most remote and wild reaches of the US. Wisconsin, Michigan, and Minnesota, like many other states, set bounties in the mid 1800s, and by the mid 1900s only a few hundred wolves remained in the lower 48 states along the border-lakes region of northeastern Minnesota.

Public attitudes toward wolves began to change as biologists and naturalists started conducting ecological research on wolves. Early conservationists and naturalists who became the pioneers of this change in attitude included Olson (1938), Murie (1944), Errington (1946), and Leopold (1949). These biologists began to paint a more realistic picture of the wolf – that wolves are highly intelligent, have a complex social structure, maintain healthy prey populations, and play important roles in ecosystems. While these researchers helped dispel myths about wolves, others also created positive images of wolves, but may have perpetuated new myths as well, such as wolves make interesting pets (Crisler 1958) or wolves are able to subsist on small mammals alone (Mowat 1963). It appeared that during the period of the 1930s and 1970s more positive attitudes toward wolves developed in the USA (Williams et al. 2002).

With these attitude changes, bounties were eliminated in the late 1950s and 1960s in the Great Lakes region. Scientists beginning research at the time, including Douglas Pimlott, Durward Allen, and L. David Mech, investigated further the ecology of wolves, and began long-term research projects in Algonquin Provincial Park, Isle Royale, and northeastern Minnesota, respectively. People like Marlin Perkins, zoologist and TV show host, and his wife Carol Perkins, formed the Wild Canid Survival and Research Center and organized and hosted the first wolf conference in 1971, with the intent to preserve the last remaining wild wolf populations. The outcome from this and other early conferences on wolves was the realization that if wild wolf populations were to be recovered, attitudes needed to be changed and education would be critical. In 1973, the federal government passed the Endangered Species Act (ESA) as part of an ecological revolution. Interest in and more positive attitudes about wolves were part of a growing environmental consciousness and realization that species such as top carnivores play an important role in ecosystems. By 1974, the remaining population of wolves in Minnesota was legally protected. Although laws granted the wolf protection, long-held fears and hostile beliefs persisted.

The message during the early years of wolf recovery was, "that 'wolves are cool!'" (Grooms 2004, p. 5). The arguments were more black and white: wolves on the landscape versus no wolves on the landscape, save them or kill them. As wolves returned and delisting was imminent, how would humans and wolves coexist? Educators found themselves in a quagmire of diverse and conflicting values. While wanting to include a biological understanding of the wolf, education was broadened to include moral and ethical considerations concerning wolves as well. Education may be able to change attitudes and specific environmental beliefs, but if it promotes attitudes that clash with people's basic ethics or values, it will not likely succeed (Gardner and Stern 2002).

In some cases, human attitudes have gone beyond appreciation and acceptance to idealization of the wolf. One example of this trend is that ownership of wolf–dog hybrids has grown in popularity, and many have caused nuisance problems (Hope 1994; Wisconsin DNR 1999). Urban residents may favor wolf protection, but not be able to understand the plight of a rancher who has lost calves to wolf depredation, and less willing to accept the need to occasionally use lethal control to remove problem wolves Educators face the dilemma of how to teach people to see wolves as wolves.

19.3   Approaches to Education About Wolves

Each program described here has its own set of goals, audiences, and methods. Educators do not agree on all the goals, appropriate methods, or the appropriateness of advocacy along with education. Our goal in this chapter is not to advocate for any particular program or philosophy of wolf education, but to illuminate programs and methods that have reached people and have thus made a difference in society's perception of the wolf.

19.3.1   An Individual Approach: Karlyn Berg

In 1968, when wolves were unprotected, Karlyn Berg began her wolf programs to expose the plight of wolves and to motivate people to take action to preserve wolves. With ambassador wolves, Karlyn was perhaps the first to travel throughout North America to schools, universities, clubs, museums, nature centers, and environmental events. In 1973, she and a team of people embarked on a wolf education effort in Minnesota, the state with the last remaining wolf population in the lower 48 states. They found that hostility against wolves was driven more by old beliefs and resentment, than by actual wolf conflicts. Berg became a resident of northern Minnesota and moved into the heart of the wolf country, where she focused her educational efforts, but continued to teach about wolves across the Midwest and in other portions of the country.

Karlyn expanded her ecological and biological programs to include history and discussions that she hoped might transform the negative attitudes and promote coexistence with the wolf. She discontinued bringing live wolves to programs when it became apparent this action was conveying wolves could make good pets, inspiring people to own one. In its place, Karlyn gathered historic, cultural, and biological artifacts creating a large traveling wolf display, a tool that would be exciting and tactile for any audience to experience.

Karlyn produced educational materials, curricula, and school programs. She created a Natural Science Museum and collaborated on numerous wolf projects and films, and in 1981, she was asked to be the exhibit consultant for The Science Museum of Minnesota's *Wolves and Humans* Exhibit. She wrote the accompanying educational materials for the exhibit using the format from her own wolf programs. These materials later served as the basis for the Wolf Education Learning Stations Box to teach about wolf ecology and biology which she designed so educators would not only have written curricula but also hands-on materials as well. Karlyn produced more than 85 boxes for state and federal agencies, and private education facilities.

Trying to coexist with wolves presents a complex challenge, and exploring this challenge remains a critical part of Berg's educational efforts. While Karlyn strongly prefers use of nonlethal methods to manage wolf–human conflicts, she accepts the necessity to sometimes use lethal controls to deal with depredating wolves, and includes this message in her talks and educational materials.

19.3.2   Wolf Park

In the early 1970s, a handful of scientists and environmentalists began to take steps to study and inform the public about wolf behavior. This included behaviorist Dr. Erich Klinghammer from Purdue University in Indiana. Klinghammer received two wolves from the Brookfield Zoo in Chicago and founded Wolf Park in Battle Ground, Indiana, a nonprofit organization dedicated to wolf conservation in 1972. Klinghammer knew the many misconceptions about wolves, so he invited curious and interested people to come and see his wolves up close and learn what a wolf is really like. Wolf Park is a 30-ha park that is home to socialized wolves, coyotes (*Canis latrans*), and red foxes (*Vulpes vulpes*) that serve as ambassadors for these animals that are difficult to observe in the wild. Klinghammer and his volunteer staff use behavioral concepts borrowed from Konrad Lorenz to socialize the captive animals so they are relaxed and not stressed around people. Through the years, students have taken advantage of the opportunity to study the captive wolves, allowing a more intimate view of the animal than is possible in the wild. Two prominent wolf ecologists, Dr. Rolf Peterson and Dr. Doug Smith, had early training at Wolf Park.

Incorporating behavioral research with public education, Klinghammer encouraged visitors to simply watch the wolves to get a sense of what this species is about. He believed that once a person could see a wolf just being a wolf, eating, howling, and socializing with members of its pack, they would see these beautiful and interesting animals as deserving respect. Each behavior a wolf displayed, whether scent marking, vocalizing, or showing pack dynamics, became an educational opportunity.

Wolf Park provides walking tours with trained guides almost daily during the park's open season. Seminars on topics including wolf behavior, ecology, and the relationship of wolves to dogs, are offered year round. Education has been geared more toward adult learners, but the Park has also expanded its education reach to include K-12 audiences. The children's program ranged from 2-h visits for busloads of students to day and overnight camps, drawing young people from the local area and adjacent states. In 10 years, Wolf Park hosted over 200,000 people. It is hoped that Wolf Park's combination of providing scientific information and the emotional impact of "getting to know" a wolf, improves people's perceptions of wolves, other wild species, and about nature, in general.

Wolf Park has modified its use of captive wolves. At one time wolves were taken off-site to do educational programs at schools, but this program was discontinued when staff discovered that the sight of a wolf on a leash led many audience members to the conclusion that wolves could make good pets. Therefore, wolves are now kept at the park in large, naturalistic enclosures. The Park now educates about the problems of raising wolf–dog hybrids and discourages their ownership.

Although a large portion of visitors are already fans of wolves, visitors' support for wolves increased after a visit to the Park (Black 2006) The Park provides wolf supporters with factual information so they themselves can speak in support of wolves and vote responsibly on regulations and policies affecting wolf conservation. Wolf Park also educates about wolf–human conflict issues, and how to manage wolf depredation of livestock. In seminars, Dr. Klinghammer expresses supports for the killing of wolves that are known to have killed livestock, and that it is in the species' best interest to do so.

Wolf Park is sometimes criticized for keeping captive wolves, but Park staffs believe that an ambassador wolf serves a worthy purpose of providing a chance for people to see this usually elusive animal up close. The Park has learned that such experience of seeing an actual wolf helps create greater respect for this animal. Wolf Park expects that education about wolves will continue to be important after delisting as more people have to learn how to live with wolves.

19.3.3 International Wolf Center

The International Wolf Center's (IWC) objective is simply to teach the world about wolves. Founded in 1985 by wolf biologist Dr. L. David Mech, the Center attempts to improve the survival of wolf populations by teaching about wolves, their relationship to wildlands, and the human role in their future. By fostering a citizenry that understands the biological and social dimensions of wolf issues, the IWC hopes people will make informed decisions about the wolf's future.

Each year 45,000 people visit the Center's interpretive facility in Ely, Minnesota and view the 550-m^2 exhibit, *Wolves and Humans: Coexistence and Conflict*, and a resident pack of ambassador wolves. The IWC provides classroom curricula, a quarterly magazine, high-tech distance learning outreach programs, symposia, an extensive Website, electronic newsletters, traveling exhibits, teacher workshops, and other educational activities.

The IWC strives to help populations of wolves to survive and help facilitate coexistence between wolves and humans. It emphasizes "populations" of wolves out of the belief that the needs of the population as a whole supersede the needs of individual animals. The Center focuses on "targeted areas," that is, states or recovery areas where there is a need for educational services that can be provided efficiently by the Center. Currently targeted areas of focus include Minnesota and southwestern USA.

The IWC focuses on providing science-based, accurate, objective information on wolves and attempts especially to share such information with teachers, the media, and natural resource professionals, who in turn are in contact with a much larger audience. The IWC estimates that a 1-day workshop that instruct, inspires, and provides resources to 30 teachers will eventually touch 1,000 students over the next 10 years. Further services for educators include several types of curriculum resources, wolf loan boxes, books, and videos suitable for the K-12 classroom, distance learning programs, a teacher newsletter, and special programs for school groups at the Center in Ely, related to the state's learning standards. Workshops include field sessions in which participants become involved in actually howling for wild wolves, tracking them in the snow, or observing them in places such as Yellowstone or the High Arctic. It is hoped that by giving teachers tools and confidence to include wolves in their curriculum, more youth will grow up with a science-based, objective view of wolves that will help them make informed decisions about wolves in the future, and support conservation of wolves.

To facilitate coexistence between wolves and humans, the IWC runs a Wolf Helpline that assists residents in northeast Minnesota to prevent and mitigate conflicts with wolves. Recently, area residents and tourists reported an increase in the number of wolves coming in close proximity to people. To encourage wolves to keep their distance from humans, Center staff members teach area residents strategies to avoid habituating wolves. These strategies include simple precautions, such as securing garbage and feeding pets indoors. Educational flyers, presentations, and radio public service announcements spread the word about ways to avoid conflicts with wolves.

The IWC works to balance its mission of advancing the survival of wolf populations with a commitment to objectivity toward wolves and wolf management issues. By respecting diverse perspectives, the Center encourages well-informed dialogue and discussion about the often volatile and value-laden topics of wolf–human conflict and coexistence. The Center promotes the message that people, both as individuals and collectively, are responsible for ensuring the long-term survival of wolves and the wildlands habitats where they best thrive.

19.3.4   National Wildlife Federation

The National Wildlife Federation (NWF) in Washington, DC worked at national levels for the passage of the Endangered Species in the late 1960s, when the gray wolf was disappearing across the lower 48 states. NWF's education efforts are geared at the national level through publications such as *National Wildlife Magazine*, *Ranger Rick*, and *Your Big Back Yard* and its Website reaching out to more than 4 million members annually. NWF's National Wildlife Week is celebrated at the end of April, and features endangered species, including the wolf. NWF's efforts keep people informed about wolf population health and viability, and encourage people to become proactive in issues that will affect the future of the wolf.

NWF works with a wide variety of special interest groups, and with state Wildlife Federation (WF) affiliates to secure a place for wolves in a healthy environment. Changing attitudes about wolves has been a very difficult process and is not always possible through education alone (Meadow et al. 2005), but attempts to improve attitudes toward wolves require a wide variety of educational tools.

To make learning come alive, NWF developed Wolf Trunks. Filled with pelts, skulls, scat, tracks, and more, these trunks are treasure chests of hands-on, real world learning for youth and adults. The trunks have been used as stand-alone resources or as tools to enhance classroom lesson plans on wolf biology, behavior and conservation, habitats, endangered species, predator–prey relationships, and more.

The NWF produced an Imax Production film, *Wolves*, for a large audience and it was shown at Imax and other giant-screen venues throughout the country for 2 years. The "Wolves Action Pack" a companion classroom activity guide to the movie was also produced. The movie *Wolves* is available for purchase on DVD.

Brochures have been developed by the NWF to teach people how to live with wolves, and to teach hunters how to minimize encounters between hunting dogs and wolves. NWF works with grassroots organizations, special interest groups, and key players to encourage conservation of wolves. Although wolves continue to be burdened with negative perceptions, NWF is developing new educational approaches to address how wolves can live with people and how this species will be managed without federal protection.

19.3.5   Timber Wolf Alliance

The Timber Wolf Alliance (TWA) was a program of the Sigurd Olson Environmental Institute at Northland College in Ashland, Wisconsin along the south shore of Lake Superior through spring 2008, when it transferred to the North Lakeland Discovery Center in Manitowish Waters, Wisconsin in the heart of the northwoods lake country. TWA focuses on educating citizens of the western Great Lakes region, especially in Wisconsin and Michigan, about wolves of the region.

In 1986, the Wisconsin Department of Natural Resources (WDNR) drafted a white paper stating facts about gray wolves and their potential recovery in the state, and responses by the general public (Thiel and Valen 1995). A majority of those who responded were antagonistic toward wolves. It was apparent that if the state was ever to see a viable wolf population, education was needed. In winter 1987, a group of people representing 11 environmental and educational organizations, as well as agency biologists, met to address this need. Over the course of several meetings, this group decided to create an organization that would disseminate information about wolves to the public, which became TWA.

TWA quickly grew, and by 1990 the Sigurd Olson Environmental Institute provided staff support and a home. TWA's mission has remained similar for the past 21 years to provide education programming based on strong science to promote sustainable populations of wolves in the region. TWA worked toward addressing all sides of controversial wolf issues, without taking specific sides except supporting use of science to address wolf management issues. TWA provided education to all age levels, and from various urban and rural backgrounds across the region. By meeting, working, teaching, and supporting people in their own communities within and outside of wolf range, TWA believes that a change in attitudes can take place.

TWA maintains a volunteer Speakers' Bureau, trained by staff, agency biologists, and volunteer coordinators with a goal to give wolf presentations in their own communities where they know and understand the particular interests and needs of the audience. The broad geographic distance across the Great Lakes region and diffused population makes community-based programming an effective tool for teaching local citizens about wolves. Currently, TWA volunteers provide programs to ~10,000 people annually in large and small communities around Wisconsin, Michigan, and northern Illinois. In partnership with federal and state agencies and private organizations (US Forest Service – Ottawa National Forest, Wisconsin DNR, and Wisconsin Trappers Association), TWA developed educational displays that travel to schools or are stationed at state parks where thousands of visitors can view them. TWA volunteers also use these displays at events tailored to specific groups such as hunters, loggers, and farmers.

In the late 1990s, Michigan DNR wildlife biologist Jim Hammill began a "hunter outreach" program during the fall firearm deer season, visiting remote hunting camps to inform hunters about wolves in their area (McLeer and Warren 2004). It has since become a program of TWA, relying on volunteers and wildlife agency personnel to share information on status and biology of wolves with deer hunters. The fall hunting season typically is the time when illegal killing of wolves is greatest. Volunteers traverse the back roads during the opening weekend of the hunting season. The intent is to greet hunters on their own turf, provide packets of factual information about wolves, answer questions, and ask if hunters are seeing wolf sign. By providing information and listening to concerns, volunteers try to respond to hostile attitudes, and hopefully discourage illegal shooting of wolves. Generally, each year's hunter outreach focuses on different areas and especially focuses in areas where illegal kills have occurred recently. A dozen or more volunteers reach nearly 1,000 hunters annually in Michigan and Wisconsin.

Training people to help survey wolves encourages stewardship toward wolves and the northern forest. In 1995, Wisconsin DNR biologists developed a volunteer tracking program for large and medium carnivores using citizens to augment data collected by wildlife professionals. TWA has contributed to this effort by hosting several training workshops around Wisconsin and Michigan for novice and advanced trackers, bringing in Jim Halfpenny to provide expert training in track identification and interpretation (Halfpenny 1986). These workshops encourage young students and retired professionals to spend time in winter searching for tracks of wolves and other forest carnivores. TWA also conducts summer workshops in which people are involved in howl surveys and searching for wolf sign in known wolf territories. As budgets tighten, using citizen volunteers as helpers in data collection not only benefits the state's pocketbook but also creates stewards for the wolf program, builds local support, and uses place-based knowledge (Nie 2003).

Since 1990, TWA has sponsored Wolf Awareness Week in the fall, highlighted by distributing engaging educational posters. Started as Wisconsin Wolf Awareness Week in October 1990, the program expanded to become a regional wolf awareness week in 1992, and expanded again in 1998 to both a regional and national Wolf Awareness Week. The regional Wolf Awareness Week focused on wolf issues in the Great Lakes region, while the national Wolf Awareness Week covered wolf issues from across the country. In recent years during Wolf Awareness Week, TWA has distributed 35,000 national educational posters around North America, and 35,000 regional posters within the Great Lakes region. The posters were supported by over 40 sponsors. Also during Wolf Awareness Week, special lectures by wolf experts were sponsored and special children's newsletters on wolves were distributed across the Great Lakes region.

TWA's work remains focused on the issues and needs of wolves in the upper Great Lakes region. Although wolves are no longer endangered in the region, teaching how to live with wolves, helping to resolve human–wolf conflicts, and teaching about the role of wolves in the ecosystem will continue to be important for maintaining long-term viability of wolf populations in the region.

19.3.6   Wildlife Science Center

The Wildlife Science Center (WSC), a private, nonprofit organization, was established in 1991 in Forest Lake, Minnesota after funding ceased for the Wolf Project, a federal program dedicated to physiological and behavioral research of wolves. Wildlife biologist Peggy Callahan, a specialist in chemical and nonchemical capture and immobilization techniques, had managed a colony of captive gray wolves at the facility. Without federal funds, Callahan and fellow biologists faced a choice of euthanizing the wolves or adopting them out to zoos.

Instead, Callahan decided to create the WSC. In 1994, after 3 years of intensive preparation and program development, the center opened to the public. In addition to hosting tours and special events, the facility offers educational outreach programs based upon the National Science Standards for K-12, research opportunities for scientists, and hands-on training for wildlife professionals.

The WSC is best known for its population of gray, red (*Canis rufus*), Mexican gray (*Canis lupus baileyi*), and hybrid wolves (*C. lupus* × *C. familiaris*). Its resident wildlife also includes other native carnivores, porcupines (*Erethizon dorsatum*), New Guinea Highland dogs (*Canis familiaris hallsrtomi*), and raptors such as hawks, owls, and falcons. Teaching ranges from elementary school students learning about scientific methods to wildlife biologists studying wolf genetics. Research opportunities are available for amateur naturalists and professional scientists alike. The center also participates in the Species Survival Plan (a program of the Association of Zoos and Aquariums) for the red wolf and Mexican gray wolf. Both species were considered extinct in the wild and relied on captive facilities such as the WSC for their survival.

In addition, the facility provides training in wildlife handling for animal control officers, zoo professionals, veterinary students, and others. They receive instruction in chemical immobilization, veterinary emergency response, and animal handling techniques.

As wolf management shifts from issues of recovery to population management, wildlife biologists and managers face new challenges and research needs. Wolf biologists and managers have an obligation to share their findings with the educators who interface with the public, and educators have an obligation to seek out the latest available scientific information. The WSC provides a facility where researchers and managers can interface with educators to make sure that findings on wolf management and research are broadly disseminated.

The WSC believes education centers are obligated to disseminate the best possible information available. Education about wolves is considered a conduit to other subjects such as ecology, literature, politics, math, and geography. The educational staff, made up of licensed educators, develops and conducts programs with a systems approach.

The wolf program at the WSC incorporates the biological, historical, and cultural role of the wolf in the environment. The student-driven curriculum involves inquiry-based learning and hands-on activities, with a focus on real-world issues. In a time where students are lacking interest in science and spending less time outdoors (Louv 2006), the WSC teaches people about their role in nature and encourages them to take part in the conservation of wolves on the landscape.

19.4   Summary and Conclusions

Attitudes toward wolves began to improve by the 1970s, probably due to education about the scientific research on wolves in the 1930s–1960s, and the environmental movement beginning in the 1960s (Williams et al. 2002). Despite better acceptance of wolves, human attitudes still seemed to be a major impediment to recovery in the Midwest in the 1980s (Thiel and Valen 1995). The 1980s–1990s, when wolf educational organizations developed and grew in the Great Lakes region, was also the period of most rapid growth of wolf populations in the region (Erb and DosCarlos, this volume; Beyer et al., this volume; Wydeven et al., this volume). While it is not possible to attribute all this population growth to educational efforts, it is reasonable to assume that educational programs were one of the factors that allowed growth and spread of the wolf populations. There appeared to be growing respect for wolves and intolerance for past management practices such as systematic removal of wolves (Jickling 1996). It appears that some of the hostile convictions about the wolf have been replaced by greater ecological understanding and acceptance of carnivores.

A challenge that educators continue to face is that despite overall improvement in attitudes toward wolves, people living close to wolves in rural are still less likely to be supportive of wolf conservation (Williams et al. 2002). If this pattern continues, this means as wolves spread out across the landscape and occupy more areas where people live, negative attitudes could spread in localized areas. Knowledge by itself does not guarantee improved attitude toward wolves and often people with diametrically opposed attitudes score highest on species knowledge; generally, personal experiences and peers have more impact on attitudes (Meadows et al. 2005). As people experience depredations or associate with people who have experienced wolf depredations, they are more likely to develop negative attitudes toward wolves (Naughton-Treves et al. 2003). Wolf educators such as Karlyn Berg and Wolf Park in the past have provided positive personal experiences by bringing wolves to people. While this may have improved attitudes for some, it also had the unexpected consequence of encouraging use of wolves or wolf–dog hybrids as pets. Facilities such as Wolf Park, the International Wolf Center, and the Wildlife Science Center continue to use ambassador wolves to encourage positive attitudes about wolves, but within highly controlled captive situations. Karlyn Berg, the National Wildlife Federation, and the International Wolf Center have produced wolf educational boxes that also try to create positive personal experiences associated with wolves. The Timber Wolf Alliance and International Wolf Center take people into wolf habitat and attempt to create such positive connections by encountering wolves through their howls, tracks, and occasional observations. While more highly educated people tend to be more positive toward wolves (Naughton-Treves et al. 2003), providing positive experiences that connect people with wolves is also critical to maintaining a public willing to tolerate wolves on the landscape.

In general, wolf education messages are likely to become more nuanced and complex (Grooms 2004). Education in support of wolf conservation will need to continue to include information on wolf biology, but people should also be taught about the need of wolf controls, options available, and other aspects of wolf management. While educators have been successful in creating more respect for and acceptance of wolves, it is critical to also strive for respect for those who have suffered depredation losses to wolves, and those who hold different attitudes and opinions about wolf management. Wolf educators may also need to learn more and share information on ethics and values (Williams 2004).

The success of wolf recovery can serve as a good example to educate people about endangered species recovery, and wildlife restoration in general. As a top predator, the wolf lends itself well to discussion of prey populations and the health of plant communities. Worldwide, wildlife is being affected by factors such as habitat destruction, global warming, and mass extinctions. The successful recovery of wolves in the north woods of the Great Lakes region can provide hope and inspiration for wildlife restorations in other locations. As a large, magnificent predator, wolves can connect people more closely to nature and create greater ecological consciousness. Creating stronger connections to nature is especially important for children that often have minimum opportunities to learn about nature in modern society (Louv 2006). As stated by Louv (2006)

> "Healing the broken bond between our young and nature is in our self interest, not only because aesthetics or justice demand it, but also because our mental, physical, and spiritual health depends on it. The health of the Earth is at stake."

As we look to the future, education will need to be less of a lesson plan on wolf ecology and more about how we value the wolf and the environment. Education will be necessary for coexistence between wolves and humans to occur. As stated by Peggy Callahan of the Wildlife Science Center, "The survival of the wolf and other organisms relies on three factors: political, social, and biological. Biologically, the wolf can survive, but politics and social perspectives revolve around knowledge. Education is the key to the wolf's future."

References

Black, P. 2006. Evaluating the effectiveness of wolf education programs at Wolf Park. MS Thesis, Tufts University, Center for Animals and Public Policy, North Grafton, MA.

Crisler, L. 1958. Arctic Wild. New York, NY: Harper and Brothers.

Errington, P. L. 1946. Predation and vertebrate populations. Quarterly Review of Biology 21: 144–177, 221–245.

Fritts, S. H., Stephenson, R. O., Hayes, R. D., and Boitani, L. 2003. Wolves and humans. In Wolves, Behavior, Ecology, and Conservation, eds. L. D. Mech and L. Boitani, pp. 289–316. Chicago, IL: University of Chicago Press.

Gardner, G. T., and Stern, P. C. 2002. Environmental Problems and Human Behavior, Second Edition. Boston, MA: Pearson Custom Publishing.

Grooms, S. 2004. We're not in Kansas anymore: The rapidly changing world of wolf education. International Wolf 14 (3): 4–6.

Halfpenny, J. 1986. A Field Guide to Mammal Tracking in North America. Boulder, CO: Johnson Publishing Company.

Hope, J. 1994. Wolves and wolf-dog hybrids as pets are big business – But a bad idea. Smithsonian 25 (3): 34–45.

Jickling, B. 1996. Wolves, ethics, and education: Looking at ethics and education through the "Yukon Wolf Conservation and Management plan." In A Colloquium on Environment, Ethics and Education, ed. B. Jickling, pp. 158–163. Whitehorse, YT: Yukon College.

Kellert, S. R., Black, M., Rush, C. R., and Bath, A. J. 1996. Human culture and large carnivore conservation in North America. Conservation Biology 10: 977–990.

Leopold, A. 1949. Thinking like a mountain. In A Sand County Almanac, pp. 129–133. New York, NY: Oxford University Press.

Lopez, B. H. 1978. Of Wolves and Men. New York, NY: Charles Scribner's Sons.

Louv, R. 2006. Last Child in the Woods. Chapple Hill, NC: Algonquin Books.

McLeer, D., and Warren, N. 2004. Hunter outreach: Face to face education. International Wolf 14 (3): 11.

Meadow, R., Reading, R. P., Phillips, M., Mehringer, M., and Miller, B. J. 2005. The influence of persuasive arguments on public attitudes toward a proposed wolf restoration in the southern Rockies. Wildlife Society Bulletin 33: 154–163.

Mowat, F. 1963. Never cry wolf. New York, NY: Dell Publishing Company.

Murie, A. 1944. The Wolves of Mount McKinley. U.S. National Park Service, Fauna Series No. 5.

Naughton-Treves, L., Grossberg, R., and Treves, A. 2003. Paying for tolerance: Rural citizens' attitudes toward wolf depredation and compensation. Conservation Biology 17: 1500–1511.

Olson, S. F. 1938. A study in predatory relationship with particular interest to the wolf. Science Monthly 66: 323–336.

McIntyre, R. (ed.). 1995. War Against the Wolf: America' Campaign to Exterminate the Wolf. Stillwater, MN: Voyageur Press, Inc.

Nie, M. A. 2003. Beyond wolves: The politics of wolf recovery and management. Minneapolis, MN: University of Minnesota Press.

Thiel, R. P., and Valen, T. 1995. Developing a state timber wolf recovery plan with public input: The Wisconsin experience. In Ecology and Conservation of Wolves in a Changing World, eds. L. N. Carbyn, S. H. Fritts, and D. R. Seip, pp. 169–175. Edmonton, AB: Canada Circum Polar Institute.

Williams, J. 2004. At the crossroads: Toward a new era in wolf education. International Wolf 14(3): 7–9.

Williams, C. K., Ericsson, G., and Heberlein, T. A. 2002. A quantitative summary of attitudes toward wolves and their reintroduction (1972–2000). Wildlife Society Bulletin 30: 575–584.

Wisconsin DNR. 1999. Wisconsin Wolf Management Plan. PUBL-ER-099 99. Madison, WI: Wisconsin Department of Natural Resources.

Chapter 20
The Role of the Endangered Species Act in Midwest Wolf Recovery

Ronald L. Refsnider

20.1 Introduction

In its 1978 Tellico Dam Snail Darter ruling, the US Supreme Court called the Endangered Species Act (ESA) of 1973 (available at http://www.fws.gov/endangered/esa/content.html) "the most comprehensive legislation for the preservation of endangered species ever enacted by any nation." For those who have been involved in its implementation for a decade or more, this statement certainly rings true. However, we are also well aware of its weaknesses, inconsistencies, the difficulty with which it can be applied to complex biological situations, and, above all, its openness to citizen involvement and litigation. In this chapter, I examine the listing, recovery, and delisting of the gray wolf (*Canis lupus*), which serves as an informative case history demonstrating many aspects of the ESA.

20.2 The Endangered Species Act – An Overview

The ESA directs the Secretary of the Interior and the Secretary of Commerce (delegated to the US Fish and Wildlife Service [FWS] and National Marine Fisheries Services [NMFS], respectively) to identify species currently in danger of extinction or likely to be so in the foreseeable future, to protect them and their critical habitats, to develop and implement recovery plans, to "delist" them when they are "recovered" or become extinct, and to monitor recovered species after delisting. I will describe how these, and other, features played a role in the recovery and delisting of the gray wolf in the Midwest.

20.2.1 Listing – Adding Species to the List of Endangered and Threatened Wildlife and Plants

The ESA specifies that species can be listed as either endangered or threatened. An endangered species is defined (ESA section 3(6)) as one "in danger of extinction

A.P. Wydeven et al. (eds.), *Recovery of Gray Wolves in the Great Lakes*
Region of the United States,
DOI: 10.1007/978-0-387-85952-1_20, © Springer Science+Business Media, LLC 2009

throughout all or a significant portion of its range." A threatened species (ESA section 3(19)) is "likely to become an endangered species within the foreseeable future throughout all or a significant portion of its range."

The ESA's definition of "species" goes beyond the spectrum of definitions used by biologists, and different aspects of this definition have been used to list and delist the gray wolf. The definition of a species (ESA section 3(16)) includes "any subspecies of fish or wildlife or plants, and any *distinct population segment* of any species of *vertebrate* fish or wildlife which interbreeds when mature" (emphasis added). Thus, the ESA allows not only the listing of the taxonomic units of species and subspecies, but also the nontaxonomic vertebrate-only entity that has come to be known as a "distinct population segment," or DPS. Under the ESA, a listed DPS must be treated the same as a listed species or subspecies.

DPSs are not defined in the ESA, so in 1996 the US FWS and the NMFS jointly developed a policy (commonly called the "DPS Policy"; available at http://www.fws.gov/endangered/policy/Pol005.html) that guides the two agencies in the listing and delisting of DPSs. It describes the two criteria that a DPS must meet: (1) it must be discrete, that is, it must be markedly, but not necessarily completely, separate from other populations of the larger biological taxon; and (2) it must be significant to the larger taxon of which it is a part. However, the DPS Policy is silent on where to place the boundaries of a DPS, other than the implicit requirement that the boundaries must be located where they do not violate the discreteness and significance criteria.

The ESA lays out the steps that are taken to list species as endangered or threatened, ensuring that the interested public and peer reviewers are involved in the process. All listing decisions must be published as proposals in the *Federal Register* (FR) and must include a public comment period of at least 60 days. If requested, the FWS must hold at least one public hearing during the comment period. The final rule (the decision) must be published in the FR within 12 months of the proposal.

The ESA requires that listing decisions be made "solely on the basis of the best scientific and commercial data available" (ESA section 4(b)(1)(A)). "Commercial data" refers to data indicating that the species may be threatened by commerce involving the species, not to the economic impacts of a listing decision. Economic impacts, social factors, or political considerations cannot be part of a listing decision.

The ESA (section 4(a)(1)) requires that the scientific and commercial data be used to determine if a species is an endangered or threatened species as a result of one or more of five categories of factors of threats. This five-factor analysis requires the FWS to consider (1) threats to habitat; (2) overutilization of the species due to commercial, recreational, scientific, or educational use; (3) disease and predation; (4) the inadequacy of regulatory mechanisms; and (5) other natural or manmade factors. These same factors are used for all listing, reclassification, and delisting decisions; however for delisting proposals, the analysis should look ahead to the threats that the species would reasonably be expected to experience once the ESA's protections are removed.

20.2.2   Critical Habitat

Critical habitat is defined by the ESA (section 3(5)) as "the specific areas … occupied by the species … on which are found those physical or biological features, (I) essential to the conservation of the species, and (II) which may require special management consideration or protection" and those areas unoccupied by the species that "are essential for the conservation of the species." In contrast to listing decisions, critical habitat designation decisions must be made "on the basis of the best scientific data available and after taking into consideration the economic impact, the impact on national security, and any other relevant impact…" (ESA section 4(b)(2)).

A designation of an area as critical habitat has a single function under the ESA – to prevent the actions of federal agencies from destroying or "adversely modifying" the critical habitat (ESA section 7(a)(2)). When a federal agency proposes an action in an area where it might adversely modify or destroy designated critical habitat, the agency must consult with the FWS to minimize the effects on the species. The designation of critical habitat has no impact on the activities of private citizens unless there is a federal nexus, for example, federal funding or the need for a federal permit or approval for the action.

20.2.3   Special Regulations Under Section 4(d) of the ESA

For threatened, but not for endangered species, the ESA allows its normal protections to be changed in any manner "deemed necessary and advisable for the conservation of such species" (ESA section 4(d)). These special regulations or rules can increase the protection beyond what the threatened species would otherwise have, or they can decrease the normal protections. In most cases, these "4(d) rules" are written to reduce protections in order to reduce the conflicts that result from the increasing numbers or expanding range of the threatened species. Being able to effectively deal with such conflicts can reduce opposition to the species' recovery, thus promoting the overall conservation of the species. Additionally, 4(d) rules that reduce or eliminate the normal federal restrictions for actions that are not impacting the species allows the FWS to focus its efforts on the truly important threats.

20.2.4   Recovery – Plans, Teams, and Criteria

Section 4(f) of the ESA requires the FWS to "develop and implement plans … for the conservation and survival of endangered and threatened species…." These "recovery plans" must describe the management actions necessary to achieve conservation and survival of the species and must contain "measurable criteria which, when met, would result in a determination ... that the species be removed from the list..." (ESA section 4(f)(1)(B)). In short, recovery plans must specify "recovery criteria" and list the tasks that are believed necessary to achieve those criteria. The ESA does not specify whether recovery plans are to be written by FWS or an outside expert or a "recovery team" of experts; the FWS has successfully used all three methods. Although not required by the ESA, FWS recovery plans contain an implementation table that prioritizes recovery tasks and lists the appropriate agency or organization to carry out each task. This brings additional parties into the recovery program even if they are not represented on the recovery team, and helps coordinate tasks and speeds recovery.

20.2.5   Protection

One of the key, and often misunderstood, components of the ESA is the protection it provides to threatened and endangered animals. (The ESA's protections for plants are significantly less than those for animals, and will not be discussed here.) The key protection can be stated as prohibition of take. The ESA prohibits the take of endangered animals within the United States or its territories or upon the high seas (ESA section 9(a)(1)(A)). By regulation the FWS has extended a similar prohibition to the take of threatened animals (50 Code of Federal Regulations (CFR) 17.31(a)).

The key word "take" is broadly defined in the ESA and is further defined in regulations. "Take" means to harass, harm, pursue, hunt, shoot, wound, kill, trap, capture, or collect, or to attempt to engage in any such conduct (ESA section 3(18)).

The regulations for implementing the ESA allow the take of endangered and threatened animals under certain circumstances (50 CFR 17.21(c), 17.22, 17.31, and 17.32). Individuals of endangered or threatened wildlife species can be taken by any person "to protect himself or herself, a member of his or her family, or any other individual from bodily harm" (ESA section 11(a)(3); see also 50 CFR 17.21(c)(2)). The FWS generally has interpreted this to mean that an individual animal can be killed if it is attacking a person in a manner that can reasonably be expected to result in the death or serious injury of the person. Additionally, certain state and federal agency personnel can take individuals "constitute a demonstrable but nonimmediate threat to human safety" (50 CFR 17.21(c)(3)).

Except for the "in defense of human life" provision, endangered and threatened animals are protected from "take" by private citizens. While designated employees of state and federal agencies can take endangered and threatened animals under restricted circumstances (50 CFR 17.21(c)(3) and (c)(4); 17.31(a) and (b)), a private citizen must have a permit from FWS to take the same animals. Such endangered species take permits can be issued only for scientific purposes, enhancement of propagation or survival, or for take that is incidental to otherwise lawful activities (50 CFR 17.22). Take permits for threatened species can be issued for the same purposes and also can be issued for economic hardship, zoological exhibition, educational purposes, or special purposes consistent with the purposes of the ESA (50 CFR 17.32(a)(1)).

The penalties to individuals for criminal violations of certain aspects of the ESA, including knowingly violating the take prohibitions for endangered animals and plants, included a fine of up to $50,000 and imprisonment for not more than a year. Fines and imprisonment were up to $25,000 and not more than 6 months for illegally taking a threatened species (ESA section 11(b)). However, the Fines Enhancement Act of 1984 doubled the fines for endangered species take violations by individuals to $100,000, with the maximum prison term remaining at 1 year. It also established a fine of up to $200,000 for organizations that illegally take an endangered species.

20.2.6   Protection from Jeopardizing Federal Agency Actions – Section 7

Section 7(a)(2) of the ESA protects threatened and endangered species, and their designated critical habitat, from excessive adverse impacts resulting from the actions of federal agencies. Federal agencies are required to consult with the FWS if their actions are likely to adversely affect threatened or endangered species so that the FWS can assess the likely impacts, estimate the extent of incidental take, and specify measures to minimize the impact of the incidental take. If the FWS determines that the action is likely to jeopardize the continued existence of a listed species, the FWS, in consultation with the agency, formulates an alternative project that will avoid jeopardizing the species but still achieve the primary purpose of the original project. The FWS also reviews the project for potential impacts to critical habitat; destruction or adverse modification of critical habitat is prohibited by the ESA. Section 7 also requires that federal agencies use their authorities to promote the conservation of listed species (ESA section 7(a)(1)).

20.2.7   Postdelisting Monitoring

When a species has been "delisted" it loses the protections previously provided by the ESA. However, the FWS is required by section 4(g) to monitor a recovered species' status for a minimum of 5 years after delisting. The FWS is directed to make prompt use of the ESA's emergency listing authority if necessary to ensure the well-being of the species. An emergency listing (ESA section 4(b)(7)) can take effect immediately upon its publication in the FR and provides the full protections of the ESA for 240 days, giving the FWS time to further evaluate the threats and to conduct the normal listing process, if appropriate.

20.2.8   Citizen Lawsuits

The ESA (section 11(g)) allows any person to initiate a civil suit to stop a person or any government agency from violating any provision of the ESA, or to force the FWS to take certain actions that are nondiscretionary, such as listing/delisting actions or critical habitat designations. Such lawsuits must be preceded by a 60-day notice of intent to sue.

20.2.9   Secondary Benefits Stemming from Listing As Threatened or Endangered

In addition to this spectrum of statutory benefits that are directly provided by the ESA and its implementing regulations, there are several important secondary benefits that a species may gain from being listed as threatened or endangered.

Listing is a stimulus for conservation actions by other federal and state agencies and nongovernmental organizations (NGOs). Agencies and NGOs often increase their public education efforts and their funding for conserving, researching, and protecting newly listed species, thereby going beyond what is required of them by the ESA. For example, some state Departments of Natural Resources (DNRs) place higher priority on acquiring land if it is occupied by a listed species. Other organizations may fund or conduct specific conservation actions on their land base, even if the organization and the action are not specifically identified in the species' recovery plan.

Increased awareness by the public and the consequent grass roots involvement are other important benefits of being listed as threatened or endangered. This awareness can result in public pressure being exerted on elected officials and other decision makers (even in private business) to protect habitat or to undertake other conservation actions. It can also supply additional funding for recovery activities and provide substantial volunteer assistance to agency and NGO recovery actions. Such secondary benefits of listing can be key components of successful recovery programs.

20.3   The ESA in Action – Guiding and Constraining Wolf Recovery and Delisting

The rest of this chapter summarizes the stages of wolf recovery in the Midwest and describes how the ESA directed and, in some cases, limited agency actions. Due to space limitations, several lawsuits, court rulings, and petitions to delist Midwestern wolves are not discussed.

20.3.1   Original Listing of the Gray Wolf and Changes in 1978

Under the ESA of 1973, the gray wolf was first listed as endangered in 1974 and 1976. Those initial listings covered the four subspecies believed to still exist in the wild in the United States. The "eastern timber wolf" (ETW, *Canis lupus lycaon*) and the "northern Rocky Mountain wolf" (*C. l. irremotus*) were both listed in 1974 (39 FR 1158). The Mexican wolf (*C. l. baileyi*) and the "Texas gray wolf" (*C. l. monstrabilis*) were listed in 1976 (41 FR 17736 and 41 FR 24062, respectively).

Upon its listing as the ETW in 1974, the remnant wolf population in northeastern Minnesota began to experience the benefits of the ESA's protection. Human-caused mortality had already declined with the ending of the state bounty on wolves a decade earlier, and the addition of the ESA's large fines and jail sentences further reduced human-caused mortality. This led to an increase in wolf numbers, an expansion in occupied range, and an upward trend in wolf attacks on livestock in northern Minnesota (Ruid et al., this volume).

Although wolf numbers were not formally quantified in 1974, by 1976 it was apparent to the FWS that the Minnesota population was no longer on the brink of extinction and did not warrant endangered status. A total of 1,235 wolves were estimated in Minnesota in 1978–1979 (Fuller et al. 1992). The endangered status did not allow for the necessary control of wolves that attacked livestock in the state.

To address these and several other issues, the FWS made four changes to the listing of the gray wolf in 1978: (1) The agency (43 FR 9607) discarded the listing of gray wolf subspecies, and instead listed all gray wolves *(Canis lupus)* across the lower 48 states and Mexico. (2) Minnesota gray wolves were reclassified from endangered to threatened, whereas wolves in the other 47 states and Mexico retained their listing as endangered (wolves were not listed in Alaska and Canada where populations remained high). (3) FWS established five management zones in Minnesota (see below), and designated critical habitat for wolves in three of these zones in northeastern Minnesota that were considered primary wolf range, as well as Isle Royale National Park, Michigan. (4) Provisions of section 4(d) of the ESA were used to establish special regulations for Minnesota wolves to deal more effectively with wolves that attacked livestock, allowing employees or agents of the Minnesota DNR or the FWS to take wolves "committing significant depredations on lawfully present domestic animals" in zones 2 through 5. In 1985, the FWS revised the 4(d) rule to restrict taking for depredation control to ½ mile (0.8 km) from depredation sites and required the release of any young of the year captured on or before August 1.

20.3.2   1978 Eastern Timber Wolf Recovery Plan

As directed by the ESA, FWS developed a recovery plan that established recovery goals and described the tasks to achieve them. As done with several other wide-ranging species (e.g., bald eagle [*Haliaeetus leucocephalus*] and peregrine falcon [*Falco peregrinus*]), the FWS ultimately developed several regional gray wolf recovery plans that addressed the unique circumstances, opportunities, and limitations found in large subareas of the listed range. In 1975, the FWS established the ETW Recovery Team, named for the subspecies believed at that time to have historically ranged from Minnesota east to Maine and southeast to the northeastern corner of Florida. Team members were chosen for their expertise in wolf biology or as a representative of the major federal land management and state natural resource management agencies whose involvement would be crucial to a successful recovery program. The plan was approved in 1978 and describes a primary objective to "maintain and reestablish viable populations of the Eastern Timber Wolf in as much of its former range as is feasible" (USFWS 1978, p. 10). It identifies two main objectives of smaller scope, and whose achievement would indicate successful recovery of this wolf entity: (1) to ensure the survival of the animal in Minnesota by highly regulated management and (2) to attempt reestablishment of at least one viable population of Eastern Timber Wolves outside Minnesota and Isle Royale, Michigan. The plan described areas having potential for wolf reestablishment in northern Wisconsin, the Upper Peninsula of Michigan, Maine, New Hampshire, New York's Adirondack Forest Preserve, the central Appalachians in Virginia and West Virginia, and the southern Appalachians extending from southwestern Virginia into northern Georgia.

The 1978 plan also recommended the establishment of five management zones in Minnesota, of which Zone 1 was to be a wolf sanctuary, and Zones 2 and 3 were to be "managed sanctuaries" with a population goal of one wolf per ten square miles (26 km^2) in Zones 2 and 3. The plan considered Zones 1, 2, and 3 to be primary wolf range and stated that attempts to maximize the wolf population should be restricted to these areas. The plan indicated the possible need for some form of wolf population reduction if the wolf population reached a level that was reducing the prey base. Zone 4, considered to be peripheral range, was to be managed at a wolf density of about one wolf per 50 square miles (130 km^2), and the plan recommended a hunting and trapping season to maintain that wolf population density. Zone 5 was not considered a wolf management area, but the plan notes that wolves occasionally would disperse into it. A wolf control program was recommended for Zones 2 through 5 in which government agents would remove wolves that had attacked domestic animals.

Thus, the 1978 recovery plan contained the first gray wolf recovery criteria – to ensure the survival of the Minnesota wolf population and to develop a second viable gray wolf population outside of Minnesota and Isle Royale. The plan did not contain reclassification criteria (i.e., to change the ESA status from endangered to threatened), nor did it describe the necessary characteristics of the second population in order for it to be considered "viable."

In a follow-up letter to the FWS (Ralph E. Bailey, ETW Recovery Team Leader, to Harvey K. Nelson, FWS Regional Director, dated September 15, 1981), the Recovery Team addressed two unclear issues of the 1978 recovery plan: (1) the apparent inconsistency between the geographically broad primary objective and the two narrower main objectives, and (2) the lack of a definition for the term "viable population." That letter said that the primary objective – to maintain and reestablish viable populations of the ETW in as much of its former range as is feasible – would be accomplished when the survival of the wolf in Minnesota would be assured and when at least one viable population was established outside Minnesota and Isle Royale (which were the "main objectives" of the 1978 plan). The letter went on to say that this one additional viable population would be achieved when (1) a self-sustaining wolf population of at least 200 was established, or (2) when a population of at least 100 wolves is established within 100 miles (161 km) of another viable population. The FWS distributed this letter to recipients of the 1978 plan on October 19, 1981 (Daniel Bumgarner, FWS Assistant Regional Director, memorandum to holders of ETW Recovery Plan).

The 1978 ETW Recovery Plan, along with the 1981 letter from the Recovery Team, established the target of the ETW recovery program, provided specific guidance on where and how wolves should be managed in Minnesota, and identified areas throughout the eastern United States that should be considered for establishing at least one additional viable wolf population. Additionally, the plan identified a series of steps to be taken if reintroduction was necessary to establish the additional wolf population. However, in the late 1970s, wolves were already moving into northwestern Wisconsin from the expanding wolf population in Minnesota (Mech and Nowak 1981) and forming the nucleus of a wolf population within 100 miles (161 km) of the Minnesota population. This natural recolonization removed the need to reintroduce wolves elsewhere in the eastern US to achieve the Recovery Plan's second recovery criterion.

20.3.3 Recovery Plan Revision in 1992

In the early 1990s, the ETW recovery program was making significant progress. The Minnesota wolf population was estimated to be 1,500 or 1,750 animals by two independent estimates (Fuller et al. 1992), Wisconsin wolves had overcome the severe impacts of canine parvovirus and the population was on an upward trend (Wydeven et al. 1995), and there was evidence of wolf pup production in Michigan for the first time since the 1950s (Hammill 1992). In addition, wolf biologists had learned a great deal from many research efforts. Finally, the southeastern states were now recognized as being outside the historical range of gray wolves (historically being occupied by a separate species, the red wolf, *Canis rufus*), and several other potential recovery areas had been excluded from serious consideration. The 1978 recovery plan had become outdated in many ways, and the FWS reconvened the recovery team to revise the original plan and to develop "reclassification criteria" that, when achieved, would indicate that wolves in Wisconsin and/or Michigan could be reclassified to threatened status.

The FWS approved a revised recovery plan in early 1992 (USFWS 1992). While it carried forward the same "primary objective" of maintaining and reestablishing populations of the ETW in as much of its former range as is feasible, it clearly specified that the ETW will be recovered when the survival of wolves in Minnesota is assured and at least one other viable population is established outside of Minnesota and Isle Royale. The definition of a viable population remained as specified in the Recovery Team's 1981 letter to the FWS, with the additional requirement that wolf numbers remain above the minimum population levels for five consecutive years. Therefore, the 1992 plan largely adopted the endpoints that were established by the 1978 plan and the 1981 letter by Bailey.

The 1992 plan identified a criterion that could trigger reclassification of Wisconsin wolves to threatened status – a population of at least 80 wolves for three successive years (USFWS 1992), which was also part of the first state recovery plan (Wisconsin DNR 1989). Although the 1992 plan did not clearly apply the same criterion to Michigan wolves, it subsequently was accepted that Michigan wolves similarly could be reclassified when they reached a population of 80 for three successive years.

The 1992 plan also contained discussions about the threats posed to wolves from high road density (increasing wolf exposure to vehicle collisions, humans, and human activities) and from diseases and parasites (USFWS 1992). Canine parvovirus and sarcoptic mange had arisen as significant causes of mortality in Wisconsin and Michigan since the 1978 plan was completed.

20.3.4 Mid-1990s Planning for Reclassification and Delisting

By 1995, both the Wisconsin and Michigan (excluding Isle Royale) wolf populations reached 80 wolves, prompting the FWS to consider the steps needed for reclassification to threatened status in the next few years. The rate of population increase indicated that the delisting criteria also would be achieved the year after the reclassification criterion would be met, so delisting was considered as well. However, the FWS was faced with a novel and complicated national recovery situation for the gray wolf – the existence of three independent recovery programs for different geographic areas of the same listed entity. It was clear that the ETW Recovery Program was well ahead of the Northern Rocky Mountains (NRM) Recovery Program, which in turn was outdistancing the Mexican (Southwestern) Wolf Recovery Program. While it did not seem reasonable to require that all three recovery programs achieve their separate recovery criteria before delisting could occur for any gray wolf entity, it wasn't clear how to delist or reclassify one gray wolf recovery program independently of the other two programs.

For several years, through 1997, biologists and managers representing several FWS regions and L. David Mech (of the US Geological Survey) discussed and drafted a national recovery strategy for the gray wolf. The driving reason for this effort was a belief that the separate recovery programs could be reclassified or delisted independently if FWS would show that separate, but not simultaneous, achievement of three sets of recovery criteria would constitute recovery of the larger listed entity, the gray wolf in the 48 states and Mexico. At the same time other FWS and NMFS biologists and managers were independently developing the DPS Policy.

Once finalized, the DPS Policy made it clear to the FWS national gray wolf strategy group that the designation of gray wolf DPSs was an appropriate way to address varying levels of threats and differing degrees of recovery progress that were being made within a single listed entity. The FWS wolf strategy group proceeded to consider how many gray wolf DPSs should be designated and where their boundaries should be drawn. While the DPS Policy provided precise criteria to be used to determine whether a population of vertebrates qualified as a DPS, it provided no guidance on where the geographic boundary around or between DPSs should be placed. Ultimately, FWS concluded that all gray wolf DPSs should be designated at one time, and they should be laid out in a way that placed the entire listed entity (the 48 states and Mexico) in one of the DPSs, leaving no portion of the listed entity in a "non-DPS remnant area." Such a coordinated DPS designation would demonstrate that the FWS' obligation to recover the species as listed in the 48 states and Mexico was being achieved by the separate recovery programs and would not be curtailed in any way by downlisting or delisting the individual DPSs as their recovery progressed.

In late 1997, the FWS switched its focus from completing a national gray wolf strategy to developing a national reclassification and delisting proposal. The proposal was also based on the DPS policy, so in many ways it was the implementation of the unfinished national strategy. Similar to previous discussions, the FWS concluded that all of the listed gray wolf range must be included in one of the DPSs that were to be designated by this national rulemaking.

FWS endangered species and wolf biologists continued debating various boundaries for gray wolf DPSs into mid-1998. While there was general agreement within the group that there should be one DPS for each of the three existing gray wolf recovery programs, there was no consensus on whether a fourth DPS should, or even could, be established for the northeastern states. Some members of the group believed that a DPS could only be designated if there was a population of gray wolves already known to exist in the area. At the time (and into 2008) there was no conclusive evidence that a wild wolf population existed in the Northeast. However, at that time there were documented occurrences of several individual wolves in the Northeast, and it seemed likely that the additional searches that would be triggered by a proposal to establish a Northeastern DPS would produce evidence of a wolf population there. Additionally, the FWS believed that any proposal of four DPSs could be scaled back to a final rule with only three DPSs if public comments or lack of data confirming a wolf population made a Northeastern DPS unjustifiable. At that point a four-DPS approach gained the support of the FWS Director, and the FWS Minneapolis Regional Office was selected to draft the proposed designation of the four DPSs.

20.3.5   1998–1999 Plans to Delist Wolves in the Midwest

On June 29, 1998, Secretary of the Interior Bruce Babbitt and FWS Director Jamie Clark traveled to Minneapolis, Minnesota, to announce an impending proposal to designate four gray wolf DPSs. The proposal would be to delist wolves in the

Western Great Lakes (WGL) DPS, reclassify wolves in the NRM and Northeastern DPSs to threatened, retain the Southwestern DPS wolves as endangered, and delist gray wolves in the 30 states outside of these four DPSs.

Because wolf numbers in the Wisconsin–Michigan population were expected to achieve the numerical delisting criterion in 1999, and the Minnesota DNR had begun preparing a state wolf management plan in accordance with recommendations produced by a citizens' wolf roundtable committee, the FWS believed the WGL DPS was ready to be proposed for delisting. The NRM DPS was to be reclassified to threatened status due to its continued progress toward recovery. However, the NRM population was still short of its delisting criteria, and there were no state wolf management plans yet in place that would indicate what threats a delisted wolf population would face, so a delisting proposal would be premature. The wolf recovery program in the Southwestern DPS was still in its early stages thus retaining endangered status was the only reasonable option for the wolves there.

The FWS chose to propose the reclassification of wolves in the Northeastern DPS to threatened status. Although the FWS recognized that the existence of only a small number of wolves, at best, argued strongly for endangered status, the agency believed that wolf recovery in the Northeast would require the full support of the state or states where the wolf population resided. It seemed that the most likely way to gain support would be by providing a way for the state(s) to take the lead in on-the-ground wolf management. The only way that could be done was via a 4(d) rule that would reduce the federal role and put more in the hands of the state(s). Because a 4(d) rule can only be implemented for a threatened species, the FWS proposed threatened status for the Northeastern DPS. Contrary to opinions expressed at the time by some wolf advocates, I remain convinced (from personal discussions with FWS biologists and Endangered Species Program managers) that threatened status and the 4(d) rule were not intended to dilute wolf recovery. Rather they were viewed within the FWS as the best way to use the flexibility of the ESA to initiate wolf recovery, and to maximize the probability of recovery success, in the Northeast. (Threatened status coupled with a 4(d) rule provides flexibility similar to that provided by the experimental population classification in the NRM that subsequently allowed wolves to recover in that region.)

The FWS also believed there were large areas of the historical range of gray wolves where wolf recovery was not necessary to achieve the goals of the ESA and that much of these areas were no longer suitable for wolf populations. Thus, FWS indicated in the proposed rule that outside of the 19 states that were in the four proposed DPSs, the remaining areas of the 48 states would be delisted because gray wolf recovery was neither feasible nor necessary there.

However, this first attempt to delist Midwestern wolves ended when the 1999 Minnesota Legislature failed to approve the Minnesota wolf management plan that was based on the 1998 consensus recommendations for wolf management adopted by the Minnesota Citizens Roundtable. Without an approved Minnesota wolf plan, the FWS had no basis for any conclusions on how Minnesota wolves might be protected and managed if delisted. Thus, any attempt at a five-factor analysis for Minnesota wolves would be based more on speculation than on fact. It had become impossible to evaluate postdelisting threats to Minnesota wolves, so the FWS abandoned delisting Midwestern wolves and substantially revised the draft proposal.

20.3.6   2000 Proposal – Four DPSs with Three Reclassified to Threatened

The wolf proposal that FWS published in 2000 for public review (65 FR 43450) retained the four-DPS approach, but it did not recommend any changes to the threatened status of wolves in Minnesota. Instead, it suggested reclassifying to threatened status the wolves in the remainder of the proposed WGL DPS, making their protection equivalent to that of Minnesota wolves since 1978. It also proposed a special rule under section 4(d) that would allow lethal control by government personnel, or their agents, of wolves attacking domestic animals, also very similar to the situation in Minnesota since 1978. The other aspects of the 2000 proposal were identical to the 1998 approach previously announced by Babbitt and Clark – four DPSs, reclassification to threatened status in the NRM DPS and the Northeastern DPS with 4(d) rules for each, endangered status for the Southwestern DPS, and delisting for all other areas of the 48 states (all or parts of 30 states).

The proposal was highly complex, and several aspects involved actions rarely, or never, previously taken by the FWS (e.g., using the 1996 DPS Policy to establish DPSs for a listed entity; specifying the "significant portion of the range" for a wide-ranging species). The 120-day public comment period and the 14 public hearings generated comments from over 43,000 individuals. Due to the magnitude of the comments and the complexity of the proposal, the FWS took nearly 3 years to analyze the issues and to publish the final rule, despite the requirement of the ESA that final rules should be published within 12 months of publication of the proposal.

20.3.7   2003 Final Rule – Three DPSs and Two Lawsuits

The April 1, 2003 final rule (68 FR 15804) was substantially different from the 2000 proposed rule in several ways. First, upon further analysis the FWS had decided that the ESA did not allow delisting a species in parts of its listed range if recovery in that area was not feasible and/or not necessary to achieve recovery. Therefore, in the final rule many of the 30 states that had been proposed for delisting in 2000 retained ESA status as either endangered or threatened by being included in one of the DPSs. Only all or parts of 16 states that were beyond the historical range of the gray wolf, and thus were erroneously included in the 1978 species listing, were delisted. The FWS also had concluded that the ESA would not allow the existence of listed areas that would not constitute listable entities (species, subspecies, or DPSs) on their own, so these areas had to be added to one of the

DPSs, rather than remain listed "non-DPS remnants." Therefore, the originally proposed DPSs were greatly expanded due to the addition of these areas.

Second, during the comment period and the ensuing 28 months the FWS received no additional data supporting the existence of a wolf population in the Northeast. Without a wolf population to designate as a DPS, the FWS decided it had to abandon the proposed Northeast DPS. Also genetic analysis suggested that the original wolf population in the Northeast may have been *Canis lycaon,* a possible conspecific of the red wolf *Canis rufus,* and not *Canis lupus,* the gray wolf (Wilson et al. 2000). Again, due to the belief that non-DPS remnant areas were not allowed by the ESA, the FWS added the Northeastern states to the nearest DPS, creating the large (and newly named) Eastern DPS. The entire Eastern DPS was reclassified to threatened status based upon the achievement of the ETW recovery criteria in the Midwestern states, but the 4(d) rule allowing lethal control of depredating wolves was limited to Midwestern wolves in order to give greater protection to any wolves that might subsequently appear in the Northeast.

Proponents of additional wolf recovery initiated litigation in federal district courts in Vermont and Oregon. The Oregon lawsuit was led by Defenders of Wildlife and included 19 organizations; the Vermont lawsuit was led by the National Wildlife Federation and included five organizations. The plaintiffs' legal arguments and allegations were similar in both lawsuits: inadequate recovery had occurred to reclassify across such large areas; the ESA required a species to be recovered throughout all significant portions of its range and the large DPSs would preclude such additional recovery; and the FWS had not adequately assessed the five factors across all significant portions of the range of the gray wolf.

Subsequent to the 2000 gray wolf reclassification/delisting proposal, the FWS had lost a separate lawsuit over a listing decision for the flat-tailed horned lizard (Defenders of Wildlife v. Norton, 9th Circuit, 2001). That ruling found that FWS must assess the threats to a species across all "significant portions of its range" (SPR), and must make the listing/delisting decision based upon that wider assess-ment, rather than focusing solely on the areas where the species is doing well. Thus, when developing the final rule for the national reclassification proposal, the FWS identified the SPR for the gray wolf and endeavored to conduct its threats assess-ment in the SPR for each of the three DPSs – those areas that the FWS deemed essential to the conservation and recovery of the wolf population in each DPS, but not all areas of suitable habitat within each DPS.

In 2005, both the Oregon and Vermont District Courts issued rulings containing views of SPR that differed markedly from the view of FWS. The courts noted that FWS recognized that there were large areas of apparently suitable habitat that were unoccupied by gray wolves in both the Eastern and NRM DPSs, and the courts found that these areas must be SPR within those two DPSs. Because these areas lacked wolves, and because FWS did not conduct a five-factor analysis for these areas, their reclassification from endangered to threatened was ruled "arbitrary and capricious" and was invalidated by the courts. Both rulings seemed to express disa-greement with the large size of the DPSs, but they did not cite specific problems with DPS size; rather, they dealt primarily with the seeming problem of a recovered population being combined with unoccupied suitable habitat, and the successful recovery being used to downlist the larger area of unoccupied suitable habitat. The Oregon District Court's ruling specifically stated that wolf populations in the WGL and the NRM had achieved their recovery goals, but that the FWS had erred in designating large DPSs that included the wolf's entire historical range (Defenders of Wildlife v. Secretary, US Department of the Interior; D. Or. 2005: 27).

Of note is the statement by the Oregon District Court that it could find no support for the FWS' contention that there must be a population present in an area for it to qualify for DPS status (Defenders of Wildlife v. Secretary, US Department of the Interior; D. Or. 2005: 28). Additionally, the Vermont District Court found no problem with the creation of non-DPS remnants (National Wildlife Federation v. Norton; D. Vt. 2005: 20).

Prior to the rulings on these two lawsuits, the FWS had moved ahead with a proposal to delist the Eastern DPS (69 FR 43664; July 21, 2004). However, the agency had not finalized that delisting decision pending the District Court rulings. The Oregon District Court ruling caused the FWS to shelve that delisting proposal, and the subsequent Vermont District Court ruling terminated work on it. The FWS and the Department of the Interior considered appealing the rulings to pursue acceptance of the narrower interpretation of SPR, but Interior ultimately decided to forego an appeal and instead focused on trying to delist the recovered wolf populations in compliance with the rulings.

20.3.8   Efforts to Control Depredating, Endangered Wolves in Wisconsin and Michigan in 2005–2006

The District Court rulings, which revoked threatened status for Wisconsin and Michigan wolves, also removed the section 4(d) special rule that allowed certain government agents to kill problem wolves after verified depredations on domestic animals. Eighty-two wolves had been removed in the two states during the 22 months they were classified as threatened, and the FWS and the two state DNRs believed that it was important to conduct an effective program to remove these problem animals. Without such a program, these agencies believed that public support for an expanding wolf population would diminish and illegal killing would increase. In February 2005, the two state DNRs applied to FWS for authority to kill a limited number of depredating endangered wolves under restricted circumstances, and FWS issued subpermits to both states in April 2005. The subpermits were issued under the FWS Midwest region's existing permit to take wolves for conservation purposes.

The FWS was promptly sued by 12 environmental and animal welfare organizations for not providing adequate public notice of the states' applications for the lethal control authority. Instead of fighting the lawsuits, the FWS revoked both subpermits, and both states immediately reapplied for full permits to take a limited number of depredating wolves. This time, FWS announced the permit applications in the FR (70 FR 54401; September 14, 2005), opened a public comment period, conducted internal Section 7 consultations on the requested level of lethal take of wolves, and jointly prepared environmental assessments of the requested take levels with each state DNR and the US Department of Agriculture–Wildlife Services program that would be conducting the lethal control actions for the states. At the conclusion of these analyses the FWS decided that the limited and closely monitored lethal control of depredating endangered wolves was an essential conservation action and therefore was authorized under the ESA's permitting provisions. Lethal depredation control permits were issued to Wisconsin DNR on April 24, 2006 and to Michigan DNR on May 6, 2006.

Again, FWS was promptly sued by seven plaintiffs, but this time mounted a defense of its rigorous adherence to the impact evaluation requirements of National Environmental Policy Act (NEPA) and ESA, and its belief that the lethal control of a small number of depredating endangered wolves was a valid conservation action that was necessary for wolf recovery to continue in Wisconsin and Michigan. However, the Washington, DC, District Court judge did not consider the conservation arguments advanced by the FWS. Instead, it ruled that the endangered status of Wisconsin and Michigan wolves precluded their lethal control, and that FWS lacked any discretion to issue such permits, and the permits were again revoked in August 2006. Although initially the Departments of the Interior and Justice chose not to appeal this second permit ruling, intervener Safari Club International defended the legality of FWS to issue such permits, and on June 3, 2008, the DC Court of Appeal vacated the lower court decision. Although this decision was moot to permits for Michigan and Wisconsin where wolves were already delisted, the decision reaffirms authority of FWS to issue special "enhancement permits" for listed endangered species.

20.3.9   The Third Midwest Delisting Proposal

In late 2005, the FWS began drafting another proposal to delist the recovered gray wolf population in the Midwest using a smaller DPS that the agency believed would be in compliance with the Oregon and Vermont court rulings. By that time, Midwestern wolves numbered nearly 3,900, with 840 in northern Wisconsin and the Upper Peninsula, excluding Isle Royale. This approach carved out a WGL DPS that contained the Midwest's core recovered wolf population plus a surrounding area whose size was based on wolf dispersal data. There was little attention given to potential problems from non-DPS remnants because the Vermont District Court ruling clearly approved the retention of endangered or threatened status for a remnant of a former larger listed entity (National Wildlife Federation v. Norton; D. Vt. 2005: 19–20). Also, by carving out a WGL DPS, the endangered status of the Northeast was retained, which preserved the full ESA protections for any gray wolves that occurred there, and maintained the opportunity for establishing a wolf recovery program there.

On March 27, 2006, the FWS published a proposal to designate and delist this WGL DPS of the gray wolf (71 FR 15266, Fig. 20.1). In contrast to the earlier, much larger 21-state Eastern DPS of 2003, the WGL DPS contained Minnesota, Wisconsin, and Michigan, and parts of six adjacent states. The legal status of gray wolves outside the DPS remained unchanged. The proposed WGL DPS is sufficiently small so that, except for some areas of potential suitable habitat in the northern Lower Peninsula of Michigan, nearly all areas of suitable habitat within the DPS are already occupied by wolves. The FWS conducted a five-factor analysis for all SPR within the proposed DPS and concluded that there were no SPR within the DPS that lack wolves, or where wolves are sufficiently threatened to as to warrant threatened or endangered status.

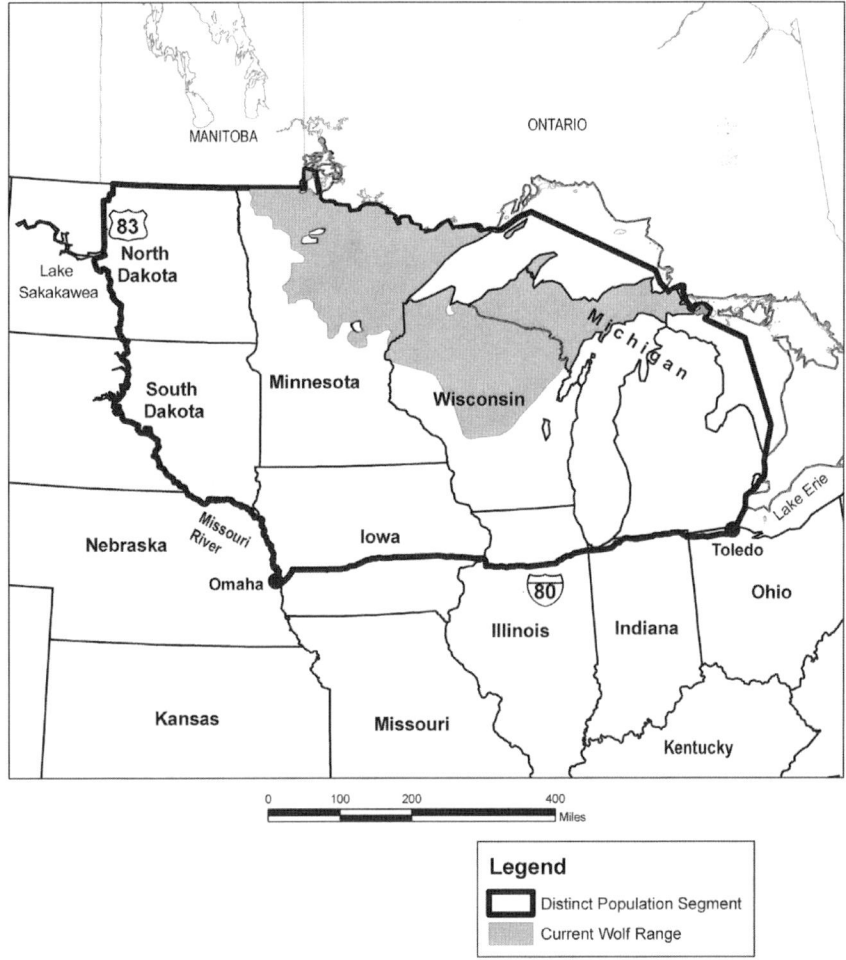

Fig. 20.1 Western Great Lakes Distinct Population Segment of gray wolves, removed from federal list of Endangered and Threatened Wildlife on March 12, 2007

In contrast to the 2000 reclassification proposal, the comments from the ten peer reviewers in 2006 overwhelmingly supported this approach, as were comments from a number of the larger environmental and conservation organizations. Comments from individual citizens were mixed, and the responses from animal welfare organizations strongly opposed the delisting proposal.

20.3.10   Midwest Delisting in 2007 and More Litigation

On February 8, 2007, the FWS published its decision to delist the WGL DPS as it had been proposed, with an effective date of March 12, 2007 (72 FR 6052). This delisting decision removed wolves of the WGL DPS from their endangered and threatened categories under the Act; ended the ESA's protection of critical habitat in northern Minnesota and on Isle Royale, Michigan; and removed the federal regulations regarding the taking of problem wolves in Minnesota. In short, the delisting returned management authority and responsibility to the respective states and tribes within the WGL DPS.

Once again, litigation was promptly initiated to undo the delisting, this time by four animal welfare organizations led by the Humane Society of the US. As of early 2008, a number of additional parties had filed motions to intervene on the side of the FWS, but legal briefs have not been completely filed with the District Court for Washington, DC. A ruling by the court is expected in 2008.

20.3.11   The ESA's Role After Delisting

Although the delisting ended the ESA's regulatory role for the WGL DPS, the FWS remains responsible for 5 years of postdelisting monitoring (PDM). At the time of publication of the final delisting decision, the FWS had prepared an advanced draft PDM plan, and the Recovery Team had reviewed several drafts of the PDM plan. The plan was subsequently published for public review and comment and was revised and finalized in February 2008. In the meantime, the FWS began implementing the monitoring program described in the draft PDM plan, recognizing that the program might change as a result of public review and plan finalization.

The PDM focuses on continuing the same sort of data review and evaluation that the FWS had been doing over the previous decade as it conducted several five-factor evaluations for the various proposed and final reclassification and delisting documents. These past evaluations assessed the threats to wolves by looking at changes in wolf population and occupied range, mortality data, and health data obtained from individual wolves. The PDM will evaluate those same types of data, and will also include a review of significant changes, or proposed changes, to state and tribal regulatory mechanisms that might increase threats to delisted wolf populations. If appropriate, the FWS can extend the monitoring beyond the required 5 years. The FWS, likely, will involve members of the former Recovery Team in the annual review of data, and the results of each evaluation will be posted on the FWS Midwest Region's web site.

In addition to the authority to relist (on an emergency basis, if necessary) the WGL DPS of the gray wolf, the ESA's petition provision remains available. At any time, any individual, organization, or agency can submit a petition containing substantial data supporting a relisting request to force the FWS to consider whether relisting may be warranted.

20.4   Summary

The gray wolves of the Midwest were listed under the ESA for nearly three and a half decades. During that period, the ESA protected wolves and provided direction and objectives for a coordinated recovery program of federal, state, tribal, and local government agencies, NGOs, researchers, and private citizens. The ESA also provided an avenue for citizen involvement and lawsuits that modified or blocked several FWS regulations and actions. Although the wolf's role as a top predator and the strong emotions evoked by the species made recovery and delisting challenging, this success story shows that the ESA can accomplish its goal of conserving species that were nearing extinction.

References

Fuller, T. K., Berg, W. E., Radde, G. L., Lenarz, M. S., and Joselyn, G. B. 1992. A history and current estimate of wolf distribution and numbers in Minnesota. Wildlife Society Bulletin 20:42–54.

Hammill, J. 1992. Wolf reproduction confirmed on mainland Michigan. International Wolf 2: 14–15.

Mech, L. D., and Nowak, R. M. 1981. Return of the gray wolf to Wisconsin. American Midland Naturalist 105:408–409.

USFWS. 1978. Recovery plan for the eastern timber wolf. Washington, DC: US Fish and Wildlife Service.

USFWS. 1992. Recovery plan for the eastern timber wolf. Twin Cities, MN: US Fish and Wildlife Service.

Wilson, P. J., Grewal, S., Lawford, I. D., Heal, J. N. M., Granacki, A. G., Pennock, D., Theberge, J. B., Theberge, M. T., Voigt, D. R., Waddell, W., Chambers, R. E., Paquet, P. C., Goulet, G., Cluff, D., and White, B. N. 2000. DNA profiles of the eastern Canadian wolf and the red wolf provide evidence for a common evolutionary history independent of the gray wolf. Canadian Journal of Zoology 78:2156–2166.

Wisconsin DNR. 1989. Wisconsin Timber Wolf Recovery Plan. Wisconsin Endangered Resources Report 50. Madison, WI: Wisconsin Department of Natural Resources.

Wydeven, A. P., Schultz, R. N., and Thiel, R. P. 1995. Monitoring of a recovering gray wolf population in Wisconsin, 1979–1991. In Ecology and Conservation of Wolves in a Changing World, eds. L. N. Carbyn, S. H. Fritts, and D. R. Seip, pp. 147–156. Edmonton, AB, Canada: Canadian Circumpolar Institute.

Chapter 21
Wolf Recovery in the Great Lakes Region: What Have We Learned and Where Will We Go Now?

Adrian P. Wydeven, Timothy R. Van Deelen, and Edward J. Heske

21.1 Introduction

When we originally wrote this chapter in July 2008, gray wolves had been off the federal list of endangered species in the western Great Lakes region of the USA for 16 months. As Ron Refsnider indicated in Chap. 20, several animal welfare organizations challenged federal delisting after delisting was completed on March 12, 2007. On September 29, 2008, a federal district judge in Washington, DC vacated the delisting, and wolves in Minnesota returned to the threatened list and wolves in the remainder of the Western Great Lakes region returned to endangered status. The judge did not indicate that wolves had not recovered in the region but questioned the use of Distinct Population Segments for designating the delisting. We expect these technicalities of the Endangered Species Act to be resolved over the next few years, and feel that biological recovery of this population has occurred in this region.

While some, including some of the authors in this volume, might argue that the federal government was premature in delisting gray wolves in this region, it is obvious to all that wolf numbers have increased drastically and the population has spread extensively across forested areas of the three states of the Great Lakes region. That this expansion provides rationale for federal delisting may be debatable, but few could argue that this is not a tremendous recovery for wolves in the region.

Winter counts for Michigan and Wisconsin in 2008 were similar to those in 2007, with about 520 (95% CI ± 144) wolves in Michigan (Dean Beyer, personal communication) and 537–564 in Wisconsin. This compares to estimates of 509 (95% CI ± 36) in Michigan and 540–577 in Wisconsin in 2007, suggesting that wolf populations have remained similar for both states during the last 2 years. Slowing growth rates were predicted by Van Deelen (Chap. 9, this volume), and Mladenoff et al. (Chap. 8, this volume) posited increasing saturation of most suitable habitat as a mechanism. In Minnesota, the wolf population in winter 2007–2008 was estimated at 2921 wolves, similar to numbers from the last survey in 2003–2004, and range expansion seems to have ceased since 1998 (J.Erb, *personal communication*). The

A.P. Wydeven et al. (eds.), *Recovery of Gray Wolves in the Great Lakes Region of the United States*,
DOI: 10.1007/978-0-387-85952-1_21, © Springer Science+Business Media, LLC 2009

only large block of wildland without wolves in the three-state area is the northern portions of the Lower Peninsula of Michigan (Gehring and Potter 2005; Mladenoff et al., this volume).

By 2008, the three Great Lakes states contained roughly 4,000 wolves, more than one-half of recent wolf numbers in Alaska (~7,000 wolves), yet occupying an area of only about 10% of the Alaskan wolf range (Stephenson et al. 1995; Boitani 2003). Wolf numbers for the Great Lakes states were similar to wolf numbers in Canadian provinces of Alberta, Saskatchewan, and Manitoba, and were exceeded in Europe only by Russia (Boitani 2003). Alaska and the Canadian provinces sustainably manage wolves as game species (Boitani 2003).

21.2   Are Wolves Recovered?

While numerical increases in wolves have been rather spectacular in the Great Lakes region, numbers alone do not indicate recovery of an endangered species. One might ask whether wolves truly are recovered and whether their populations are secure in the region. The symposium held in 2005, the genesis for this book, occurred because of a widespread feeling among conservationists that gray wolf recovery in the Great Lakes region was making something of a transition from a population at risk of extirpation to a population whose risk of extirpation was becoming trivial. This perception is mirrored almost perfectly by the changing regulatory status of wolves under state and federal endangered species laws (Beyer et al., this volume; Erb and DonCarlos, this volume; Wydeven et al., this volume; Refsnider, this volume). In this context, it is useful to consider whether the perceptions about population security and changes in legal status support a rigorous definition of recovery.

Specifying conditions that must be met for a species to be considered recovered is a normative decision (Vucetich et al. 2006). This means that acceptable levels of population risk and the spatial extent of the population under consideration are essentially judgment calls. The role of science is to quantify population trends, extinction risks, and occupied range to inform those judgments.

Vucetich et al.'s (2006) argument that recovery requires normative and scientific decisions anticipates what might be the point in this book. Relative to the normative judgments and supporting science behind the federal and state endangered species laws, wolves in the Great Lakes region are now recovered (no longer endangered). Whether the current size, growth, and extent of their population are sufficient or optimal and whether conservationists should act to change them remain open questions. The story of gray wolf recovery took place against a backdrop of shifting sociocultural, ecological, and legal milieus as described by our authors. Given that wolf recovery was ultimately successful, the themes that emerge from our chapters may be useful for future wolf conservation (because recovery may be temporary) and for recovery of similar endangered carnivores.

21.3 Broad Themes

21.3.1 The Importance of Shifting Sociocultural Influences

Although it is naive to talk about "a Native American attitude toward wolves" because Native Americans probably have as diverse attitudes toward wolves as other Americans, native peoples of the Great Lakes region do have a unique relationship with wolves, as described for the Ojibwe (Anishinabe) by David (this volume). Along with the Ojibwe, several other tribes occur in wolf range in the Great Lakes region, including the Menominee, Dakota, Ho-Chunk, Potawatomi, Stockbridge-Munsee, Odowa (Ottowa), and Oneida. The wolf plays a dominant role in the origin story of the Ojibwe, and as do many of the tribes in the region, the Ojibwe have a wolf clan. The Ojibwe are also unique in having maintained their hunting and fishing rights outside of their reservations at the time they signed treaties with the US Government in the 1800s. David (this volume) describes the struggle of the Ojibwe reestablishing those treaty rights in parallel with the reestablishing of the wolf population in the region. In Wisconsin, the Ojibwe maintained hunting rights across much of the wolf range within the state, and in any future discussions of possible public harvest, could request up to one-half of allowable harvest in the ceded territories. Clearly, future management of wolves within the region will need to include perspectives of Native Americans.

The broader society has gone through a transition from treating wolves as pests (including bounties) to protecting them as endangered species (aided by an emergent status as iconic, charismatic megafauna; Schanning, this volume; Troxell et al., this volume). This improved attitude toward wolves probably allowed wolves to recover in the Great Lakes region. While general attitudes toward wolves have improved, persistent antiwolf sentiments continue to occur among some stakeholders including bear hunters and livestock owners in Wisconsin (Naughton-Treves et al. 2003). In northern Wisconsin, 13% of residents indicated a willingness to shoot wolves when encountered (Naughton-Treves et al. 2003). Attitudes toward wolves may have started to decline in the Upper Peninsula of Michigan (Beyer et al., this volume), and also remain fairly negative among livestock owners in Minnesota (Chaves et al. 2005). In general, attitudes toward wolves tend to decline in areas occupied by wolves (Williams et al. 2002), especially in rural areas with livestock production (Naughton-Treves et al. 2003; Chavez et al. 2005). Thus, it will be an ongoing challenge for educators and managers to maintain public acceptance in these areas occupied by wolves.

Management of wolf depredation problems is led by state Department of natural Resources (DNRs) and United States Department of Agriculture-Animal and Plant Health Inspection Service (USDA-APHIS)-Wildlife Services (Ruid et al., this volume). Depredation control methods that are generally supported by the public include lethal control by governmental agents and landowners (Schanning, this volume). Minnesota and Wisconsin have provisions that allow landowners to participate in controlling problem wolves in some cases (Erb and DonCarlos, this volume; Wydeven et al., this volume), and Michigan will propose similar regulations in its new management plan (Michigan DNR 2008). In general, wolf depredation has increased with growth of the wolf population, but in Minnesota depredations have declined since the late 1990s (Ruid et al., this volume). Thus, expanded control of wolf populations, especially in agricultural areas, may be an important component of wolf management in the future.

All three states have minimum population goals, but currently only Wisconsin has a state management goal. Population-level management using government trappers or private individuals potentially can be used to reduce wolf depredation at lower costs and removal of less wolves than required for traditional reactive controls by government trappers (Haight et al. 2002).

21.3.2   Changes in Understanding of Wolf Ecology

The expansion of wolves across wildlands of the Great Lakes region has demonstrated that wolves do not need wilderness to survive (Mladenoff et al., this volume). Wolves do benefit from areas of low road density that perhaps provide refugia from human-caused mortality. In a highly protected population, with improved human attitudes and highly productive prey populations (DelGiudice et al., this volume), roadless refugia are probably less important, but attitudes and prey populations are likely to change over time. While maintaining wolves is not contingent on creating new wilderness, maintaining areas of low road density on the landscape evidently benefits wolves by providing sufficient habitat to support a source population, especially where poorer-quality, densely roaded habitat tends to be a population sink (Mladenoff et al., this volume). Although wolves were historically considered habitat generalists, in the Great Lakes region wolves seemed to be most successful in forested wildland areas (Chaps. 4–6, this volume; Mladenoff et al., this volume). Future developments in these areas, especially second-home developments, are likely to reduce and further fragment forest areas of the upper Midwest (Radloff et al. 2005). Thus, large blocks of wildland areas with low road densities are likely to remain important to wolves.

21.3.3   Changes in Legal Status

While wolves in the Great Lakes region demonstrated tremendous recovery for a population of gray wolves, *Canis lupus*, it is less clear whether this population may also contain eastern wolves, *Canis lycaon* (Kyle et al. 2006; Nowak, this volume). Eastern wolves appear to be the same species as red wolves, currently listed as *Canis rufus* (Kyle et al. 2006). It is possible that Great Lakes wolves may include combinations of both species and perhaps includes hybrids of both species. Wolves in Minnesota are intermediate in size between gray wolves of northern Canada and Alaska, and eastern wolves of southern Ontario (Mech 1970). Federal delisting of wolves in the Great Lakes states was predicated on recovery of a gray wolf population segment (Refsnider, this volume). If eastern wolves or hybrids of gray wolves/eastern wolves are also found in the area, it is not clear how they should be managed. From an ecological standpoint, a large canid that regularly hunts deer and larger ungulates has been restored to the area. The current management systems for the three states will likely preserve this large canid, despite its genetic background, but additional research should be done on the genetics of this population to insure that any unusual genotypes that might exist in the region are adequately protected.

Although wolves could eventually become designated game species in the Great Lakes region, wolves are currently managed as nongame mammals in all three states. This contrasts with the Northern Rocky Mountain region where the states of Idaho, Montana, and Wyoming planned to hunt wolves during fall 2008 (Morell 2008), after federal delisting on March 28, 2008 for 1,500 wolves in that region (USFWS 2008) but on July 8, 2008 these wolves were placed back on the Endangered Species List by a Federal judge. In North America, wolves are managed as game species in Alaska and in all ten Canadian provinces and territories with wolves (Boitani 2003).

Regulatory status and socially mandated goals for wolf conservation in Michigan, Minnesota, and Wisconsin are evolving, and conservationists will be confronting the issue of whether to limit further growth in the states' wolf populations. In Wisconsin, the Conservation Congress, a group that advises the Wisconsin DNR on issues of fishing and hunting regulations, held hearings in all counties in spring 2008, and 86% of people attending these meetings recommended that efforts begin to develop public harvest regulations for wolves in the state. Thus, apparently there is a strong interest among hunters and anglers to begin developing the framework for wolf hunting and trapping seasons in Wisconsin. The Michigan DNR is developing a new wolf management plan and found that among people surveyed, 67% support hunting and 60% support trapping for reducing wolf abundance, and most opposed using government agents for reducing wolf abundance (Michigan DNR 2008). In Minnesota, public harvest of wolves would not be considered until 5 years after federal delisting (Erb and DonCarlos, this volume). The questions about public harvest bears directly on issues related to the usefulness of population goals that were established during early wolf recovery (Mladenoff et al., this volume; Van Deelen, this volume; Wydeven et al., this volume), the appropriateness and means of a public harvest, and a backlash against wolves on the part of important stakeholders. Implementing public harvests is likely to be controversial, and as stated by Nie (2003, p. 59), "The issue of hunting and trapping wolves - a public take - after they become delisted is perhaps the most divisive and potentially explosive issue in the entire wolf debate." David (this volume) also notes that many Native Americans oppose hunting of wolves for cultural and spiritual reasons.

21.4   The Future of Wolf Conservation in the Great Lakes Region

The complexity and dynamism of the combined social, cultural, ecological, and legal landscape that enabled wolves of the Great Lake region to recover during the last half decade strongly suggests that continued complexity and dynamism will influence the region's wolf population into the future. Human impacts to core portions of wolf range in northeastern Minnesota, northern Wisconsin, and upper Michigan will undoubtedly increase. These areas are popular tourist destinations, and the conversion of former wildlands into smaller parcels is being driven by the desire for more vacation homes, more rural homes, and more retirement homes on the part of an expanding regional human population (Radeloff et al. 2005). While it is encouraging that, given protection from human-caused mortality, wolves seem more tolerant of human influence on the landscape (Kohn et al., this volume; Thiel et al., this volume), it is also true that core habitat is increasingly influenced by human encroachment. Conservationists will need to be vigilant about monitoring wolf population trends and conservation educators will likely be especially important in training rural residents to live with and appreciate wolves (Troxell et al., this volume).

Expanding human populations will also generate expanding depredation problems and the public support for maintaining the wolf population in its recovered state may depend critically on management of depredating wolves that is professional, responsive, and effective (Ruid et al., this volume). This will be especially true if wolves are allowed to expand into the mixed agriculture-forest regions of central Wisconsin and the northern Lower Peninsula of Michigan (Beyer et al., this volume; Thiel et al., this volume; Wydeven et al., this volume; Mladenoff et al., this volume).

Wolves have shown that they can be resilient. When not treated as an enemy to be extirpated by humans, wolf populations can recover in abundance, recolonize parts of their former range, and resume important ecological roles as top predators in ecosystems. However, humans now dominate the world to an ever-increasing extent, and even the ability of flexible, intelligent animals such as wolves to acclimate to anthropogenic changes of the landscape has its limits. Humans now must make deliberate decisions about our relationships with large predators such as wolves in a new context. In regard to large predators, the questions have become, what will we allow to share the world with us and how much of it are we willing to share? Rather than confront wildness as something to fight and conquer, we need to develop at least détente if not respect and appreciation. The recovery of wolves in the Great Lakes region is a source of great optimism for these relationships.

Few species have been as reviled, or as admired as wolves. Wolves inspire us. A world without wolves would be a pale and impoverished place.

References

Boitani, L. 2003. Wolf conservation and recovery. In Wolves: Behavior, Ecology, and Conservation, eds. L. D. Mech and L. Boitani, pp. 317–340. Chicago, IL: University of Chicago Press.

Chavez, A. S., Gese, E. M., and Krannich, R. S. 2005. Attitudes of rural landowners toward wolves in northwestern Minnesota. Wildlife Society Bulletin 33:517–527.

Gehring, T. M., and Potter, B. A. 2005. Wolf habitat analysis in Michigan: an example of the need for proactive land management for carnivore species. Wildlife Society Bulletin 33:1237–1244.

Haight, R. C., Travis, L. E., Nimerfro, K., and Mech, L. D. 2002. Computer simulation of wolf-removal strategies for animal damage control. Wildlife Society Bulletin 30:844–852.

Kyle, C. J., Johnson, A. R., Patterson, B. R., Wilson, P. J., Shami, K., Grewal, S. K., and White, B. N. 2006. Genetic nature of eastern wolves, past, present and future. Conservation Genetics 7:273–287.

Mech, L. D. 1970. The Wolf: Ecology and Behavior of an Endangered Species. Garden City, NY: Natural History Press.

Michigan Department of Natural Resources. 2008. Michigan Wolf Management Plan, Wildlife Division Report No. 3484. Lansing, MI: Michigan Department of Natural Resources, Wildlife Division.

Morell, V. 2008. Wolves at the door of a dangerous world. Science 319:890–892.

Naughton-Treves, L., Grossberg, R., and Treves, A. 2003. Paying for tolerance: rural citizens' attitudes toward wolf depredation and compensation. Conservation Biology 17:1500–1511.

Nie, M. A. 2003. Beyond Wolves: The Politics of Wolf Recovery and Management. Minneapolis, MN: University of Minnesota Press.

Radeloff, V. C., Hammer, R. B., and Stewart, S. I. 2005. Sprawl and forest fragmentation in the U.S. Midwest from 1940 to 2000. Conservation Biology 19:793–805.

Stephenson, R. O., Ballard, W. B., Smith, C. A., and Richardson, K. 1995. Wolf biology and management in Alaska 1981–92. In Ecology and Conservation of Wolves in a Changing World, eds. L. N. Carbyn, S. H. Fritts, and D. R. Seip, pp. 43–54. Edmonton, AB: Canadian Circumpolar Institute.

U. S. Fish and Wildlife Service. 2008. Final rule designating the Northern Rocky Mountain population of gray wolves as a Distinct Population Segment and removing this Distinct Population Segment from the Federal List of Endangered and Threatened Species. Federal Register 73(39):10514–10560.

Vucetich, J. A., Nelson, M. P., and Phillips, M. K. 2006. The normative and legal meaning of *endangered* and *recovery* in the U.S. Endangered Species Act. Conservation Biology 20:1383–1390.

Williams, C. K., Ericsson, G., and Heberlein, T. A. 2002. A quantitative summary of attitudes toward wolves and their reintroduction (1972–2000). Wildlife Society Bulletin 30:575–584.

Index

Color Plates

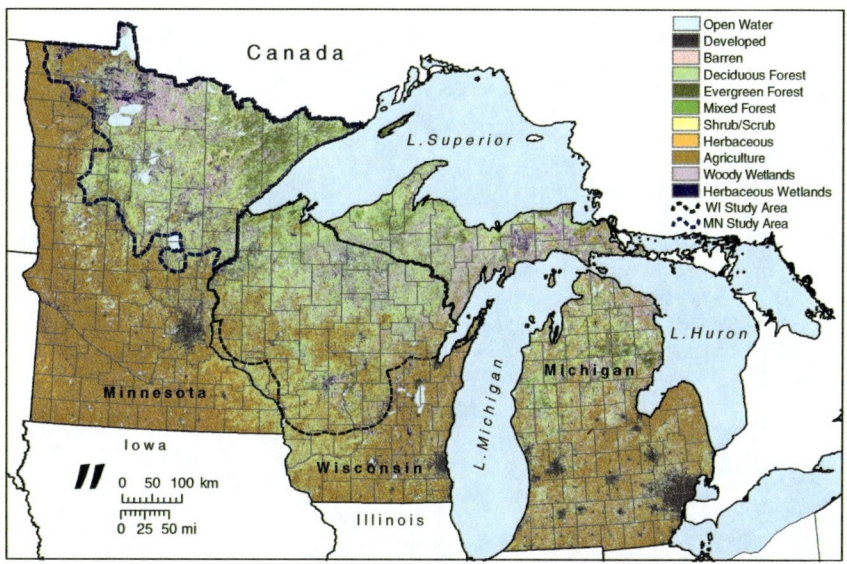

Plate 1 Current land cover in Minnesota, Wisconsin, and Michigan. Data source: USGS National Land Cover Database 2001 (2006). Map created: Forest Landscape Ecology Laboratory, Department of Forest and Wildlife Ecology, University of Wisconsin-Madison. © David J. Mladenoff

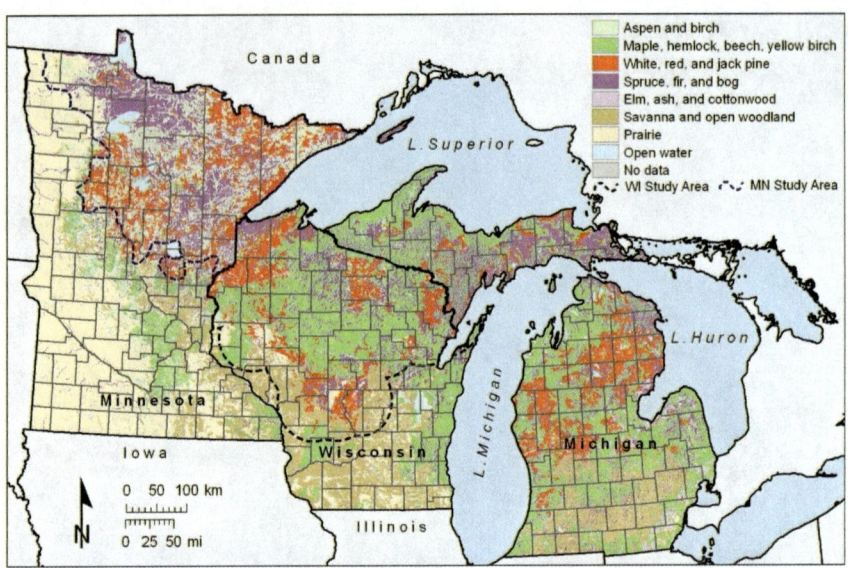

Plate 2 Pre-European land cover (1800s) in the Great Lakes region. Data compiled for the three states from the US Government Land Office survey of the 1800s. Data source: US Forest Service Great Lakes Assessment. Land cover classification and map: Forest Landscape Ecology Laboratory, Department of Forest and Wildlife Ecology, University of Wisconsin-Madison. © David J. Mladenoff

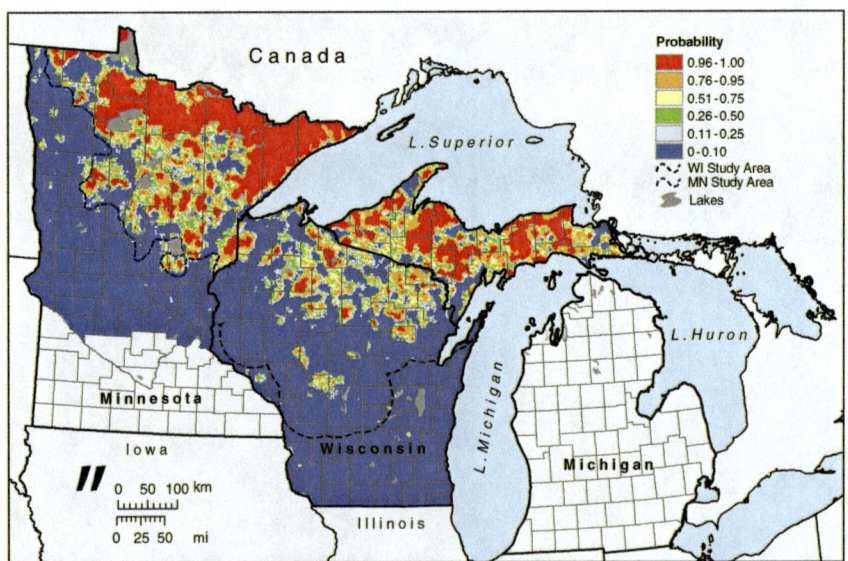

Plate 3 Map of northern Great Lakes states wolf habitat classes from Mladenoff et al. *(1995)*. Map created: Forest Landscape Ecology Laboratory, Department of Forest and Wildlife Ecology, University of Wisconsin-Madison. © David J. Mladenoff

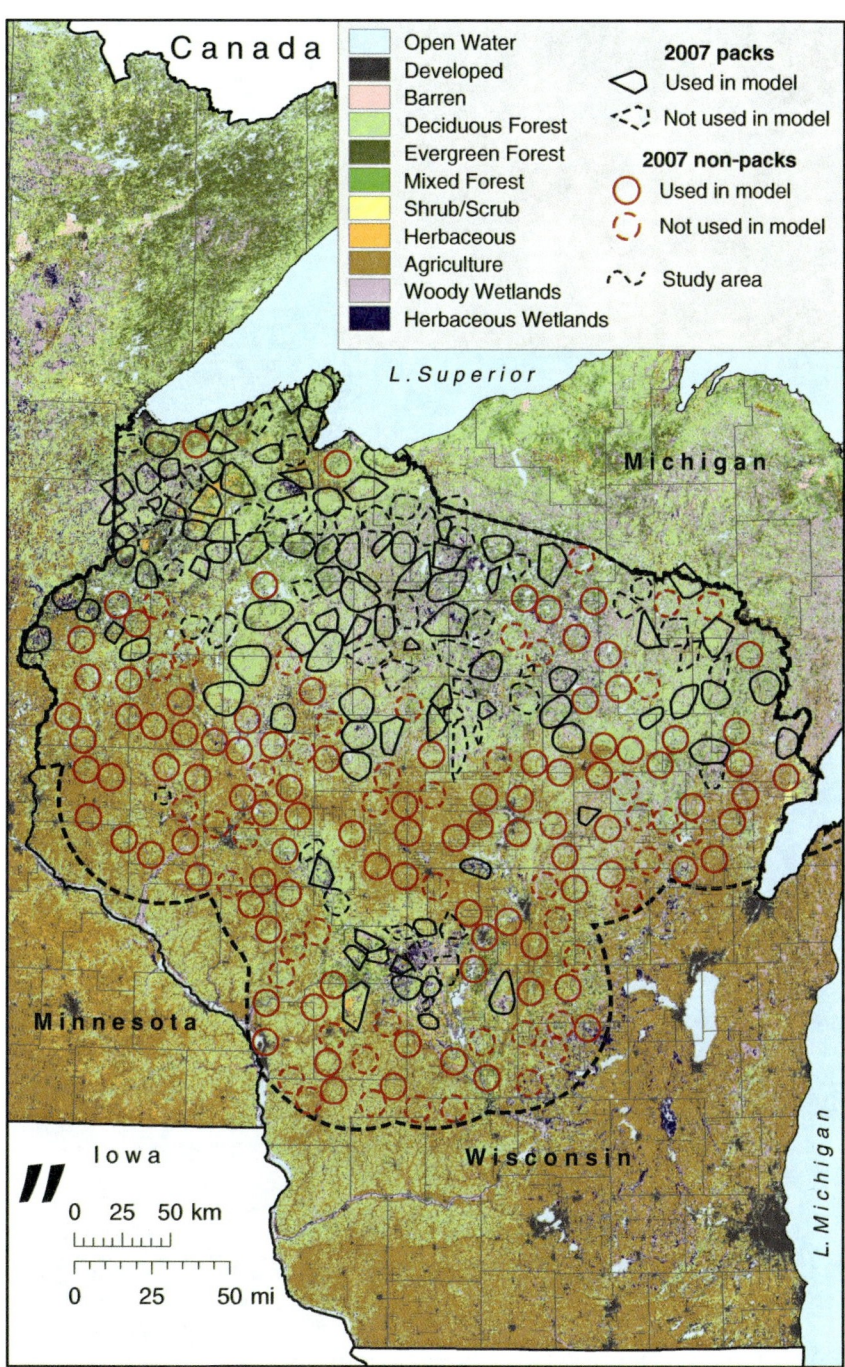

Plate 4 Map of 2006–2007 wolf pack areas and randomly selected non-pack areas used in the analysis for Wisconsin. Map created: Forest Landscape Ecology Laboratory, Department of Forest and Wildlife Ecology, University of Wisconsin-Madison. © David J. Mladenoff

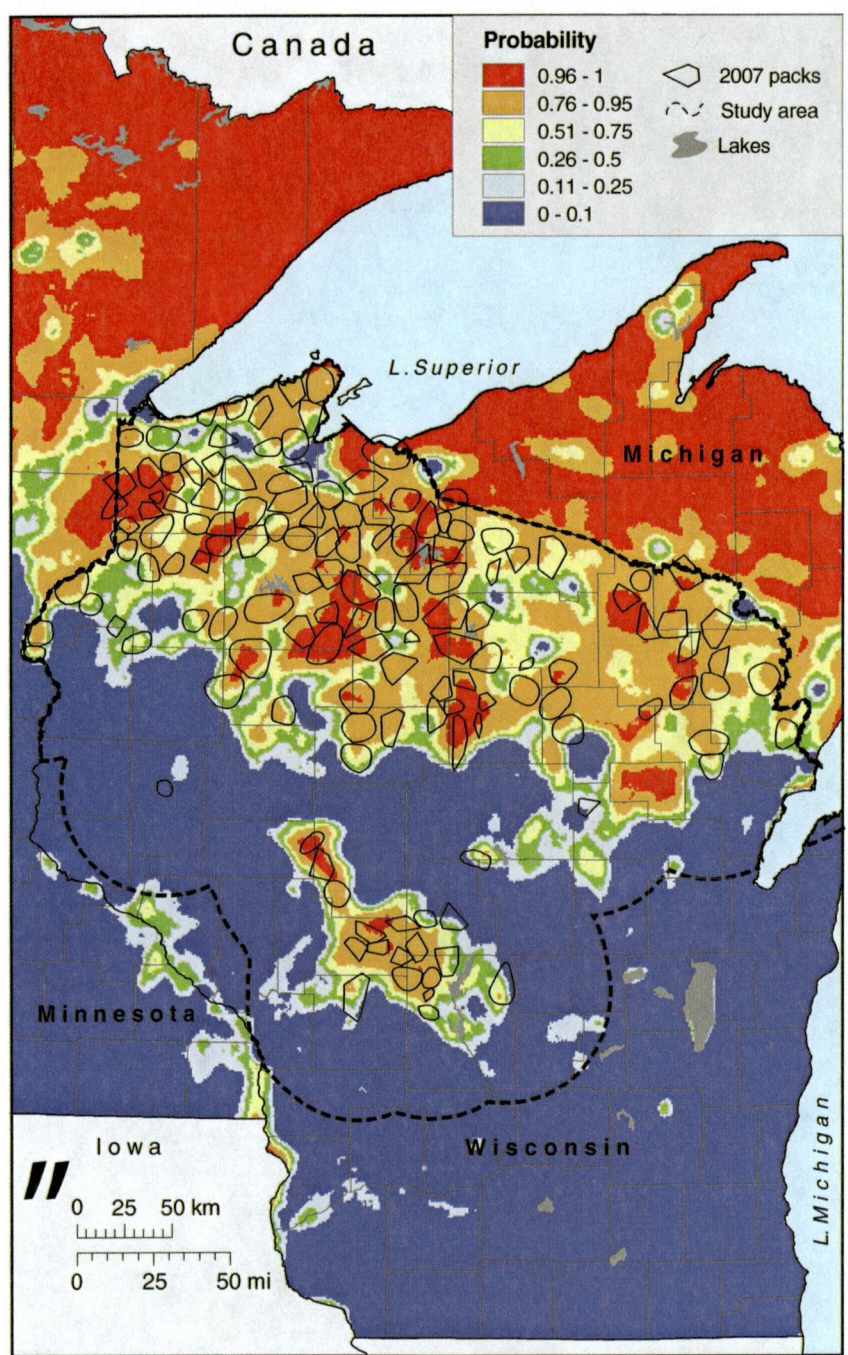

Plate 5 Map of wolf habitat probability classes in Wisconsin based on new analysis and wolf pack occupancy in winter 2006–2007. Map created: Forest Landscape Ecology Laboratory, Department of Forest and Wildlife Ecology, University of Wisconsin-Madison. © David J. Mladenoff

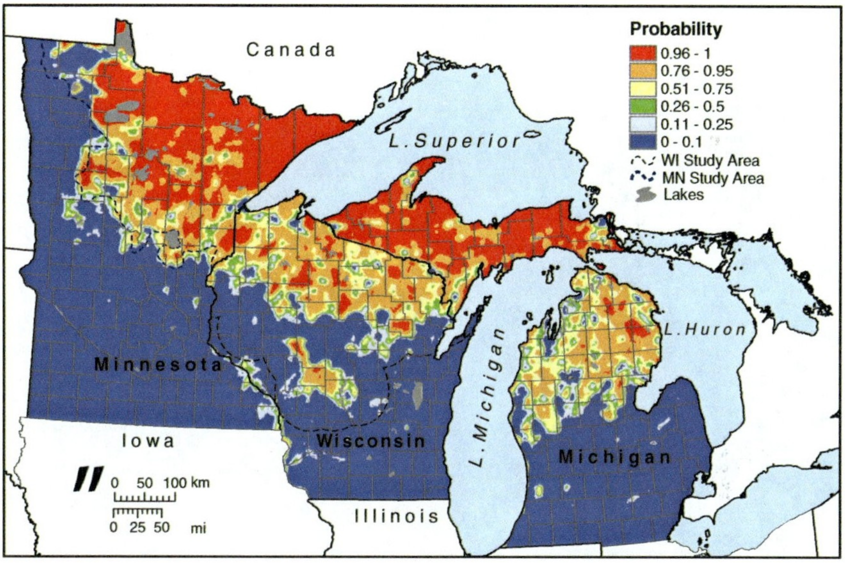

Plate 6 Mapped extrapolation of the new Wisconsin model to the three states in the Great Lakes region. Map created: Forest Landscape Ecology Laboratory, Department of Forest and Wildlife Ecology, University of Wisconsin-Madison. © David J. Mladenoff